Monographs on
Theoretical and Applied Genetics

20

Monographs on Theoretical and Applied Genetics

Volume 1 **Meiotic Configurations**
A Source of Information for Estimating Genetic Parameters
By J. Sybenga (1975)

Volume 2 **Pollination Mechanisms, Reproduction and Plant Breeding**
By R. Frankel and E. Galun (1977)

Volume 3 **Incompatibility in Angiosperms**
By D. de Nettancourt (1977)

Volume 4 **Gene Interactions in Development**
By L.I. Korochkin (1981)

Volume 5 **The Molecular Theory of Radiation Biology**
By K.H. Chadwick and H.P. Leenhouts (1981)

Volume 6 **Heterosis**
Reappraisal of Theory and Practice
Editor: R. Frankel (1983)

Volume 7 **Induced Mutations in Plant Breeding**
By W. Gottschalk and G. Wolff (1983)

Volume 8 **Protoplast Fusion**
Genetic Engineering in Higher Plants
By Y.Y. Gleba and K.M. Sytnik (1984)

Volume 9 **Petunia**
Editor: K.C. Sink (1984)

Volume 10 **Male Sterility in Higher Plants**
By M.L.H . Kaul (1988)

Volume 11 **Tree Breeding: Principles and Strategies**
By G. Namkoong, H.C. Kang, and J.S. Brouard (1988)

Volume 12 **The Wheat Rusts – Breeding for Resistance**
By D.R. Knott (1989)

Volume 13 **Population Dynamics and the Tribolism Model**
Genetics and Demography
By R.F. Costantino and R.A. Desharnais (1991)

Volume 14 **Genetic Improvement of Tomato**
Editor: G. Kalloo (1991)

Volume 15 **Evolution and Function of Heterostyly**
Editor: S.C.H. Barrett (1992)

Volume 16 **Distant Hybridization of Crop Plants**
Editors: G. Kalloo and J.B. Chowdhury (1992)

Volume 17 **Cytogenetics in Plant Breeding**
By J . Sybenga (1992)

Volume 18 **Cytogenetics of the Festuca-Lolium Complex**
By P.P. Jauhar (1993)

Volume 19 **Breeding Oilseed Brassicas**
Editors: K.S. Labana, S.S. Banga, S.K. Banga (1993)

Volume 20 **Oenothera**
By C. Harte (1994)

C. Harte

Oenothera

Contributions of a Plant to Biology

With 30 Figures, 3 Color Plates, and 2 Tables

Springer-Verlag
Berlin Heidelberg New York
London Paris Tokyo
Hong Kong Barcelona
Budapest

Prof. Dr. C. Harte
Institut für Entwicklungsbiologie
Universität Köln
Gyrhofstraße 17, D-50931 Köln

ISBN-13:978-3-642-84288-7 e-ISBN-13:978-3-642-84286-3
DOI: 10.1007/978-3-642-84286-3

Library of Congress Cataloging-in-Publication Data.
Harte, C. (Cornelia), 1914- Oenothera : contributions of a plant to biology / C. Harte.
p. m. -- (Monographs on theoretical and applied genetics ; 20) Includes bibliographical references.
ISBN-13:978-3-642-84288-7
1. Oenothera--Genetics. 2. Oenothera--Development. I. Title. II. Series. QH470.035H37 1993
583'.44--dc20 93-29300

Softcover reprint of the hardcover 1st edition 1994

Typesetting: Best-set Typesetter Ltd, Hong Kong
31/3145- 5 4 3 2 1 – Printed on acid-free paper

Preface

When I was asked to write a book on *Oenothera* because I had personally known several of the classic *Oenothera* geneticists and was occupied for a great part of my scientific research with *Oenothera* myself, I immediately liked the idea of summarizing the knowledge about this plant genus accumulated during more than a century by a great number of outstanding biologists. Several books and summaries on the subject already exist. Some are outdated by newer publications by the same author (Gates 1928; Cleland 1962b), others were already old-fashioned at the time of printing (Gates 1958), and some are still modern in spite of their publication date, for example the excellent book by Cleland (1972) and the summary by Steiner (1975).

I did not intend to write in competition to Cleland's classical text. There are chapters which I cut short because I could not do it better than Cleland who wrote from his own experience and that of his students on the evolution of the genus. Instead I chose my topics to cover fields in which some progress has been made since then, and I did not hesitate to bring in critiques of odd results or interpretations and to present a personal view on some points.

It was most interesting for me to discover the interactions between the observations and the interpretations given to them, on the one hand, and the general scientific background in biology of the time, on the other hand. Still more interesting was the opportunity to use my experience in developmental biology to find out how much is really known about developmental genetics in *Oenothera* but is hidden in publications with titles which do not mention these problems.

I hope the reader will become fascinated by this extraordinary plant genus, its contributions to biology in the past and the prospect of its scientific future which I tried to point out.

It is a pleasure for me to thank the editor of this series, Prof. Dr. H.F. Linskens (Nijmegen/Netherlands) and Prof. Dr. D. L. Mulcahy (Amherst/USA) for their valuable criticism and corrections of my sometimes funny English. Thanks are due to all colleagues who provided me with reprints of their papers on *Oenothera*, facilitating the literature search, and to Prof. Dr. Kehr, director of the university library in Freiburg i.B./FRG, for providing access to the reprint collection of Prof. Dr. F. Oehlkers where I found, due to his worldwide scientific relations, which even continued through two world wars, most of the old publications on *Oenothera*. Gratitude also to my *Oenothera* friends, Dr. Ines Noher de Halac (Cordoba/Argentina) and Dr. Renata Sniezko (Lublin/Poland), who managed to be part of a scientific cooperation spanning half the globe through both cold and hot wars and contributed photomicrographs and drawings for this book.

Finally, thanks are due to Dr. Eberhard Rudlof (Cologne) for the computer drawings of chromosomes and to those colleagues who provided original photographs or drawings from their collections or gave permission for reproduction of pictures from their publications: Prof. Dr. Gertrud Linnert (Berlin) (chromosomes), Prof. Dr. Wilfried Stubbe (Düsseldorf) (plastid mutations and chimaeras), Prof. Dr. Franz Schötz (Munich) (EM of plastids). Photographs of the classical *Oenothera* geneticists came from the archives of several universities: Botanical Laboratory of the University of Amsterdam (de Vries), Botanisches Institut der Universität München (Renner), Botanisches Institut der Universität Freiburg (Oehlkers) and University of Indiana in Bloomington (Cleland).

C. Harte

Contents

Chapter 1 Introduction 1

Chapter 2 Some Things to Know Before Starting Experiments on *Oenothera* 3

2.1 A Plant Enters Science ... 3
2.2 A Short History of Research on *Oenothera* ... 4
2.3 Who is Who in *Oenothera*? or: What's in a Name? ... 9
2.4 Pedigree of Genetically Used Stocks of *Oenothera* ... 13
2.5 Geographical Distribution of the Genus ... 14

Chapter 3 Morphology 18

3.1 Vegetative Parts ... 18
3.2 Flowers ... 19

Chapter 4 Genetics of *Oenothera* 21

4.1 Investigations Using the Methods of Classical Genetics in *Oenothera* ... 21
4.1.1 Genetic Analysis of Oenothera ... 21
4.1.2 Introduction of a Special Terminology ... 23
4.1.3 The Renner Complexes ... 26
4.1.4 How To Be a Constant Hybrid ... 30
4.1.5 Renner Complexes and Genes in *Oenothera* ... 32
4.1.6 Gene Loci and Linkage in *Oenothera* ... 38
4.1.7 Variability of Crossing-Over Frequency in *Oenothera* ... 40
4.1.8 Unexpected Segregation Ratios and Competition Between Genotypes ... 43
4.1.9 Incompatibility ... 46
4.1.10 Selective Fertilization ... 50

4.1.11	The Genetic Model	52
4.1.12	Limitation of the Genetic Method	54
4.2	Karyology of *Oenothera*	55
4.2.1	General Considerations	55
4.2.2	Mitosis	57
4.2.3	Meiosis	58
4.2.4	Facts and Artefacts	60
4.2.5	*Oenothera* Specialities	66
4.2.6	The Basis of Chromosome Catenation	68
4.2.7	Chromosomes and Translocations, or: Chromosomes Unlimited	69
4.2.8	Pairing Behaviour of Translocated Chromosomes	75
4.2.9	Physiology of Meiosis and Chiasmata	82
4.2.10	Meiosis in Polyploids	86
4.2.11	Meiosis in Aneuploids	87
4.2.12	Some Additional Observations	88
4.2.13	The Karyological Model	88
4.2.14	Limitation of the Karyological Method	89
4.3	Cytogenetics of *Oenothera*	89
4.3.1	Combination of Two Methods	89
4.3.2	Renner Complexes and Chromosome Circles	90
4.3.3	Genetic Segregation and Chromosome Catenation	93
4.3.4	Chromosome Catenation and Crossing-Over	94
4.3.5	Trisomics	96
4.3.6	Evolution of Translocations	98
4.3.7	The Cytogenetic Model of *Oenothera*	99
4.3.8	The Limitation of the Cytogenetic Methods	100
4.4	Mutations and Mutation Theory	100
4.4.1	The Start of Mutation Research	100
4.4.2	The Mutants of Oenothera	102
4.4.2.1	Polyploids	102
4.4.2.2	Trisomics	103
4.4.2.3	Chromosomal Translocations and Lethals	106
4.4.2.4	Gene Mutations	109
4.4.2.5	Summary: The Role of Spontaneous Mutations in *Oenothera*	109
4.4.3	Experimental Mutagenesis	110
4.4.3.1	Gene Conversion or Multiple Mutable Alleles?	110
4.4.3.2	Radiation and Temperature as Mutagens	112
4.4.3.3	Chemical Mutagenesis	118
4.5	Genes Without Chromosomes: Extrachromosomal Inheritance	121
4.5.1	The Plastome	121
4.5.2	Plastome Mutations	127
4.5.3	Molecular Biology of the Plastome	128
4.5.4	Evolution of the Plastomes	129
4.5.5	Mitochondria	130

4.5.6 Still More Extrachromosomal Factors 130
4.5.7 The Model of Extrachromosomal Genetics in *Oenothera* 131

Chapter 5 The Species Concept in *Oenothera* and Evolution of the Genus 132

5.1 Genetics and Taxonomy 132
5.2 Some Aspects of Taxonomy of the Genus *Oenothera* 133
5.3 *Oenothera* in Europe 134

Chapter 6 Developmental Biology of *Oenothera* 140

6.1 General Aspects 140
6.2 The Sporophyte 140
6.2.1 Embryogenesis 140
6.2.2 Development of the Seed 142
6.2.3 Germination 145
6.2.4 Development of the Seedling 145
6.2.5 Development of Shoot and Leaves 146
6.2.6 The Flowering Process 149
6.2.7 Development of the Flower 150
6.2.7.1 General Aspects 150
6.2.7.2 Sepals and Petals 151
6.2.7.3 Hypanthium 153
6.2.7.4 Stamens and Anthers 154
6.2.7.4.1 Normal Development 154
6.2.7.4.2 Male Sterility 155
6.2.7.5 Ovary and Ovules 156
6.2.7.6 More Mutants 159
6.2.7.7 Stigma 161
6.2.7.8 The End of Flower Development 162
6.2.8 Still more Genes 162
6.2.9 Summary: Development of the Sporophyte 163
6.3 Plastomes and Development 163
6.3.1 Plastides in Gametophytes, Embryos and Meristems 164
6.3.2 The Pale Plants 168
6.3.3 The Pale Plastids 169
6.3.4 Plastid Development in Plastome Mutants 172
6.4 Developmental Biology of the Haploid Generation 177
6.4.1 Why Are the Haploids So Important? 177
6.4.2 Development of Megaspores and the Female Gametophyte 178
6.4.2.1 Ovules and Embryo Sacs 178

6.4.2.2 Competition Between Megaspores and Embryo Sacs (or Catch as Catch can) 181
6.4.2.3 Genetics of the Development of Polarity 184
6.4.2.4 Polarity in the Megaspore Tetrad and Genetic Competition 190
6.4.3 Development of the Pollen Grain and the Male Gametophyte 193
6.4.3.1 From Meiocyte to Pollen Grain 193
6.4.3.2 Development of Pollen in Male-Sterile Anthers 196
6.4.3.3 Metabolism in Developing Pollen Grains 200
6.4.3.4 Pollen Grains and Embryos, Comparable Developmental Stages? . 201
6.4.3.5 Germination of Pollen Grains and Growth of Pollen Tubes in the Style and Ovary of *Oenothera* 203
6.4.3.6 Elimination of Male Gametophytes 207
6.4.3.7 Differential Growth Rate of Pollen Tubes and Competition Between Pollen Tubes 210
6.4.3.8 Physiology of Incompatibility 211
6.5 Fertilization, Selective and Otherwise 215
6.6 The Developmental Model of *Oenothera* 219

Chapter 7 Propagation System and Evolution in Oenothera 221

Summary 222

Postscript: The End is a Start 223

References 225

Author Index 251

Species Index 254

Subject Index 257

1 Introduction

When we talk about the contributions that a plant makes to biological science, we should first say what we mean. Biology is a way of thinking, starting with observations, making hypotheses for possible explanations of the phenomena, designing and performing experiments to test the hypotheses and ending up, if we are lucky, with a theory that puts our first observations, and the observer too, at the start of a new chapter of biological knowledge. This is exactly what *Oenothera* has done. Some people still consider this genus as the most perfect collection of exceptions to what a decent plant should be. I disagree with them and look at *Oenothera* not only as a plant presenting the most fascinating questions and challenging scientists to find the solutions, but at the same time providing the ideal system for studying phenomena found in many other, less convenient plants.

For more than 200 years, *Oenothera* was looked at only by taxonomists, especially those who were interested in the discovery of new species in Europe. Nearly 100 years ago a new era began. *Oenothera* was introduced into experimental biology by the Dutch botanist Hugo de Vries, professor of botany at the University of Amsterdam. Since that time, the genus is of interest not only to taxonomy but especially to genetics and all branches of biology which are connected with this field, e.g. cytology and karyology, cytogenetics, evolution, developmental biology of the sporophyte and the gametophytes, and molecular biology too.

As a consequence of the great number of investigations in all these fields on different species of this genus, *Oenothera* belongs to the best-known genera of plants not used economically. Considering the great variety of problems which this genus presented to biologists, not many plants will give more. The results are published in many different journals, which are often not present in modern libraries, the older papers being mainly in French or German, sometimes even in Gothic characters, in any case long, papers with 50 pages as normal, and 100–200 pages not uncommon. This means extensive documentation of the experiments and an elaborate discussion of all aspects related to the problem. Discussing many different problems in one paper under an uninformative title means putting up high barriers to young scientists, who are accustomed to having everything presented briefly, in small portions, with keywords and in modern scientific English. So it seems appropriate to bring together some of the knowledge on *Oenothera* in a compact book as an introduction to the genetics of this extraordinary plant genus.

Plate 1. Flower colours in *Oenothera*.
The photographs are from large-flowered plants of the genetic constitution *Oe.*^{h}*franciscana velans* which originated from the F_2 of the hybrid *Oe. franciscana sulfurea* dwarf × *Lamarckiana* de Vries mut. *vetaurea*. Both parents were from the cultures of F. Oehlkers. The four types are maintained by self pollination.
A. Wild type yellow flowers, genotype $+^s\ +^{vet}/+^s\ +^{vet}$
B. Old gold flower, vetaurea, genotype $+^s$ *vet*/$+^s$ *vet*
C. Pale yellow flower, sulfurea, genotype *s* $+^{vet}$/*s* $+^{vet}$
D. Golden center flower, genotype *s vet*/*s vet*

2 Some Things to Know Before Starting Experiments on *Oenothera*

2.1 A Plant Enters Science

What should you do if you are a plant and want to become famous? Behave like a starlet who wants to play a great role in a big movie: first, look for a man who wants to become famous too and needs a star to help him, put yourself in his front garden, wear a bright dress and look as if **you** can solve all **his** problems. Don't worry about how to do that, **he** will forget about it and soon be busy solving **your** problems. That is the story of *Oenothera* and de Vries in a nutshell.

De Vries was interested in heredity. He had developed the theoretical concept of pangenesis as the genetic basis for characters of the individual (de Vries 1889). In our context it does not matter that the pangenes are not identical with the modern concept of genes as units of heredity. Furthermore, he has the same bright idea as Mendel, that you need **differences** in characters when you want to investigate the mode of transfer of the information about these characters from one generation to the next. He decided that abnormalities in plants which were observed repeatedly could be suitable for his purpose and started looking for morphological differences of one character between plants in otherwise homogeneous populations. He found many species with characters that seemed promising, annual plants with fasciations or abnormal leaves or malformations of the flowers (see de Vries 1927; Harte 1953a). Then he collected seeds and grew the plants in the experimental garden and when the abnormality proved to be inherited, i.e. the characteristic appeared again in the progeny, he began with controlled self-pollination of the "abnormal" plants and their "morphologically normal" relatives, made crosses between them and followed them up in progenies from controlled pollinations for several generations. He tried to establish selection lines of high vs. low frequency or strong vs. weak expression of the character, and, most important of all, he counted the frequency of both the normal and the abnormal plants in the different progenies. In between, he laid the foundation for two of the most modern aspects of developmental biology: (i) looking for the interaction between environmental influences mainly nutrition, and the supposed genetic basis in the expression of the character, and (ii) working with figures and tables, using the possibilities presented by the new methods of quantitative biology and statistics. He had only bad luck in working for many years with characters of highly variable expressivity and low penetrance of the genes.

Then, in a neglected garden, he found *Oenothera*, a species which he called *Oe. Lamarckiana*. He became interested in the species because of the great variability of this population, and observed the difference in the length of the style between the plants. This fitted nicely into his project. He included *Oe. Lamarckiana brevistylis* in his experiments and found the laws of heredity, confirming Mendel's observations on segregation ratios in controlled progenies.

Although based on sound reasoning, this was good luck. Short style was the only character in his whole collection of abnormalities which could give reliable segregation ratios. The allele for normal development of the style and stigmas is completely dominant in the heterozygotes. The penetration of the character "short, multilobed style" vs. "long quadrilobed style" is 100% in the recessive homozygotes, the variability of expressivity of the abnormal character is low and does not overlap with the normal phenotype. The phenotype of each plant can be easily determined and there is an absolute correlation between phenotype and genotype of the plants.

Furthermore, de Vries had more than just ordinary good luck. Until now, *brevistylis* is the only known character in *Oe. Lamarckiana* de Vries that behaves according to the laws of classical Mendelian genetics. Isn't it funny? A character that led to the rediscovery of the general laws, describing the fate of genes on their way from parents to progeny, behaves at the same time in the most abnormal manner according to the rules of the species in which it is observed. The laws of heredity of genes in controlled progenies, derived from the behaviour of this character, are generalized and proven to be true and to form a sound basis for the genetics of all plants and animals, and the species which led to the discovery behaves as an exception. The genetic behaviour of *brevistylis* is an exception to an exception! De Vries could not see this from his own experiments. Later it will be shown that these exceptions can be fully integrated in the framework of modern genetics, but that was not known to de Vries either.

This is not the only time that *Oenothera* played a trick. It is repeated in a similar way, but that is a different story and will be discussed later. For now let it be as it is, *Oenothera* was introduced as the star in the first performance of the big new show of experimental biology: **Genetics**.

2.2 A Short History of Research on *Oenothera*

In more detail: in March 1900 Hugo de Vries, professor of botany at the University of Amsterdam, sent to the Academie des Sciences in Paris a short communication which was printed in the *Comptes rendues de l'Academie des Sciences*. Shortly afterwards he submitted a longer paper to the Deutsche Botanische Gesellschaft in Berlin. This appeared in the Berichte der *Deutschen Botanischen Gesellschaft*. Both articles came out in the same year in which they were submitted to the editors. The titles of these papers are: "Sur la loi de disjonction des hybrides" and "Das Spaltungsgesetz der Bastarde". In these papers he published the results of crossing experiments which were performed over many years. The F_2-generations, in which he found the segregation of phenotypes, had been raised since 1882. This means that he started the experiments at least 2 years earlier, in 1880. His material consisted of controlled progenies up to the F_2-generation of nine species. Six other species were investigated after free pollination of the F_1, most probably these were spontaneous self-pollinations or sister-plant-pollinations of self-incompatible species. With three species he made backcrosses of the F_1-hybrid with one parent which was, in modern terminology, the recessive homozygote.

His list of experimental material contains *Oe. Lamarckiana* de Vries. In this species he observed many aberrant types which he called "mutants". He crossed one of them, *Oe. Lamarckiana* de Vries mut. *brevistylis*, with the type he considered to be the pure species. The *brevistylis* differs from the normal type in some quantitative characters (leaf form, leaves shorter and rounded at the tip, shoot shorter, flower buds a bit thicker, delay of shoot formation and flowering by a few days) so that the mutants can be recognized at a glance by anybody who is trained in looking for small differences between plants. The most prominent characters, however, are the form of the style and stigmas and the nearly complete sterility of the ovary. In *Oe. Lamarckiana* de Vries the style is longer than the anthers; in *brevistylis* it is considerably shorter and ends with an irregular number (1 to 8) of stigmas. In most flowers the tips of the stigmas with characteristic malformations can just be seen at the upper end of the hypanthium.

The F_1 has normal styles. In the F_2-families 22% of the plants had the *brevistylis* phenotype. In the backcross of the F_1-hybrid with pollen of the abnormal parent, 55% *brevistylis* plants were observed. The papers contain neither data about the number of F_1-families or plants in the individual progenies, nor do they mention whether a "family" consisted of the seeds of one fruit, or one mother plant, or whether the seeds of one pollen parent were pooled. Only the summarizing statement is given that in general the families consisted of several hundred plants, and in some cases up to 1000 plants were grown. From the point of view of modern genetic statistics, this is a perfect example of incomplete documentation of the experiments, but who could know at the start of a new chapter in biology and from the state of biometry in 1900 that variability in the frequency of types in segregating progenies could be of importance? Not de Vries himself and neither the editors nor the referees of the journals in which his papers appeared! Nobody should blame one of them for not considering methods of scientific reasoning that were developed half a century later. In this respect, de Vries is in a different situation from that of Mendel, who published all the details of his experiments and now some people raise the suspicion that his results were too good.

The list of publications by de Vries, which appeared before and after the crucial ones, demonstrates that *Oenothera* was not just by accident included in the experiments. The genus was of special interest to him, because in his population of *Oe. Lamarckiana* many aberrant types were found, not only *brevistylis*. These different types were later described in detail in several publications (e.g. de Vries 1929) and were also observed in other *Oenothera* species (de Vries 1918, 1925b). They were used by him as material for many crossing experiments which were the basis for his mutation theory. This will be discussed later.

Soon after the introduction of *Oenothera* and the new science of genetics, several botanists shifted their interest to the new problems. Besides the students of de Vries in the Netherlands, the first scientists to become interested in *Oenothera* after the first genetic results were published were Otto Renner, who was at that time "Privatdozent" (lecturer) at the university in Munich, who started experiments in 1910 and published his first results in 1915, and Davis (1909) and Shull (1921) in the USA. They must have started their research at least 4 years earlier before the results could be printed. Soon they were followed by other scientists.

The first result of Renner was his discovery that the occurrence of "mutants" was not the only peculiarity of the genus. Renner found the interpretation of the odd results of species crosses in *Oenothera*, presented in a series of publications and two voluminous papers on "Versuche über die gametische Konstitution der Oenotheren" (Experiments on the gametic structure of the Oenotheras) (1917) and "Untersuchungen über die faktorielle Konstitution einiger komplexheterozygotischer Oenotheren" (Investigations on the factorial structure of some complex-heterozygotic Oenotheras) (1925). How revolutionary the ideas of Renner were and why they caused such confusion among geneticists can only be understood when first the facts are presented.

At the same time the chromosome theory of inheritance was developed and many cytologists looked for the chromosomes of *Oenothera*. After many errors in observation and interpretation of cytological preparations, Ralph E. Cleland became interested in the cytology of the genus (nowadays it would be called karyology) and found, first in cooperation with Albert Blakeslee and later with Friedrich Oehlkers, the solution of the problems of *Oenothera* chromosomes. By combining the results of genetics and cytology to the interesting cytogenetics of the genus, an interpretation of the unconventional genetic behaviour of *Oenothera* was presented.

From the first stages of genetic research it was clear that the solution to some of the problems could not be found in repeated hybridization, but that investigations on development and reproduction had to be included. Among all other problems, developmental biology of the gametophytes, fertilization and embryology in *Oenothera* became the most interesting field.

Parallel to a growing interest in genetics and cytology of *Oenothera*, a differentiation of scientific interests took place. One line was the evolution in the genus *Oenothera*, which developed naturally where the greatest diversity of types was found, namely in the USA, by Cleland. Another line was developed in Europe, following the ideas of Renner, who had already suggested in his first papers on *Oenothera* the hypothesis of plasmatic or plastidal (now extra-chromosomal) inheritance as an explanation for some results of *Oenothera* genetics which did not fit into the usual scheme. For all these fields, it turned out that the haploid phase of the life cycle is crucial to understanding the genetic behaviour of *Oenothera*, so the developmental biology of the haploid phase had to be investigated in more detail than for any other plant.

At the same time as the formal genetic analysis was going on, observations by de Vries on *Oenothera* introduced another important field of modern genetics, mutation research. In the population of *Oe. Lamarckiana* de Vries, in which the *brevistylis* plants were seen, aberrant plants were observed every year. Following self-pollination, their characters were not constant in the progenies. Some were sterile, others segregated for the parental type and normals, and still others segregated new types beside those resembling the parent and the normal *Oe. Lamarckiana* de Vries. However, in progenies of self-pollinated *Oe. Lamarckiana* de Vries these types were observed repeatedly. De Vries called the new types "mutants". His observations on the sudden appearance of new types were the basis for his mutation theory, which was soon published in two thick volumes *Die Mutationstheorie* (Mutation theory) (1901, 1903). The ideas were further developed and presented in another book with the title *Gruppenweise Artbildung* (Groupwise evolution of new species) (1913). The explanation which de Vries presented for his observations does not hold for *Oenothera*, but this did not prevent new contributions of the genus to mutation research. The continued cytological experiments with *Oenothera* again puzzled scientists by yielding unexpected results which led to the discovery of chemical mutagenesis by Oehlkers (1943a).

The mutation theory of de Vries was the starting point for an important branch of genetics, the investigation of mutagenesis. What had happened? A correct observation, the sudden appearance of new phenotypes in both wild and controlled progenies, was given a wrong interpretation: mutation. This interpretation was the basis for a theory on the occurrence of genetic changes as the cause for an increase in biological variability and evolution, that proved to be true, except in those cases of *Oenothera* on which the theory was developed originally.

Was this just another dirty trick to stay in the headlines, played by *Oenothera* on her discoverer and most famous investigator? Or is it an

example that scientific fantasies of men and nature do not always correspond to one another? Which of the two partners is to blame for the discrepancy?

With the exploration of a new field in genetics, molecular biology, *Oenothera* is there already present as a research object and we have to wait for the next surprise.

2.3 Who is Who in Oenothera? or: What's in a Name?

The purpose of giving a name to an object, living or not, is identification and communication about this identity between persons interested in it. A name that does not fulfill this purpose is useless. If one object has different names and the same name is used for different objects, the result is complete confusion and worse than giving no name at all.

Before we start to discuss the experiments, we have to introduce our object. As is the problem with all plants, various taxonomists became interested in the genus at different times. The taxonomy of the genus has a history of variable interests and results and is a source of confusion to young geneticists who want to know whether the plants with which they are working belong to the same species as that described in some interesting but rather old papers, written by authors who died long ago and can no longer be asked to have a look at the plants to identify them.

The genus *Oenothera* belongs to the Dicotylendons, Choripetalea, Dialypetalae, Myrtales, Onagraceae, tribe Onagreae (Raven 1964). The family name in the older literature was Oenotheraceae. Once the genus was named *Onagra*, later renamed *Oenothera*. The orthography is variable, *Onothera* or *Oenothera*, and even the pronunciation of the now valid name

←

Fig. 1. Four leading geneticists who contributed the important steps in *Oenothera* research. **a** Hugo de Vries, one of the founders of modern genetics, rediscoverer of Mendel's rules of inheritance and the first in mutation research, who introduced *Oenothera* in genetic experiments (photograph from 1932, persbureau Lenarts, published by de Veer 1969; courtesy of the Archives of Biologisch Centrum Anna's Hoeve, Hugo de Vries Laboratorium, Amsterdam/Netherlands). **b** Otto Renner, who interpreted the odd genetic behaviour of *Oenothera* as being caused by gene complexes of many loci with variable linkage relations and by plasmatic inheritance (photograph from 1930 by Alfred Bischof, Jena/Germany, courtesy of Prof. Dr. Mägdefrau and the Archives of the Botanical Institute, University of Munich/ Germany). **c** Ralph E. Cleland, who found the solution for the chromosome catenation in meiosis by reciprocal translocations of chromosome arms and the relation between the genetic Renner complexes and the chromosome complexes, and who also investigated the evolution of the genus in North America (courtesy of the Archives of the University of Indiana, Bloomington/USA). **d** Friedrich Oehlkers, who pushed forward the cytogenetics of *Oenothera* by investigations on the physiology of meiosis, connected cytological chiasma formation to genetic crossing-over, and detected chemical mutagenesis (photograph from ca. 1950, from the Archives of the Botanical Institute of the University of Freiburg i.Br./Germany) is not generally accepted: either the Latin form *Oenothéra* or English *Oenóthera*, both combined with the "th" as Latin "t" or English "th", as you prefer. In any case it is wise to make sure that your partner in a discussion follows the same rules as you do.

The remaining taxonomy is confusing too. The genus was at one time divided into several genera. This led to difficulties in the definition of the important, distinguishing morphological characters. On the other hand, heterogeneous material is sometimes combined into one species, a method which generates other difficulties in the morphological delimination of the types. More on this problem is found in the books of Cleland (1972) and Gates (1958) and the publications of Munz (1949, 1965), Raven (1964), Dietrich (1977) and Raven et al. (1979).

Nowadays the genus is subdivided in several sections. Only three of them are of importance in the genetic literature, and so only these will be discussed. Here too, confusion is normal. The geneticists first used the names which were acknowledged in taxonomy at the time. For nearly 60 years this was a sound basis for information to other geneticists about the objects of their study. Then, taxonomists started reviewing the genus and searching for the names, valid according to the meanwhile accepted rules of botanical nomenclature, but without taking into consideration the widespread use of certain names in the genetic literature. In the older genetic publications, up to about 1955, a section *Eu-Oenothera* is distinguished from the *biennis* group. Then it was found that the name *Eu-Oenothera* should be appointed to that section of the genus which contains the species, described first by the name *Oenothera*, and that is what was known up to that time as the *biennis* group. Later the sections were revised again (Fischer 1963; Dietrich 1977). The current valid version is:

old name	temporary	new name
biennis group	*Eu-Oenothera*	*Euoenothera*
Eu-Oenothera	*Raimannia*	*Raimannia*
		Renneria

There are more sections recognized by taxonomists, but they are not investigated genetically and will therefore not be discussed here. The sections are characterized by several morphological characters, the most striking being the form of the ovary and the seeds. In the sect. *Euoenothera* the ovary is nearly square in cross section, while in the sect. *Raimannia* it has, in some species, outgrowths at the four corners, called "wings". The seeds of the sect. *Euoenothera* have sharp edges, those of the sect. *Raimannia* are smooth and egg-shaped. In the sect. *Renneria* the form of the ovary, like the *Euoenotheras*, is combined with oval seeds (Fischer 1963). *Raimannia* and *Renneria* are considered by some taxonomists to be separate genera (Munz 1935; Raven 1964). Other genera of the Onagreae, e.g. *Clarkia*, *Godetia* (now sometimes included in *Clarkia*) and *Gaura*, have some similarities with *Oenothera* in their cytogenetic behaviour.

As most of the genetic experiments were done with species belonging to the section *Euoenothera*, in this book the section will be mentioned only when the species belongs to another section, in most cases *Raimannia*.

Concerning the species names, similar difficulties arise. In this book the valid taxonomic names will be used whenever possible, but besides this the historical names, the classical *Oenothera* nomenclature, will be used for two reasons: first to facilitate comparisons with the literature; and second, because the genetic and karyological relationships of the acknowledged taxonomic types with the phenotypes (genotypes) of *Oenothera* used since de Vries are not established. In fact in some cases a disagreement is evident. For example, the taxonomic *Oe. villosa* Thunb. (syn. *Oe. strigosa* (Rydb.) Mack and Busch) (Munz 1965; Dietrich and Raven 1976) is not identical with the "*Oe. strigosa*" used in the genetic publications of de Vries, Renner, Oehlkers and others, even when de Vries called it "*Oe. strigosa* Rydb.". Comparison of the two types, grown side by side in the experimental garden under the same conditions, revealed such morphological differences that they cannot be considered to belong to the same taxonomic species. Furthermore, the cytogenetic differences are to be considered. Therefore, as already proposed by Cleland (1944), the name "*Oe. strigosa* de Vries" will be maintained for the type used by de Vries and those plants which are derived directly from seeds given by him to Renner and other geneticists.

Other name problems arise with *Oe. suaveolens* Desf. Its name cannot be "*Oe. biennis*" as proposed by Raven et al. (1979) because the two species, known in genetics as *Oe. biennis* L. and *Oe. suaveolens* Desf., are not only morphologically different, but are genetically different in terms of their Renner complexes, the chromosome translocations and known genes, to such an extent that this difference cannot be neglected in experiments where the genetic constitution of the plants is important. For this reason, the name "*Oe. suaveolens* Desf." is maintained for the type of known pedigree, used under this name in the experiments since the publications of de Vries, who received the seeds from colleagues in France.

Still more difficult are the cases of the genetic types known as *Oe. Lamarckiana* and *Oe. Hookeri*. The cases are discussed in detail by Cleland (1972). According to the rules of nomenclature, the names should be *Oe. glazioviana* Michelli (syn. *Oe. Lamarckiana* Seringe and *Oe. erythrosepala* Borbas) and *Oe. elata* H.B.K. (syn. *Oe. Hookeri* Torr. and Gray). But already Davis (1915, 1927) showed by comparisons of herbarium specimens with the descriptions in the genetic literature, that the plants used as *Oe. Lamarckiana* by de Vries are not identical with those described by Lamarck and later named *Oe. Lamarckiana* Seringe. That was a type belonging to *Oe. grandiflora* Solander. According to the proposition of Davis, the type used in genetics and whose pedigree can be followed to the garden of de Vries, should retain its name "*Oe. Lamarckiana* de Vries" and the old taxonomic type "*Oe. Lamarckiana* Seringe" should be called *Oe. glazioviana* Michelli according to the valid rules of nomenclature. The type introduced to genetics by de Vries as "*Oe. Hookeri*" is not identical with the species of this name as found by Davis (1934), when he again compared herbarium specimens and descriptions, but without stating the identity of the species used in genetics. It will therefore be named "*Oe. Hookeri* de Vries", as proposed by Cleland (1944). The change of name of the taxonomic type *Oe. Hookeri* Torr. and Gray to *Oe. elata* H.B.K. cannot involve the genetic literature, because the taxonomic type and the one used in genetic investigations are not identical.

For *Oe. grandiflora* three authors are used: *Oe. grandiflora* Solander by Davis (1915, 1927), *Oe. grandiflora* Ait. by Renner and his students in several papers and at last the now valid name *Oe. grandiflora* l'Her. (Steiner and Stubbe 1984, 1986; Schumacher et al. 1992). The identity of the stocks investigated cannot be verified with the information available. For genetic reasons, the stocks introduced by Steiner, which are mainly homozygotes, cannot be identical with the heterozygotic strain with the same name investigated by Renner. Therefore, in citations the names used by the authors will be given.

Other difficulties are created in those cases where a new type, behaving genetically like a species, appeared in experiments, was described (not in Latin as required by taxonomists), characterized by its genetic and cytological peculiarities, given a name, mentioned a few times by its discoverer and afterwards disappeared completely from the records. In these cases the given name will be used in the proper context with citation of the publication and without any further taxonomic comments.

In the species of the section *Raimannia* still greater difficulties arise. Some taxonomists recognize a genus "*Raimannia*", others include the same species in *Oenothera* as sect. *Raimannia*. This is usual in the genetic literature. The type introduced in genetic research by Schwemmle as *Oe. affinis* has, according to Tandon and Hecht (1956), to be included in *Oe. mollissima*. On the other hand, *Oe. affinis* and *Oe. mollissima* are both investigated in species hybrids. In this subgenus more name-changing took place and now the old *Oe. berteriana* "Erlangen" Haustein has been renamed *Oe. villaricae* Dietrich and *Oe. odorata* "Erlangen" Haustein became *Oe. picensis* Philippi ssp. *picensis* (Dietrich 1977). These names will be used in new publications on genetic problems. The old names were used in numerous publications by Schwemmle and his students.

The species *Oe. campylocalyx* Koch et Bouche used in genetic experiments is now included in the subgenus *Renneria* as *Oe. versicolor* Lehm. (Dietrich 1977).

In the genetic literature the species names are used mainly without addition of the name-giving author, in contrast to what is usual in taxonomy. In new monographs on the taxonomy of *Oenothera* it is clear that in the sect. *Raimannia* the same name was used for what are now recognized as different species and a new species may formerly have had many different names (Dietrich 1977). For example, 15 synonyms are mentioned for *Oe. odorata* and a considerable number of other species have *odorata* in their list of synonyms. The genetic identity of the stocks used by geneticists and the new taxa with the same name is not always proven.

To make things no more complicated than is necessary, in discussing the results of experiments with species of the sect. *Raimannia* the names familiar in the genetic literature will be used without a taxonomic author. How would you feel if you had learned something interesting about *Raimannia mollissima* and, even with a computer search of the literature, you could not find anything about this species but, instead, found some papers describing experiments on the same problems and with exactly the same results, but performed with *Oenothera affinis* which was not mentioned in the text?

In all other cases, the valid taxonomic names will be used and reference to synonyms will be given whenever necessary. This proceeding seems to be the only possible way when the names of the Renner complexes and the gametic constitution of the *Oenothera* species are discussed. We have to choose that nomenclature that will enable us to minimize the possibility of confusion and errors of understanding in the genetics of the genus.

The result of this apparently unavoidable name-changing is that, in contrast to what is generally expected and what is intended in giving a name, in *Oenothera* the name of a species, genus or section does not by itself give any information about the identity of the object in question. For this purpose you need so much additional information that the name becomes the most unimportant item.

As the old papers will not be reprinted with a revised taxonomy and new names, nor will the experiments be repeated with stocks of known taxonomic identity, the only way out of the problem for the geneticist, or the non-taxonomist, is to look for the author of the paper, the pedigree of his stocks, the year of publication and to remember the genetic structure of the plants used in the experiments instead of the names.

Remember: What's in a name? (Shakespeare).

2.4 Pedigree of Genetically Used Stocks of *Oenothera*

To identify the types of *Oenothera* used in the experiments, it will be useful to know the pedigree of the stocks. The information given in this section is mostly received by personal communication from the scientists. Only in a few cases is the pedigree printed in the first or a later paper of the authors working with *Oenothera* (Sturtevant 1931; Renner 1942a; Schötz 1954).

De Vries collected the type called by him *Oe. Lamarckiana*, other stocks like *Oe. Hookeri* and *Oe. strigosa* were given to him by colleagues as described in his book *Gruppenweise Artbildung* (1913). His stocks are used by other research workers in the Netherlands. A second type of *Oe. Lamarckiana* with red veins in the leaves was collected by Heribert-Nilsson in Sweden and given to Renner (Schötz 1954). The pedigrees of all species and genetic types used by Renner and his students were summarized later (Renner 1942a). He received some types directly from de Vries, others from several colleagues and collected European species himself. The stocks of the German geneticists are all derived from the cultures of Renner. Oehlkers had his seeds from this source and passed them to Harte. Schötz and W. Stubbe, who gave some of his cultures to Herrmann, also acquired their stocks from Renner.

In Great Britain, R. Ruggles Gates collected plants and seeds himself, but at least had *Oe. Lamarckiana* from de Vries, and Catcheside worked with the same stocks. In the USA the first American geneticists working with *Oenothera*, Shull, Davis, Sturtevant and Emerson, collected species themselves from the wild or from botanic gardens, but it seems that they all used the *Oe. Lamarckiana* of de Vries. Cleland received some of his stocks from the German scientists Renner and Oehlkers, with whom he collaborated, but he also collected many species himself. The stocks used by his students and collaborators Steiner, Stinson and Hecht are the same as those used by Cleland, but all collected from wild habitats too. The origin of the species of the sect. *Raimannia* investigated by Julius Schwemmle and his students could not be traced, but most probably they came from botanical gardens, while those used by the American authors Hecht and Tandon were collected from the original localities. The *Oe. organensis* Munz. is in all cases derived from the collection of Emerson.

As the names were transferred together with the seeds, for the papers of scientists who used the types of known pedigree it is certain that the results of experiments with plants of the same name can be directly compared without there being a danger of mixing uncomparable facts.

2.5 Geographical Distribution of the Genus

The genus *Oenothera* is indigenous only in North and South America. A few species were introduced as weeds to other parts of the world. The evolution of the North American species is described in detail by Cleland (1972) and will therefore not be discussed here. The greatest variety of species is found in Mexico and the northern part of Middle America. In South America the sect. *Raimannia* is prevalent, but the genetics of all these species is not know. In our context they are therefore of no interest. In their original region of distribution *Oenothera* species are found in many different habitats, varying from deserts to disturbed places and from plains to mountains.

It is evident that the European species, or their ancestors, were introduced from North America. Wein (1930) wrote a very interesting history of the genus *Oenothera* in Europe, based on herbarium samples and descriptions of plants in the literature. The first species mentioned as cultivated in gardens of continental Europe in 1633 is one with small flowers, probably *Oe. syrticola* Bart. (syn. *Oe. muricata* AUT.). Later from the same botanical garden a species with larger flowers, which could be *Oe. biennis* L., is described. Both are mentioned as exotic garden plants. Soon the first variety, the *Oe. biennis sulfurea* with sulphur-yellow flowers, appeared between 1668 and 1686 in the botanical garden at Leiden (Netherlands). *Oe. biennis* L. started its "Grand Tour of Europe" probably as an escapist from gardens in France, where the plants were grown as early as 1623. In 1683 the species was already seen in the Netherlands. In this year, J. Commelin found *Oe. biennis* L. as an intruder in the dunes between Haarlem and Den Haag. Shortly afterwards the species was found growing in large numbers farther

324 PROSPERI ALPINI

Hyoſciamus Virginianus.

Ab hinc

DE PLANTIS EXOTICIS. 325

De Hyoſciamo Virginiano. Cap. LI.

AB hinc annos duos mihi nata eſt planta ex ſeminibus nomine Lyſimachię Virginianæ ad me miſſis à Ioanne Moro medico, & philoſopho anglo eruditiſſimo. Quæ planta fert caulem craſſum, rotundum, veluti ſtrijs actum, rectum, non durū, bicubitalem, & tricubitalem, qui poſtea paulò ſupra radicem profert duos alios, aut tres caules parum primo minores, bicubitales, rectos, rotundos, fere ad craſſitiem digiti minoris, ſolijs multis, longis, non tenuibus, extrema acuminata habentibus, atque circum leuiter crenatis, & ambitum inequalē poſſidentibus caulibus inordinate ab imo ad cacumen vſq. inhærentibus, Ex ſingulis verò alarum foliorum cauis exibat petiolus digitali longitudine ferè rotundus, gracilis, mollis, qui in cacumine fert florem luteum quadrifolium leucoijs ſimilem, ſed maiorem in medio illius exeunt multa capillamenta tenuia, lutea, quibus deſloreſcentibus ſuccedūt ſiliquæ paruæ, craſſæ, oblongę, in extremis tenuiores, & in medio craſſiores quæ matureſcētes indureſcunt, & circum ſtrijs aguntur, colore viridis dilutioris figura, & magnitudine ad hyoſciami albi ſiliquas non parum inclinant: ad ſaporem ſubdulcem accedentes & inodorę etiam ſunt. Tota planta nititur radice craſſa in multis diuiſa craſſa alba, & abſq. odore & ſapore. Credidi ego hanc plantam eſſe hyoſciamum nigrum Verginiæ familiarem, folia enim ſiliquæ, & ſemina hyoſciamo Aegyptio à nobis deſcripto ad hyoſciamum accedunt, ad quem ſemina magnitudine, & colore maximè accedunt, & ex ſapore inſipido ſubdulci conijcitur eſſe frigidioris temperamenti, cuiuſmodi eſt Italicus hyoſciamus.

Bellis

Fig. 2. The first published drawing of a species of *Oenothera* in *De Plantis Exoticis* by Prospero Alpino (1627), probably *Oenothera syrticola* (according to Rostanski 1985). Courtesy of the Library of the Herbarium Hausknecht, University of Jena/Germany north in a cemetery in Harderwijk, all these places are situated in the Netherlands. In 1711 it was seen in Germany. During the following years, the species is mentioned as growing on riverbanks and near the sea coast. The first species becoming a real immigrant was, however, *Oe. muricata* (now *Oe. syrticola* Bart.), found already before 1612 in England. Wein states that other dates as mentioned by de Candolle and Gates cannot be confirmed. The first picture of a plant that cannot be anything other than an *Oenothera*, probably *Oe. syrticola*, even when the artist granted it five petals, is published by Prospero Alpino in a book "*De Plantis exoticis*" (1627) (Fig. 2).

An interesting hypothesis about the distribution of *Oe. biennis* L. is presented by Rostanski (1968c, 1975, 1982). This species is a long-established member of the flora not only in Europe but also in northern Asia. It is found on islands off the Pacific coast between Asia and Alaska, on Sachalin and in Siberia, growing on riverbanks and in other habitats similar to those occupied by the genus in Europe. Specimens were collected early and the distribution is documented in old Russian herbaria (Rostanski, pers. comm.). Cleland did not find in his collection of North American *Oenothera* species a type like the European *Oe. biennis* L. Therefore it seems possible that this species, now extinct in its original site, started a few thousand years earlier, travelled the other way round the globe, over the northern Pacific islands and then crossing Asia to the Ural and into Europe, leaving isolated populations in suitable habitats and arriving in Europe, at nearly the same time as those species that came across the Atlantic. Anyhow, hybridization started immediately when different species met in western Europe.

For other species, e.g. *Oe. Lamarckiana* de Vries and *Oe. suaveolens* Desf., it can be suspected that they originated in Europe from introduced American species by hybridization, maybe repeatedly. The suspected ancestors are *Oe. biennis* L. with a type similar to *Oe. grandiflora* Solander in the case of *Oe. suaveolens* Desf. and *Oe. biennis* L. with *Oe. franciscana* for *Oe. Lamarckiana* de Vries. Evidence for this will be presented later. Starting from their place of origin, these species were distributed as weeds to other parts of the world.

The places where *Oenothera* species mainly grow wild in Europe, namely as ruderal flora, in bare, abandonned places, along stony riverbanks, the sides of canals, roads and railroad tracks, can be seen as evidence that they are not indigenous. The same holds true for the immigrant species in other countries. *Oe. Lamarckiana* is known in Japan as the "railroad plant".

When Otto Renner, at the time professor of botany at the University of Jena, described the species of *Oenothera* growing in northern and eastern Germany, he could find the plants very easily, doing "Kleinbahnbotanik" (small-train-botany). As he once told, he just travelled by the slow narrow gauge trains which were used in the area at that time connecting villages and small towns (the slogan, not invented especially for botanists, was: "It is forbidden to jump out and pick flowers while the train is moving"), and made notes about the species he saw at the side of the track. Then he had only to inspect the stony banks of canals and rivers and he could be sure of having found all the species growing in that particular region (Renner 1937a, 1938a).

The appearance of new species in Europe, as they are described repeatedly by botanists from different countries, can be understood only on the basis of the genetic specialities of the genus and this will be discussed later.

3 Morphology

It is not the intention to present here a complete description of the morphology and morphogenesis of *Oenothera*, but only to describe some characters which are important for the experiments. A monograph on the morphology and anatomy of the genus does not exist, but some details are to be found in most handbooks on plant taxonomy and in old texts on flower morphology (e.g. Eichler 1878). The peculiarities of developmental biology of the genus, as revealed by experiments, will be discussed in a separate chapter.

3.1 Vegetative Parts

The species of the sect. *Euoenothera* and most of those in the sect. *Raimannia* are annuals or biennial plants of the hapaxanth type (a term created by Troll for plants flowering only once in their life, irrespective of the time needed to finish the cycle), growing, flowering and dying after maturation of the seeds. Only a few species, like *Oe. organensis* or some in the sect. *Raimannia*, are perennials. Many details of the morphology of the plants are described by Hegi (1935). In most cases, after germination a rosette of leaves develops. When ca. 20–50 leaves are present in the rosette, the growth of the shoot starts, which becomes the main inflorescence. Side branches can develop from the axils of rosette leaves or on the stem. They all end in an inflorescence. The inflorescence is a spire with potentially unlimited growth. The flowers are single, without a peduncle in the axils of bracts.

The shoot is herbaceous, with a closed cambium ring, from which secondary woody tissue with intraxyllary phloem is formed. In the parenchyma of all vegetative parts, long files of elongated cells are present which contain bundles of raphides, needle-like crystals of Ca-oxalate, whose fine structure and development is described by Schötz et al. (1970). The root system is allorhizous. The main root becomes a more or less fleshy structure which is considered to be edible (Hall et al. 1988, who do not rely on their own experience but on other publications according to citations).

Author's comment: I first heard that the root was potentially edible long ago when I was still a student, but could not discover who had invented the story. I tried once myself, just out of curiosity to learn if the claim was true, and do not recommend repeating the experiment. A proverb says: When the devil is hungry he will even eat flies. Maybe a botanist near starvation should try to eat *Oenothera* roots, but I know things which taste better, even considering war-time conditions in Germany. However, I am still alive, so it is certain: *Oenothera* is not poisonous.

3.2 Flowers

A description of the morphology of the flowers is given by Eichler (1878) and Planchon (1896). The flowers are pentacyclic, tetrameric, eucyclic and zygomorphic (five circles of flower organs, each containing four equal members). In the flower bud the sepals are valvate, not overlapping, but soon after initiation they are connected in the very young bud at the margins by protuberances of the epidermis. Only the tips of the sepals are free. More details on the histology, especially of sepals and petals, are described by Kowalewicz (1956a).

The petals in the flower buds are flat and convolute (type *Oe. Hookeri* de Vries and *Oe. biennis* L.) or crooked and corrugate (type *Oe. Lamarckiana* de Vries). The anthers are diplostemonous, versatile with four loculi and open at the sides with two longitudinal splits. The ovary is inferior, syncarpic, tetracarpic, tetrameric and has a central placenta with many anatropic ovules. These number several hundred but, in some species, can reach more than 1000. Many of them degenerate before fertilization. Up to 500 seeds per fruit can be found. The fruit is a capsule which opens at maturity, starting from the tip, by four longitudinal splits, one in the middle of each carpel. At the same time the partitions between the four compartments are separated from the central placenta which remains free in the middle of the mature, open fruit.

The characteristic morphological feature of the Myrtales, the hypanthium, is seen in all *Oenothera* species. It is a continuation of the flower axis over the ovary, forming a tube. At the upper margin the sepals, petals and anthers are inserted. In the middle the free style, with four lobes at the top, is found. The style ends at the level of the anthers, allowing self-pollination in the mature bud, or is longer, placing the stigmas above the level of the anthers and requiring pollination by insects or artificially by the experimenter.

In the four loculi of each anther many pollen mother cells (PMC, or microspore mother cells) are found which, after meiosis, give rise to four microspores, each developing to form a three-lobed pollen grain. In the ovules one or, rarely, two megaspore mother cells (MMC) are present, giving rise to a linear tetrad each after meiosis. In most cases the spore at the micropylar end of the tetrad undergoes two haploid mitoses and develops into a monosporic embryo sac with four nuclei, containing the egg cell, two synergids and one pole cell. A second pole nucleus and the antipodal cells are missing.

The development of the spore mother cells, meiosis, and the haploid generation including embryo sac mother cells, embryo sac and pollen, are of great importance for understanding the genetics of *Oenothera* and will therefore be discussed later in more detail.

4 Genetics of *Oenothera*

4.1 Investigations Using the Methods of Classical Genetics in *Oenothera*

4.1.1 Genetic Analysis of *Oenothera*

Soon after *Oenothera* was introduced into genetics, it was discovered that there were some peculiarities in its genetic behaviour. The laws of Mendelian inheritance, which were within a few years of their rediscovery known to describe the genetic behaviour of hybrid progenies in many cases in plants and animals, seemed not to be applicable to some results observed in *Oenothera*.

The Mendelian rules of heredity are concerned with the behaviour of hybrids and hybrid progenies from controlled crossings. The progenies consist of the F_1-hybrid, the F_2-generation obtained by self-fertilization (in higher plants self-pollination) or by fertilization between individuals of an F_1-family, eventually subsequent generations obtained in the same way, and backcrosses of the F_1-hybrids with the parents or parental types, in case the parents did not survive until the F_1-hybrids had reached maturity.

Soon after the first encouraging experiences of de Vries with the gene brevistylis, more genetic experiments were carried out by Renner, Shull and Davis, as mentioned in Section 2.2 on the history of *Oenothera* research. Mendelian segregation of some characters was discovered, but it seemed as though these were rather rare exceptions. To demonstrate the problems, the observations on F_1-hybrids must first be considered. According to the first two rules of Mendelian genetics, uniformity of the phenotype of all individuals in a family of F_1-hybrids and similarity of the hybrids from reciprocal crosses is expected, irrespective of which of the two parental types provided the male and the female gametes for the cross. However, in the progeny of most species crosses in *Oenothera*, the rules of uniformity and reciprocal identity of the hybrids seem to be not valid.

The result of a certain cross A × B can be a progeny of similar plants, corresponding to the expected uniformity of the F_1-generation, but also two, three or even four different types can appear. The F_1-hybrids from crosses of the type A♀ × B♂ should be identical with those from the reciprocal cross B♀ × A♂, but in *Oenothera* this does not hold in every case. Irrespective of the number of different hybrids resulting from a cross, and even when there was homogeneity in an F_1-family, the reciprocal cross does not necessarily produce the same types. In both F_1-progenies, resulting from the reciprocal crosses, together never more than four different hybrid types are observed. Furthermore, even if the same types are found, their numerical relations are not necessarily the same in reciprocal crosses, and in any case, they do not fit into any formal description.

However, the number of types found in the progeny of each species does not vary at random. After crossing of a species which gives rise to multiple hybrid types, with several other species, always two types of hybrids from one parent can be recognized and identified on the basis of morphological characters. This phenomenon was described as the occurrence of twin hybrids (de Vries 1907; Andrews 1910; Renner 1917).

When only one hybrid type arises from a cross between two species, then in the hybrid family resulting from the reciprocal cross, either the same type, or one or two different types, or the same type with one, two or three additional new phenotypes can be observed. When two types originate from the first cross, then the reciprocal cross can yield one, two, three or four different types of F_1-hybrids.

At first sight the whole picture is very confusing, but when enough different species crosses are considered, the species of the sect. *Euoenothera* can be allocated to one of only four types of reaction. A similar grouping can be found in the sect. *Raimannia*.

Case 1, type Oe. Hookeri de Vries

In reciprocal crosses with other species of this type, the F_1 behaves according to Mendelian expectations with uniformity of the F_1 and reciprocal similarity of the hybrids. When crossed with any other species, either as male or female parent, never more than two types of F_1-hybrids are observed. When a given species is crossed with *Oe. Hookeri* as the female parent and two types of hybrids appear, then the reciprocal cross produces either one or both of the hybrid types, but never new types. In Europe there are no species behaving like this. The results prove that the functional gametes are of only one type which is the same in both sexes.

Case 2, type Oe. biennis L.

When used as the female parent, two or four hybrid types are observed. When used as the male parent never more than two types are generated and often only one type is observed which is always identical with one of those from the egg cell progeny.

Case 3, type Oe. syrticola Bart. (*syn. Oe. muricata* Aut.)

When used as male or female parent, they normally give rise to only one type of hybrid, the F_1-family being homogeneous, but the type is never the same as in the reciprocal cross. However, the hybrid type which is found in the pollen progeny can be seen in the egg cell progeny at very low frequency, usually less than 1%, and rarely up to ca. 5%. Nearly all European and most of the North American species belong to this group.

Case 4, type Oe. Lamarckiana de Vries
In the F_1-generation of crosses with species of this group, irrespective of the sex of the gametes provided by them, never fewer than two or more than four types appear. When crossed with a species corresponding to case 1, then only two types occur, and these are found in the reciprocal cross as well. With a partner corresponding to case 2 as the pollen parent, two different hybrids are observed, but with the same species as the egg cell partner, four different hybrids are found. When the second parent is a species corresponding to case 3, then in each of the two reciprocal hybrid progenies two types are found, but different ones in each of the two families. Summarizing the reactions between species belonging to these four cases we find:

Number of different hybrid types in species crosses

		Male parent Case 1	Case 2	Case 3	Case 4
	Case 1	1	1	1+	2
	Case 2	2	2	1+	4
Female parent	Case 3	1+	1+	1	2
	Case 4	2	2	2	4

+ Hybrid type not observed from reciprocal cross.

The first conclusion from these observations is that in the *Oenothera* species belonging to the cases 2–4, the gametes cannot all have the same genetic constitution, as is expected in homozygous species. Either there are genetic differences between the gametes in the pollen grains and embryo sacs, or in one or even both sexes gametes of two different genotypes occur. In some cases, both types of gametes can be active in both sexes; in other cases one type is inactive in one sex, or each type is active only in one sex and inactive in the other, and is transmitted to the next generation either by the male gametes through the pollen or by the egg cells through the embryo sacs, but never by both ways (Fig. 3).

4.1.2 Introduction of a Special Terminology

A terminology adapted to this situation will help to describe the facts. When all functional gametes of one sex are genetically identical, they are called monotypic. When differences are observed, the gametes are called ditypic.

Fig. 3. Twin hybrids in *Oenothera*. Young inflorescences of two plants from the F_1 of the cross *Oenothera suaveolens* Desf. × *Oe. Hookeri* de Vries, written as Renner complexes this cross is *Oe. albicans·flavens* × *Oe.* ^{h}*hookeri*. The genotype of the F_1-plants is **a** *flava, flavens·*^{h}*hookeri*, characterized by a pointed apical bud; **b** *albata, albicans·*^{h}*hookeri*, characterized by a flat top of the inflorescence

When a species gives the same types in the progeny, irrespective of whether it is used as male or female parent, the species is called isogametic. When, however, the genotypes transferred by the male and female gametes are different, the species is called heterogametic. If, on the other hand, one sex is monotypic and the other ditypic, the constitution is designated as semiheterogametic. This condition is presently recorded only as female ditypic + male monotypic. The other combination, female monotypic + male ditypic, should be possible, but has not yet been observed. When this nomenclature is applied to some of the *Oenothera* species frequently used in experiments, their gametic behaviour can be described as follows:

Oe. Hookeri de Vries is isogametic – monotypic (genotype of pollen and eggs identical),

Oe. syrticola Bart. (*muricata* Aut.) and *Oe. strigosa* de Vries are heterogametic while both sexes are monotypic (pollen and egg cells genetically different, but in each sex all the functional gametes are of the same genotype),

Oe. Lamarckiana de Vries is isogametic – ditypic (the same two genotypes of gametes in both sexes),

Oe. biennis L. and *Oe. suaveolens* Desf. are semi-heterogametic, female ditypic and male monotypic (egg cells of two genotypes, only one of these genotypes in the pollen).

In the case of ditypic egg cells, both types can be found, either with nearly equal frequency and with variability only as random deviation from a 1:1 distribution, or, alternatively, one type appears statistically significant more often in the progeny than the other. It is suspected that strictly heterogametic species, in which the genotype transmitted by the pollen is never found in an egg cell, do not exist. For practical purposes, the term heterogametic is therefore used to describe those cases in which the rare type is transmitted in less then 10% of the egg cells. Using this terminology, the behaviour of *Oenothera* species in crosses can be properly described.

Investigations on the phenotype of the F_1-generation of species hybrids lead to the conclusion that many species are in fact heterozygotes. During the formation of the gametes, a segregation of two types of gametes is observed (Renner 1917). Such experiments, which were performed in order to evaluate the genetic differences between gametes, were called by de Vries (1913) "gamolysis". In the following years, however, this term was used by only a few investigators, the last being Gates and Catcheside (1932), and it then disappeared completely.

According to the behaviour of the gametes, the species must be heterozygotic. In contrast to this statement, species are constant when propagated by self-pollination. The segregation of different phenotypes, as is expected following self-pollination of a hybrid, is not observed. The species can therefore be designated as "constant hybrids" (Renner 1917).

Observing the different types which appear from crosses with a certain species it is found that the variability is not at random but that groups of characters appear together, irrespective of the other parent of the hybrid. This is demonstrated by an example with *Oe. Lamarckiana* de Vries. This species is isogametic ditypic. Functional gametes of two different genotypes are found in both sexes. In other words, after crossing with another species, two types of hybrids are produced from pollen as well as from egg cells.

Type 1: leaves of the rosette and on the lower part of the shoot are broad with an irregular surface of the lamina in which the growth of the parenchyma exceeds that of the vascular bundles; bracts short, in the top of the inflorescence of the same length as the flower buds; short internodes in the upper part of the inflorescence, which causes a crowded appearance of buds and bracts at the top of the shoot; flower buds short and thick; sepals with short tips; flowers medium-sized; petals in the mature buds corrugate; relatively small amount of anthocyanin in the epidermis of shoots (red spots at the hair bases); sepals and fruits green with interrupted narrow red stripes.

Type 2: leaves of the rosette and the shoot are long, narrow and flat; bracts in the top of the inflorescence considerably longer than the young flower buds; internodes, especially those in the upper part of the inflorescence, are more elongated than in type 1, leaving the inflorescence less crowded; flower buds long and slim; sepals with long free tips; flowers large; petals in mature buds flat and convolute; dark anthocyanin in the epidermis of the shoot; sepals and fruits with broad red stripes.

From crosses between any species and *Oe. biennis* L. also two types emerge. One hybrid type is nearly identical with type 1 of *Oe. Lamarckiana* de Vries and is found in progenies of both pollen and ovules. The other type is different from the type 2 in the *Oe. Lamarckiana* progenies: dark green, narrow leaves with white veins, shorter than in type 2 of *Oe. Lamarckiana* de Vries; whole plant covered with dense, short hairs, especially at the short and thick flower buds; petals small, ca. 2 cm long; a few side branches develop from buds in the axils of rosette leaves, only short ones develop from the stem; main stem very high and considerably longer than the side branches; bracts longer than the flower buds, curved outwards, giving the top of the inflorescence a flattened appearance. This type is only observed in the female progeny.

When *Oe. suaveolens* Desf. is crossed with any other species, two types arise from the egg cells. One is identical with the last one described from *Oe. biennis* L., the other is different from all the types described above: stem slightly shorter than in the other type, leaves broader, flat and light yellowish-green; in the leaves at the shoot the broadest part is at approximately one-third of the length measured from the base; bracts longer than the flower buds, erect, giving the top of the inflorescence a slender and pointed apical bud; flower buds slender; sepals with long tips; flowers larger than in the other type; petals in the buds convolute. The most striking character is the distribution of anthocyanin, which is found only in the epidermis of the stem, sepals and ovaries as small dots at the base of the hairs.

The types found in progenies from crosses of other species can be described in a similar way. Soon the scientists working with *Oenothera* agreed that differences in the genetic constitution between the species must be responsible for the characters observed in the hybrids. The differences between the investigators are to be found in the interpretation which is given to the observations.

Some research workers tried to fit *Oenothera* into the meanwhile established scheme of Mendelian genetics by the identification of single genes and their linkage relations (e.g. Shull 1921, 1925, 1926, 1927; Davis 1926b; Brittingham 1931b; Emerson and Sturtevant 1932).

4.1.3 The Renner Complexes

A different approach to the problem was provided by Renner. He did not stress single genes, although he did describe some, but he emphasized the fact that groups of characters, which he called "complexes", did appear together in the progeny of a cross between "species". He accepted this fact and concluded that each species contains the genetic equivalent of two such complexes. Soon he used the same term not only for the combination of certain characters but also for the genetic basis of this phenomenon. Nowadays they are referred to as "Renner complexes" in honour of the scientist who introduced this concept.

The Renner complexes are identified by the characters observed in the hybrids and can be treated as genetic unities. Furthermore, they can be given names to simplify the description. Transferred by the egg cells only, they are called "egg cell complex" and when transferred mainly by the pollen (rare exceptions possible) they are accordingly termed "pollen complex".

In agreement with the rules of botanical nomenclature and making good use of his profound classical education, Renner chose Latin names, based on characteristic features of the hybrids, to identify the types observed in the progenies and the "participium praesentiis activis" of the appropriate verb with the ending -ens or -ans as the name for the genetic complexes. Lists of species and their Renner complexes are provided by Renner (1917, 1942a). This name-giving facilitates a formal description of experimental results. It has to be kept in mind, however, that it does not include or suggest any hypothesis about the genetic basis of the complexes of characters behaving as a hereditary unit. For some of the species mentioned as examples, the names of types and complexes are:

Species	Phenotypes		Renner complex	
	Type 1	Type 2	of type 1	of type 2
Oe. Lamarckiana de Vries	*laeta*	*velutina*	*gaudens*	*velans*
Oe. biennis L.	*rubata*	*albata*	*rubens*	*albicans*
Oe. suaveolens Desf.	*albata*	*flava*	*albicans*	*flavens*
Oe. strigosa de Vries	*depressa*	*stricta*	*deprimens*	*stringens*
Oe. muricata Aut.			*rigens*	*curvans*

To understand this system you really need enough Latin to know that *gaudere* and *laetare* are synonyms.

Some special consideration is necessary for those species which are isogametic monotypic, with gametes of only one genotype, and the same in both sexes. These are considered to have two identical Renner complexes, named as *haplo-* plus the name of the species, e.g. *haplo-hookeri*, written as ^{h}*hookeri* (Renner 1919a). At first sight this seems superfluous, but it is necessary when the genetics of the whole genus is considered.

When more species, especially those from North America, were included in the analysis, the rules of nomenclature introduced by Renner soon became too complicated. Instead of individual Latin names, Cleland named the complexes *alpha*- and *beta*- plus the species name. This system proved to be clear, simple, effective, unlimited for introduction of new species and avoided the search in dictionaries for words that could be used as new names by those scientists who did not know Latin.

In the sect. *Raimannia*, the system of Renner complexes could be applied to describe the observations on hybrids and hybrid progenies. But to identify the Renner complexes and describe the genetic constitution of the species, confusing combinations of small and capital letters were introduced by Schwemmle (1938) without presenting an explanation of the system. For the Renner complexes of the homozygotic species in the sect. *Raimannia* a different nomenclature was chosen, with *h* as the abbreviation for *haplo*- and addition of the first letter of the species name, e.g. *ha* for *haplo-argentinea*.

As soon as the description of the gene complexes by this nomenclature was introduced by Renner in his analysis of F_1-hybrids of *Oenothera* (1917), it became possible to describe the genetic constitution of interspecific hybrids in the same terminology as is used for the species. Some examples are:

Parents Female	Male	Hybrids	Renner complexes
Oe. Lamarckiana	× *Hookeri*	{ *hookeri-laeta*	*gaudens*·^{h}*hookeri*
		{ *hookeri-velutina*	*velans*·^{h}*hookeri*
Oe. Hookeri	× *Lamarckiana*	{ *hookeri-laeta*	^{h}*hookeri*·*gaudens*
		{ *hookeri-velutina*	^{h}*hookeri*·*velans*
Oe. suaveolens	× *strigosa*	{ *albistricta*	*albicans*·*stringens*
		{ *flavistricta*	*flavens*·*stringens*
Oe. strigosa	× *suaveolens*	*flavidepressa*	*deprimens*·*flavens*

{ *connects "twin hybrids"*

Let us look at the results of the species hybridizations in detail. When considering the phenotypes of the hybrids, which lead to the concept of Renner complexes, it can be suspected that some species contain identical, or at least very similar, Renner complexes. As examples the European species *Oe. biennis* L., *suaveolens* Desf. and *Lamarckiana* de Vries, used in the classical *Oenothera* literature, are presented here. Others are described by Cleland (1972).

From both crosses a hybrid type originates that cannot be distinguished from the female parent. In the second case even both parents appear. Is this the consequence of an experimental error, or can another explanation be offered? A decision can be made when the hybridizations are repeated with genetic markers, in this case genes for flower colour. By using the variety of *Oe. Lamarckiana* de Vries with old-gold flowers, *vetaurea*, and of *Oe. biennis* L. with light sulphur-yellow flowers, *sulfurea*, experimental errors can be detected.

Parents	Expected hybrids	Characteristics
Oe. Lamarckiana × *biennis*	*velans·rubens*	Indistinguishable from *Oe. Lamarckiana*
	gaudens·rubens	Not observed
Oe. biennis × *Lamarckiana*	*albicans·gaudens*	Indistinguishable from *Oe. biennis*
	rubens·gaudens	Not observed
	albicans·velans	New type
	rubens·velans	Indistinguishable from *Oe. Lamarckiana*

The cross *Oe. Lamarckiana vetaurea* × *biennis* (irrespective of yellow or *sulfurea* flowers) gives typical *Oe. Lamarckiana* de Vries with yellow flowers, and in the progeny of the cross *Oe. biennis sulfurea* × *Oe. Lamarckiana* (irrespective of whether yellow or *vetaurea* flowers) only typical *Oe. biennis* L. with yellow flowers are seen. In both cases the types arising by hybridization are not identical with the female parent and cannot originate from uncontrolled self-pollination before the artificial crossing. The appearence of the type corresponding to the female parent cannot be the consequence of inaccurate work by the experimenter.

They are real species hybrids but look like one of their parents. This observation leads to the conclusion that the Renner complexes *gaudens* from *Oe. Lamarckiana* and *rubens* from *Oe. biennis* are very similar, if not identical. They had just been given names before this similarity was discovered.

In another case, the unexpected types cannot be distinguished from the male parent. In the progeny from the cross *Oe. biennis* L. × *suaveolens* Desf., in addition to a new hybrid type, as expected from hybridization, another type appears which looks like *Oe. suaveolens* Desf. The reciprocal cross gives rise to the same new type and one similar to *Oe. biennis* L. Written with Renner complexes, these crosses should produce the following types:

Parents	Expected hybrids	Description
Oe. biennis × *suaveolens*	*albicans·flavens*	Indistinguishable from *Oe. suaveolens*
	rubens·flavens	New type
Oe. suaveolens × *biennis*	*albicans·rubens*	Indistinguishable from *Oe. biennis*
	flavens·rubens	New type, the same as in the reciprocal cross

On the basis of these results, the egg-cell complexes from both species, *Oe. biennis* L. and *suaveolens* Desf., cannot be distinguished. In this case the identity was discovered in the course of the genetic analysis, so the names given to the two Renner complexes had to be the same.

4.1.4 How To Be a Constant Hybrid

The investigations on the phenotypes of the F_1-generation of species hybrids led to the conclusion that many species are heterozygotes. In the process of formation of the gametes a segregation of two types of gametes is observed (de Vries 1907, 1913; Renner 1917). However, following self-pollination, the segregation of different phenotypes, which would be expected in a hybrid progeny, is missing. The species are constant hybrids (Renner 1917).

This contradiction between the observation of undoubtable heterozygosity, segregation of the gametes and constancy of the species has to be solved. The direction in which the solution must be found is clear: there must exist very efficient mechanisms which prevent segregation in the progeny and maintain heterozygosity. An additional question is whether these mechanisms will still work after hybridization.

It can be expected that the genetic basis of such mechanisms is different, depending on the gametic structure of the species, To achieve the goal of constant heterozygosity, two possibilities have to be considered: either the homozygotes are eliminated after fertilization or their formation is prevented. The tests for investigating these possibilities are independent of whether the heterozygosity depends on one gene locus or on Renner complexes.

In the ditypic-isogametic species homozygotes should be possible. If fertilization occurs at random, i.e. the combination of the gametes is independent of their genotypes, then the meeting of two types of gametes should give rise to a segregating progeny with two types of homozygotes in addition to the heterozygotes produced in Mendelian proportions. When the raised progeny is homogeneous, then the possibility that the expected homozygotes escape observation because they are eliminated before the progeny is investigated, has to be tested. To discuss the analysis of the questions arising here, an interesting example, analyzed by Renner, will be presented.

Oe. Lamarckiana de Vries is a constant hybrid but gives rise, as mentioned already, to the twin hybrids *laeta* and *velutina* when it is used as male or female parent in a hybridization. The result, as expected from self-pollination when both types of gametes are active and fertile, is described in the following familiar scheme of Mendelian inheritance:

		Male gametes	
		gaudens	*velans*
Female gametes	*gaudens*	*gaudens·gaudens*	*velans·gaudens*
	velans	*gaudens·velans*	*velans·velans*

The expected frequency for each of the four cases is 25%. The first task is to look for the missing homozygotes, together 50% of the zygotes. If good care is taken the death of seedlings is very rare. In *Oe. Lamarckiana* de Vries only ca. 50% of the seeds germinate. The remaining 50% is a proportion that equals that of the missing cases. Inspection of the non-germinating seeds reveals that they are neither empty nor contain dead embryos but that they are filled with a tissue of undifferentiated cells with delicate thin walls, eventually with a very small globular structure that could be the remnants of a degenerated embryo. The proportion of the two types is approximately 1:1. Comparisons with the progenies from hybrids which contained only one of the two Renner complexes of *Oe. Lamarckiana* de Vries in combination with a pollen-inactive Renner complex (e.g. *albicans·gaudens* and *albicans·velans*) revealed that these produced only one of the two types of non-germinating seeds, either with or without a degenerated embryo. This is proof that these represent the missing homozygotic genotypes (Renner 1916, 1918). The homozygotes die at early stages of embryogenesis. The interpretation given to this result is that the Renner complexes *velans* and *gaudens* act as recessive lethals. The lethal action of the two Renner complexes involved cannot be identical, because the heterozygote is viable. The two Renner complexes *gaudens* and *velans* together form a system of balanced lethals that maintains heterozygosity by eliminating the lethal homozygotic combinations. This system is characteristic of the ditypic-isogametic species in *Oenothera* (Shull 1923).

In the semi-heterogametic *Oe. biennis* L. non-germinating seeds are also present and the expected homozygotes *rubens·rubens* are missing after self-pollination. When crossed with pollen of *Oe. Lamarckiana* de Vries, in addition to viable seeds also non-germinating seeds of the same type as *gaudens·gaudens* are found. The types observed in the germinating seedlings are *Oe. albicans·gaudens*, *albicans·velans* and *rubens·velans*, as described in the preceding section. This result is in agreement with the assumption that the lethal action of the Renner complexes *gaudens* and *rubens* is the same and that the combination of the Renner complexes *Oe. gaudens·rubens* is eliminated in the early embryonic stage, an additional proof of their identity (Renner 1918).

In *Oe. Lamarckiana* de Vries and in other cases of the same type, the population of constant hybrids is maintained by sacrificing half the progeny in every generation. This method is effective but not very efficient. This does not matter in a species that produces such an abundance of seeds, but the question arises whether other species have devised other ways to avoid the trap of producing homozygotes that have to be discarded afterwards.

The possibilities to be discussed are based on the assumption that the unwanted gametes have to be eliminated by death or by inhibition of their proper function or even that they do not exist at all. Continuing research with the genetic method can give more information on the facts of missing genotypes but cannot provide a causal analysis of the phenomena. Anyhow, we first need to know the facts in order to ask the proper questions for the next steps.

In higher plants the gametes are not generated on the sporophyte but on an independent generation, the gametophytes, which exist in two sexually differentiated forms: the embryo sac with the egg cell, and the pollen grain–pollen tube with the generative male cells. When we look for methods of elimination of certain genotypes in the gametes, then in any case developmental processes in the haploid phase of the life cycle must be considered. That means a change in the methods of investigation from genetics to developmental biology and looking for events between meiosis and fertilization which could be related to the problems. The results of these experiments will be presented later, after other questions which can be answered by applying the genetic method have been discussed first.

4.1.5 Renner Complexes and Genes in *Oenothera*

The investigations on the gametic constitution of *Oenothera* species by means of observations on the F_1-generation of species hybrids have revealed several peculiarities of this genus. Two main rules of Mendelian genetics, uniformity and reciprocal identity of hybrids, are not confirmed in *Oenothera* species hybrids. The gametes are not all genetically identical. Considerable differences between egg cells and pollen, and even between gametes of one sex, are observed. The important conclusion is that the species are constant hybrids.

In their progeny, however, another rule of Mendelian genetics does not hold. There is no free recombination of characters in the progeny, as is expected when the alleles of many genes segregate. Instead, groups of characters stay together and their genetic equivalent is inherited in the species and passed to the F_1-hybrids as a unit. During development of the gametes, the two Renner complexes are separated and each gamete contains only one. In this respect the Renner complexes behave as genes. But what is their real nature? The observations of constant heterozygosity of the species and the complexes of characters appearing in their progeny can be explained by several hypotheses.

There can exist a pleiotropic gene which influences all the characters through branching of developmental pathways following the translation of the gene. This gene can exist in may alleles. Each species would be a combination of two of these alleles. Comparing the combination of characters in a species with those in its twin hybrids, it is evident that the dominance relations of this hypothetical giant gene must be complicated. The hybrids of the type *stricta* from *Oe. strigosa* de Vries resemble in many characters the male parent, but the twin hybrid with the Renner complex *deprimens*, originating from egg cells of the same species, exhibits characters that cannot be recognized in the parent. The Renner complex *stringens* seems to be dominant over *deprimens* for all characters. On the other hand, the twin hybrids *laeta* and *velutina* from *Oe. Lamarckiana* de Vries each have some characters which are seen in the parent and some others that are new. Taking many pairs of twin hybrids, the dominance relations differ for each character involved. Looking at the consequences of this hypothesis, it turns out that there is no such case known from other species, but this cannot be a reason for rejecting the possibility without proof. Pleiotropic genes exist, but in the cases that have been analyzed in detail in insects and vertebrates, there is only one primary expression of the gene and the different phenotypic characters are connected by developmental pathways where one wrong step has consequences for the whole organism (Hadorn 1955). Such connections cannot even be suspected between the characters influenced by the Renner complexes of *Oenothera*.

A second hypothesis is that there exists a group of genes with absolute linkage of many loci. Each species would contain two such gene complexes as the genetic equivalent of the complexes of characters. Each gene complex would contain a special combination of alleles of the individual genes. Although this assumption avoids some of the difficulties arising from the first hypothesis, especially the dominance relations for different characters, there are other objections against it. No other case of absolute linkage of so many genes as must be assumed for the number of characters involved, is known from investigations of other organisms. In this case too this cannot be a reason for avoiding clarification of the problem by experiment.

Neither of the two hypotheses which can be proposed as an explanation of the Renner complexes can be accepted without additional assumptions. The decision must be made between the two hypotheses "pleiotropic gene" and "complex of genes" as the genetic equivalent of the Renner complexes, the definition of which is primarily based on morphological observations. The investigation of more F_1-hybrids can only confirm the facts but cannot provide new ideas for their interpretation.

However, as only one of the two possibilities can be realized, one of them has to be rejected. The first hypothesis is falsified when it is demonstrated in at least one case that the Renner complex can break down and set free at least one character for which the genetic basis is a single gene with Mendelian inheritance.

The experimental way to solve this problem is to extend the experiments to the next generation, obtained by self-pollination of the species hybrids, and observe whether or not the Renner complexes are maintained.

The hybrids are made up of two Renner complexes and are therefore heterozygotes, irrespective of the hypotheses which are to be tested. A "genetically normal" hybrid is expected to give rise to a segregating progeny after self-pollination, with Mendelian segregation for each individual locus and either free recombination or partial linkage between loci. The expectation for the F_2 of *Oenothera* hybrids is different, depending on which of the two hypotheses mentioned above is being considered. When the combination of two Renner complexes, which originated as a species hybrid, behaves like the hybrid species, then the gametes of the hybrid should be either monotypic or ditypic in each sex, and the hybrid should be isogametic, heterogametic or semi-heterogametic, just as a species. Furthermore, it can be expected that the Renner complexes in the hybrid will behave as an egg-cell complex or as a pollen complex according to their origin.

If the genetic complexes are alleles of one locus, then the hybrid should behave as a species when used as the parent in another hybridization. The progeny of crossing the hybrid with a third species should consist of the twin hybrids corresponding to its two Renner complexes. A recombination of the characters cannot occur on the basis of this hypothesis. An implication is that the mechanisms maintaining the heterozygosity of the species are still active after hybridization. If this is not the case, the progeny of a hybrid should segregate and exhibit three types, the two homozygotes and the hybrid. If, however, a segregation of a single character is observed in the F_2-progeny, then the first hypothesis of the pleiotropic gene as the genetic equivalent of the morphological complex of characters is falsified.

Before discussing this further, the consequences of the second hypothesis, of a gene complex of many loci with absolute linkage, must be considered. When the hybridization does not influence the behaviour of such a gene complex, then the genetic consequences for the possibilities of the F_2 are the same as in the first case. It is impossible to decide between the two. However, when the mechanisms for the maintenance of the gene complexes do not work in a new hybrid in every case, then the genetic make-up of the F_2-generation is expected to be different. In a complex of linked genes the possibility of crossing-over should be considered as well as a complete breakdown of linkage with the consequence of free recombination.

When, in the progeny obtained by self-pollination of a species hybrid, new phenotypes are observed, which do not correspond to the combinations of characters attributed to the two Renner complexes combined in the hybrid, then the conclusion must be that a mechanism exists in the species which keeps the genes of the Renner complexes together. It must also be concluded that such a mechanism does not exist or is not active during the developmental processes leading to the formation of the gametes in those hybrids in which a recombination of the genetic material is observed.

When this happens, the existence of individual gene loci in the gene complex is obvious and the second hypothesis has to be accepted on the basis of positive evidence and not merely as the alternative of a falsified assumption. The genetic equivalent of a Renner complex will be a complex of gene loci with absolute linkage. In this case, the absence of recombination and crossing-over in the species remains to be explained, but that is a different problem.

The design of experiment to reach a decision between the two hypotheses is now clear. The hybrids have to be self-pollinated. Backcrosses with the two parents in both directions and test-crosses with other species, which are suitable for recognizing the gametic constitution of the hybrid under consideration, have to be carried out. Whenever possible, the progenies have to be raised under the same environmental conditions (same year and same laboratory). As an additional test, more generations obtained by self-pollination and also by repeated backcrossing can be added.

Summary of the design for experiments for determining the nature of the genetic equivalent of the morphological Renner complexes.

Observations on hybrid progenies	1st hypothesis pleiotropic gene	2nd hypothesis gene complex
No new combinations of characters	Decision not possible	
New combination of characters of the parental Renner complexes observed	Falsified	Verified

The progenies of many F_1-species hybrids were investigated by several scientists, but the decision was reached by Renner (1925) and was later confirmed by work on other hybrids (Renner 1933a, 1942a, and others). In advance: the results depend on the hybrid. The problem became clarified in one direction and new problems arose at the same time. Did you expect anything else from *Oenothera*?

Following self-pollination of some species hybrids, there are no signs of recombination of characters that would allow the observer to assume the occurrence of any kind of segregation, either by free recombination or by crossing-over, of individual genes in the progeny. The hybrid is constant, just like a species. When such a hybrid is used as the parent in other crosses, the Renner complexes appear unchanged in the plants of the next generation. It must be concluded that the genetic equivalent of the Renner complexes is not changed in these hybrids but is maintained and transferred to the next generation in the original form, following the same rules. These cases are interesting, but do not give any information about the decision to be made.

Examples of this behaviour include the following hybrids:

Oe. Lamarckiana × *biennis* hybrid *velans·rubens*
Oe. biennis × *muricata* hybrid *albicans·curvans*
Oe. biennis × *suaveolens* hybrid *albicans·flavens*
Oe. strigosa × *suaveolens* hybrid *deprimens·flavens*

Some interesting cases of breakdown of the Renner complexes were described by Renner (1918) soon after he had introduced the concept of complexes. As an example the hybrid *Oe.* (*muricata* × *Lamarckiana*) *rigens·velans* is presented here. Both Renner complexes break down, but only to make two new Renner complexes which seem to be combinations of characters of the parental Renner complexes and which were immediately given new names: *fere-velans* and *fere-rigens*. The two original Renner complexes and the two new ones are transmitted through the egg cells, resulting in quadri-typic egg cells, while only *velans* and *fere-velans* are found in the pollen progeny. In hybrids including *velans* from *Oe. Lamarckiana* de Vries this breakdown with formation of new complexes was particularly clear, producing *subvelans*, *paene-velans*, *sub-curvans* and so on. Soon the creation of new names in this way became very complicated and was later abandoned when the phenomenon was understood.

In the progeny of other hybrids a variability of phenotypes is observed and can be explained as genetic segregation and recombination. However, there are only a few cases in which the familiar numerical ratios between the different phenotypes of one character, that could be interpreted as Mendelian ratios, can be found. Often, phenotypes are observed which do not resemble one of the parents and which also were not seen in the F_1-plants. Many such cases were found by Renner (Oehlkers 1922, 1933; Renner 1925, 1941, 1942a; Hoeppener and Renner 1929; Harte 1942, 1948).

As an example the hybrid *Oe.* (*suaveolens sulfurea* × *strigosa*) *flavens·stringens* will be discussed. In the F_2-generation resulting from self-pollination of the F_1-plants, segregation can be observed for many characters: flower colour yellow vs. sulfurea, male fertile vs. male sterile (empty anthers without pollen), anthocyanin in epidermal cells at the base of hairs vs. no anthocyanin, straight stem vs. malformations, normal apex vs. degenerating apex, bracts straight and upright (forming an apical bud) vs. bracts bending away from the stem (forming a flat apex), main stem long vs. main stem only slightly taller than the branches ("Zwerge", as they were called by the German authors, are not identical with the "dwarfs" from other segregating progenies), normal growth and leaves vs. very small plant "Knirps" (little boy) with small leaves and flowers and with many parenchymal cells containing calcium oxalate as needle-like crystals (Oehlkers 1922; Harte 1942, 1948). In this case, Mendelian segregation ratios are found, and self-pollination of the different types confirmed the monogenic basis of the observed segregations. For other, mainly quantitative characters, like smaller differences in growth of the main stem, size and form of the leaves, size of the petals, quantitative differences in the distribution of anthocyanin in the epidermis of stem, sepals and ovaries and colour of the leaves, the number of different phenotypes indicates a polygenic inheritance.

In an F_2-family only a few plants resemble the F_1. All others exhibit different combinations of the above-mentioned characters and, in addition, show a continous variation over a great range of size in quantitative characters. Backcrosses with the parental species as the pollen parent (bringing the Renner complex *flavens* from *Oe. suaveolens* or *stringens* from *Oe. strigosa*) revealed that the alleles for yellow flowers, male sterility,

"Knirps", malformations of the main stem, anthocyanin in the epidermis of sepals and ovaries, red spots at the hair bases on the stem, narrow leaves and a broad apex come from *stringens*, and the alleles for the alternative characters sulphur-coloured flowers, fertile pollen, normal growth (not Knirps), straight stem, lack of anthocyanin, broad leaves, pointed apical bud and degeneration of the apex are contributed by *flavens*. The frequencies of the new recombinations of the characters show clearly that the two Renner complexes break down in the F_2 into five linkage groups, each containing several loci (Harte 1948).

Similar observations on a partial or complete breakdown of the Renner complexes have been made on many hybrid progenies in *Oenothera* by Renner and his students and by others. In the course of the processes leading to the formation of gametes, the Renner complexes in hybrids can break down into parts or into many individually segregating loci. In the hybrid progenies these genes behave in just the same way as is known from Mendelian genetics for other species (Renner 1925, 1928, 1942a, and several papers of his students; Oehlkers 1922, 1933; Harte 1948). It is no longer possible to recognize the two Renner complexes in the plants of the F_2-family. The number of linkage groups which are set free after hybridization does not depend on the individual Renner complex but on the combination of complexes in the hybrid. In any case, it is certain that the number of linkage groups to be found in the hybrid progeny varies between one and seven.

The behaviour of some other hybrids can be described as a mixture of both cases. Some of the characters, which are characteristic of a Renner complex, stay together in the progeny of a hybrid, and others are separated for free recombination. The original Renner complex breaks down into a remaining basal complex and free loci. When several hybrids of a particular Renner complex are compared, the basal complexes are not identical but contain different loci.

When we combine the results of crossing experiments with different hybrids of the same Renner complex and compare them with the expectation on the basis of Mendelian inheritance, two main differences become apparent:

1. A recombination of alleles by free segregation or crossing-over can occur but seems rather to be the exception. Instead, constancy or at least partial constancy of the hybrid is observed.
2. When segregation and recombination of loci do occur, then the linkage relations of the loci are not constant but differ between hybrids of a given Renner complex.

By the results obtained from F_2-hybrid progenies, the hypothesis of a pleiotropic gene as the genetic equivalent of the Renner complexes is falsified and the alternative hypothesis has to be accepted.

The Renner complexes consist of genes for individual characters on many loci, with absolute linkage in the species but behaving according to the rules of Mendelian genetics in appropriate hybrids. All Renner complexes from heterozygotic species contain recessive alleles that can only be recognized in segregating hybrid progenies and which have no influence on the phenotype of the species or the F_1-hybrids.

Now there are too new problems: Why does this breakdown of the linkage not occur in all hybrids? And: What is the mechanics of linkage without crossing-over in the constant hybrid species? Both questions are concerned with genetic linkage and cannot be answered by classical genetic methods. A shift to karyological investigations is necessary to look for the chromosomes as the structures bearing the genes.

4.1.6 Gene Loci and Linkage in *Oenothera*

The problems that remain to be solved are the absolute linkage of the loci to the Renner complex in the species themselves and the behaviour of some other hybrids.

The hybrid *Oe. suaveolens* × *strigosa*, *Oe. albicans·stringens*, behaves in a different way from its twin hybrid *Oe. flavens·stringens*, as has already been discussed. In test crosses the Renner complex *albicans* is transmitted only through the egg cells, and *stringens* is transmitted only through the pollen, as in the species which contributed them to the hybrid. The phenotype of the plants in the F_2-family, obtained by self-pollination of the hybrid, resembles their parent and most of them have yellow flowers and fertile anthers. A segregation which can be interpreted as recombination of alleles from the two complexes is only observed for two characters, flower colour $s/+^s$ and male sterility $fr/+^{fr}$. Only some plants have sulphur-coloured flowers and some others have male sterile anthers, like the segregants from the twin hybrid. Both characters appear independently of one another. These phenotypes occur in a frequency of not more than 5%, often considerably lower. This is in contrast to the behaviour in the progeny of the twin hybrid *Oe. flavens·stringens* where both characters segregate independently in Mendelian ratios with free recombination. This suggests that in *Oe. albicans·stringens* there is linkage of both loci to the Renner complexes and independent transfer of the *fr* allele from *stringens* to *albicans* and of *s* from *albicans* to *stringens* by crossing-over (Oehlkers 1922, 1924, 1926, 1933; Renner 1942a). In the hybrid progeny of *Oe.* ^{h}*hookeri·stringens*, however, both loci are closely linked and free from other segregating loci (Harte 1948). In *Oe. biennis* L. and *Oe. suaveolens* Desf. there is a very close but not absolute linkage of *s* to the Renner complex *albicans*. Crossing-over, with transfer of the allele *s* to the partner complexes *flavens* and *rubens*, is observed as a very rare event, with a frequency much lower than

1%. The interpretation of the occasional appearance of *sulfurea* plants as due to crossing-over was given by Oehlkers (1924).

This phenomenon of variable linkage relation was detected by Renner (1928, 1933a). More details were added by Oehlkers (1933) and it was summarized with many additional facts in an elaborate paper by Renner (1942a). He called it "Koppelungswechsel", changing linkage relations. Some alleles can be present in one of the Renner complexes in a species and in some hybrids without ever segregating, linked with crossing-over in another case, or exhibiting free recombination with the Renner complexes as demonstrated above for the gene *fr*. The relation of two loci to one another is variable too. Both can be linked to the Renner complexes, one can be linked and the other free, both can be free from the Renner complexes but linked together, or they can undergo completely free recombination, always depending on the combination of the two loci and the combination of Renner complexes in which the relation is tested.

The following table (after Oehlkers 1956, using data of Cleland and Renner) demonstrates some cases of changes in linkage relations of gene loci in hybrids. The bars connect the loci linked in the different hybrids.

Hybrid	Chromosome configuration in meiosis	Gene loci R	mP	B	Sp	Cu
flavens·curvans	6 6 2	?	—	—	—	—
flavens·percurvans	8 6	?	—	—	—	—
flavens·flectens	6 4 2 2	?	—	—	—	—
flavens·velans	4 4 2 2 2	—	—	—	—	?
rubens·flavens	12 2	—	—	—	—	?
rubens·curvans	14	—	—	—	—	—
curvans·velans	?	—	—	—	—	—
rubens·velans	12 2	—	—	—	—	?

Only a few loci give free recombination with the Renner complexes in all combinations tested. The most famous of them is *brevistylis*, $br/+^{br}$. The segregation for flower size in the progeny of otherwise constant hybrids was described by de Vries (1913) and Davis (1918, 1921), but the interpretation as due to crossing-over with high frequency of the gene for flower size *Co*/*co* between the Renner complexes of the hybrids was proposed by Renner (1917, 1942a). In extensive cultures of *Oe. Lamarckiana* de Vries some genes (*vetaurea*, *supplena*, *bullata*) were detected which were supposed to give free recombination in hybrids. This led to the construction of three linkage groups in *Oenothera* (Shull 1925, 1926, 1927, 1928a, 1930; Davis 1926b, 1937; Emerson and Sturtevant 1932), but this assumption was later discarded.

When the basic allele for anthocyanin synthesis, *P*, is present, several loci influence the pattern of anthocyanin distribution in the plant: broad, narrow or absent red stripes on sepals and/or ovaries; epidermis of the stem red or green with red dots at the base of hairs; red veins in the leaves; reddening of the first rosette leaves; red dots on rosette leaves. Some of these genes are closely linked to *P* and were first described as multiple alleles of *P* (Shull 1923; Emerson 1930, 1931a; Renner 1942a). The great variety of phenotypes appearing in F_2-families of segregating hybrids cannot occur by recombination of only two alleles (Shull 1934), but can be explained when, by crossing-over within a region of closely linked genes, new combinations of alleles of several loci are created (Harte 1948). Certain combinations of the phenotypes of these genes, describing the anthocyanin pattern characteristic of a genotype, were used in describing the "mutants" of *Oe. Lamarckiana* de Vries and are discussed in Section 4.4.2. Considering the great number of phenotypes in hybrid progenies it became clear that there cannot be only one gene locus with multiple alleles influencing both anthocyanin synthesis and pattern formation. There exists a group of closely linked genes which influence the pattern of anthocyanin distribution in the epidermis of the plant, and especially in the green parts of the flower, and which can generate new combinations by crossing-over.

Lethals, which are responsible for the elimination of homozygotes, can be treated as gene loci in absolute linkage with the Renner complex. Crossing-over of the lethals is not observed. The lethals of different complexes are not alleles. In the species with a balanced lethal situation, each of the Renner complexes contains the normal allele of the lethal of the other Renner complex. When in *Oe. Lamarckiana* de Vries the lethal alleles are designated $l(v)$ and $l(g)$, then the constitution of the two Renner complexes in relation to the two loci involved can be described as *velans* $l(v)\ +^{l(g)}$·*gaudens* $+^{l(v)}\ l(g)$. A change of linkage relation of a gene locus to free recombination as well as crossing-over to the Renner complex means a change of its linkage relation to the lethal which remains in the basal complex and keeps this in the state of constant heterozygosity. Rare cases of segregation of genes in the progeny of an otherwise constant hybrid are explained by the assumption of crossing-over between the gene in question and the site of the lethal in the giant linkage group called the Renner complex.

4.1.7 Variability of Crossing-Over Frequency in *Oenothera*

As soon as linkage and crossing-over had achieved their place in the genetics of *Oenothera*, the next problem arose. It was generally assumed that the crossing-over frequencies are constants. This assumption is the basis for establishing the order of genes in linkage groups. In *Oenothera* there were some difficulties in the experiments related to this problem, but mainly because the number of genes was small, the size of the families that could be raised in most laboratories was limited and there was a lack of appropriate multiple recessive homozygotes for backcrosses to estimate exact crossing-over values. In spite of this it became clear that in *Oenothera* not only the alternative "absolute linkage or crossing-over" needs to be discussed, but that in cases of undoubted linkage the frequency of crossing-over was variable beyond the range of random variability (Oehlkers 1933). A large series of experiments was described and discussed by Renner (1942a).

When in the species a mechanism is present which prevents crossing-over or keeps it at such a low level that the probability of observing it in a progeny of usual size is negligibly small, then it could be possible that this mechanism breaks down in a new hybrid. In this case an increase in the frequency of crossing-over should occur and a recombination of characters of the two Renner complexes in the hybrid should be possible. This is exactly what is observed, but, at the same time, it raises the question whether the restoration of the absolute linkage is possible too.

As an example, the behaviour of the locus *fr* will be presented. In *Oe. strigosa* de Vries the linkage between the Renner complex *stringens* and the locus *fr* is absolute. In the first generation of the progeny of the constant hybrid *Oe. albicans·stringens* pollen-sterile segregants were found whose appearance can only be explained by transfer of *fr* from *stringens* to *albicans* on the basis of crossing-over between the locus *fr* and the rest of the Renner complex. A few generations of self-pollination later, no more segregants were observed. Similar observations were made on several loci in other hybrids (Renner 1942a).

The interpretation given by Renner to the results of his experiments was that in new hybrids the absolute linkage to the Renner complexes broke down for several loci and crossing-over became possible at low frequency. But because after a few generations no more segregants were observed, that could be witnesses for crossing-over, the absolute linkage of all loci to the Renner complexes seemed to be restored. If this should be really the case, then it would give a plausible explanation for the lack of crossing-over in the heterozygotic species.

This conclusion, however, cannot be drawn from the experiments because in the later generations the heterozygosity at the *fr*-locus, or the other test loci, of the otherwise constant hybrids was never proved. This could have been tested by using the self-pollinated plants also for back-crossing to others which were either homozygotic recessive for the test genes or had, in one sex, only an active gametes with the recessive allele at the test locus. When the allele *fr* can be transferred from *stringens* to *albicans* by crossing-over, then this requires that, with exactly the same frequency, $+^{fr}$ will be transferred to *stringens*. When a male gamete of this constitution fertilizes a normal *albicans* $+^{fr}$ egg cell, then the result is an *Oe. albicans·stringens* homozygotic for $+^{fr}/+^{fr}$. This genetic constitution cannot be seen in the phenotype of the hybrid, but a family derived from it can never again segregate pollen-sterile *fr/fr* plants and cannot be distinguished from a heterozygote $+^{fr}/fr$ with absolute linkage of the locus to the Renner complexes without additional tests.

The present author became interested in this issue in relation to her own research long ago and made the design for an experiment to test the hypothesis of decreasing crossing-over of the *fr*-locus. It turned out that such an experiment would last at least 10 years. The number of generations, families per generation, individual plants per family, self-pollinations, back-crossing of each self-pollinated plant to homozygote recessives to test the heterozygosity and determine the exact numerical value of crossing-over in both sexes in every generation, necessary for valid conclusions was so large that it would need greenhouses and an experimental garden of immense size, plus a staff of trained and reliable gardeners and technicians appointed with longterm contracts to take care of the plants, seeds and documentation of the results. Once started, the experiment would have to be carried through to the end without interruption. Stopping half way would make every effort up to that point worthless. To carry out the programme would be beyond the scope of any laboratory, irrespective of the fact that no source of funds anywhere in the world would be willing to guarantee money for such a long time with the nice prospect that maybe in the end the hypothesis to be tested could not be verified. Imagine the possibility that one of the referees had the irresistible idea to ask for confirmation with another gene! You would have to ask your employer for an unlimited contract and heaven for a guaranteed prolongation of your life until the manuscript on the experiment was written. The result of half a day of calculations was the decision **never** to start **that** experiment. This most interesting question of the crossing-over frequency in a new hybrid decreasing to zero in a limited sequence of generations as a factor in the evolution of Renner complexes based on absolute linkage must remain unsolved.

The problem of variability of crossing-over frequency was investigated by Oehlkers from a different starting point. The simple question was, whether crossing-over frequency between two loci is really a constant in development or whether it is the result of a process that can be influenced by external factors to a considerable extent. The crucial point in the design of the experiments was to use plants which were heterozygotic for linked test loci, cultivating them under different conditions and then using the pollen of flowers which developed during the treatment for backcrossing to recessive homozygotes. The frequency of the different phenotypes in the progeny was counted and the relative frequency of crossing-over calculated. Experiments with the hybrid *Oe. albicans*·^{h}*hookeri*, a constant hybrid with the gene for flower size *Co*/*co* as the test locus, demonstrated that the frequency of crossing-over depended on temperature and hydration of the plants (Oehlkers 1940b,c). Similar experiments performed with the hybrids *Oe. albicans*·*stringens* (test loci complex *kr-fr-p-s*) and *Oe. flavens*·*stringens* (test loci *fr-p-s*) and using temperature shock and decrease in hydration of the plants as external influences demonstrated that the frequency of crossing-over depends on physiological conditions in the plants. Moreover, the direction of the change, increase or decrease, depended on the test loci, indicating differences in the reaction of the linkage groups (Harte 1942).

4.1.8 Unexpected Segregation Ratios and Competition Between Genotypes

The expectation of Mendelian segregation ratios in the progeny of hybrids depends on several assumptions. The most important are that the two alleles of one gene locus are each represented at the frequency of exactly 0.5 in the gametes, that all gametes have the same probability for performing fertilization and that all possible combinations of two gametes occur with the same probability. The first assumption means that no loss or multiplication of gametes occurs in the population of gametes. The second means that the genetic make-up of a gamete does not have any influence on its probability for fertilization and that no selection or competition occurs. The last one is based on the assumption that the gametes cannot recognize the genetic constitution of the partner and therefore meet at random. In the case of unexpected segregation ratios it is necessary to determine whether these assumptions, which are usually taken for granted, are realized. Deviations from the expected segregation ratios for single genes or linkage groups can give information about processes involved in fertilization.

Renner observed such cases where there could be no doubt about the monogenic basis of a character although Mendelian segregation was not found. The most dramatic are the monotypic egg cells in the heterogametic species. The lack of one type of gametes in the egg cells was interpreted as being the consequence of competition between cells of different genotype in the female line of sexual propagation. The chances of being represented in an egg cell are unequal for different genotypes arising from a hybrid (Renner 1917, 1921; Harte 1975). With the genetic method it cannot be decided whether the competition occurs between megaspores, female gametophytes (embryo sacs) or between gametes. To solve this question, investigations on the development of the female structures are needed, but the phenomenon as such can be analyzed without this knowledge.

It was necessary to construct an experimental design that provided a quantitative look at the problem and which not only proved the existence of competition but also allowed estimates to be made of the causes of variability (Harte 1958a,b,c). Species were crossed to make hybrids which were heterozygous for complexes and marker genes in such a way that the frequency of certain phenotypes in the progeny would give estimates for the frequency of genotypes in the fertilizing gametes. If one of the Renner complexes can pass through the egg cells but contains a pollen-inactivating gene and the other complex, which is active in pollen and egg cells, as well, carries a zygotic lethal, then this can be used as a test locus and the frequency of vital embryos is an estimate of the frequency of the "egg cell complex" in the egg cells; the remaining seeds with lethal embryos indicate the presence of the "pollen complex" in the egg cell of the ovule. This frequency is rather constant in species, but very variable in hybrids. There are significant differences between ovaries which develop under different conditions, either in different parts of the inflorescence or on different plants. The nature of these influences, varying between plants and during the vegetation period, is unknown.

In this experiment there are, besides the determination of the variability (Harte 1958a), two main hypotheses to be tested (summary in Harte 1975). If both Renner complexes of the species or hybrid have the same probability of appearing in the egg cells, then the frequency of vital seeds should be 50%. A significant deviation from this expectation in the direction of more seeds with vital embryos gives proof of a competition in the ovule in favour of the egg cell complex. The data confirm this hypothesis and, furthermore, provide the information that the "strength" of an egg cell complex, determining the chance to win the competition in the ovule, is not a constant but is influenced by variable environmental factors (Harte 1958b).

In the second hypothesis the crucial question is whether the competitive value of an egg cell complex depends only on its own genetic constitution or whether the other Renner complex in the hybrid has any influence. It could even be that the deciding factor is not the strength of the egg cell complex, but the reduced ability of the pollen complex to appear in a functioning egg cell. In this case those cells with the egg cell complex would have only to take the place left open by the inability of cells with the other Renner complex to look like winning the competition, but the goal would be reached without any effort on the part of the winner. The question can be answered by comparing the results of hybrids in which one egg cell complex is combined with different pollen complexes and vice versa. The result of the experiments was that the competitive ability of a Renner complex does not depend only on its own genes, regardless of whether it is the winning "strong" or the losing "weak" complex. The data demonstrate that both partners in a hybrid act together to determine the outcome of the competition. Irrespective of whether in its species of origin it is the egg cell complex or the pollen complex, the frequency with which a certain Renner complex appears in the egg cells of the hybrid depends also on the partner with which it is combined. The interaction between the Renner complexes in a hybrid is most important. The outcome of the competition between two Renner complexes in a certain case cannot be predicted from the known behaviour of both in other hybrids (Harte 1958c).

Analysis of the variability of the competition phenomena revealed that the genetic factors involved in the competition also determine the range of variability. This is rather narrow in the established species and is extremely enlarged in new hybrids. In a species, the Renner complexes are in balance with one another with respect to their competitive ability. In the hybrids such a balance does not exist. From this observation the conclusion must be drawn that there is a genetic basis for the competition in the female line, but this cannot be a simple one. A polygenic inheritance should be considered.

Now the question arises whether the lack of competing ability in the female line, which is observed in the pollen complexes of female monotypic species, depends only on this interaction with a much stronger partner complex. Therefore, the rare cases of transfer of a pollen complex through the egg cells are of special interest for the investigation of competition in the fertilization process. When two female monotypic species are crossed, then hybrids which contain the pollen complex of the mother plant in combination with the pollen complex of the father are observed at low frequency. Renner (1940a) investigated the rare cases of combinations with two pollen complexes, called diplarrhenic hybrids (from Greek dipl- = double and arrhenic = male). On these hybrids with two Renner complexes, which act as pollen complexes each in a different species, the seed set after self-pollination is very scarce, indicating a deficiency in fertile ovules. This demonstrates a general defect of the pollen complexes that does not completely inhibit their appearance in a fertile egg cell but considerably diminishes their ability to do so.

Experiments including backcrosses of the hybrid *Oe. flavens*·^{h}*hookeri* with appropriate multiple recessive test stocks demonstrated that the transmission of alleles through the egg cells depended on the Renner complex which had contributed them to the hybrid. The first linkage group contains the loci for flower colour (sulfurea, $s/+^{s}$) and for the form of the apex of the inflorescence (deplanata $de/+^{de}$). Those alleles which were introduced together with the Renner complex ^{h}*hookeri* appeared preferentially in the progeny arising from the egg cells. The cause is the gene locus *ga*, influencing the competition in the ovules between the "strong" allele ga^{+} in ^{h}*hookeri* and the "weak" allele ga^{-} in *flavens*. The same preference of alleles from ^{h}*hookeri* is observed in the hybrid *Oe.* ^{h}*hookeri*·*stringens* (Harte 1953b, 1969a).

A special case is the Renner complex *stringens* from *Oe. strigosa* de Vries. As a very rare exception this pollen complex can pass through the egg cells. The hybrid *Oe. stringens*·*flavens*, from the cross *Oe. strigosa* × *suaveolens*, is fertile and in the progeny the same segregation for many genes is observed as in the reciprocal hybrid. Only the pollen-sterile plants are missing. This was observed first by Oehlkers (pers. comm.) and then confirmed by Harte (1943). Backcrosses of the hybrid as female parent with pollen from *Oe. strigosa* de Vries and of the pollen on appropriate genotypes as female partner revealed that the allele *fr* is missing. It seems that the Renner complex *stringens* can only be transferred in the female line when it has lost the allele *fr*. This can only be an indication of some other change in the genetic make-up of the Renner complex *stringens*, as in other hybrids *fr* is transmitted through both sexes in Mendelian proportions. When a genetic factor, influencing the competition in the female line, is assumed, then this is linked to the locus *fr* in *stringens* in combinations of this Renner complex with *deprimens* in *Oe. strigosa* de Vries or with ^{h}*hookeri*, exhibiting a small increase in crossing-over, but shows free recombination in *Oe. flavens*·*stringens*. This behaviour is similar to that described by Renner as change of linkage relations.

A similar experimental design could be used to investigate the question whether competition also occurred in the male line. In test crosses with pollen of the same hybrids it could be shown that a locus in the same linkage group influences the competition between pollen, again with a preference of the alleles of ^{h}*hookeri*. At first sight it seemed possible that the same gene could influence the competition in both sexes. More progenies of the hybrid *Oe.* ^{h}*hookeri·stringens* gave proof that here even pollen with that part of the Renner complex *stringens* with the allele *fr* as marker, which belongs to the clearly defined pollen complex of *Oe. strigosa* de Vries, has only very limited chances for fertilization (Harte 1948, 1961, 1972b). The same loci used as markers in the hybrid *Oe. flavens·stringens* do not give any indication of competition (Harte 1942).

From the results of the genetic experiments it cannot be decided whether there is one gene locus influencing competition in both sexes or whether there are two loci close together in one linkage group, but investigations on development of the haploid generation, which will be discussed in a separate chapter, indicate that the mechanisms of competition in both sexes are so different that the activity of only one gene locus cannot be assumed. The results from other hybrids make clear that in other parts of the genome of *Oenothera* there must be more loci influencing the competitive ability of cells with different genotypes, at least in the female line.

It has to be kept in mind that in constant hybrids the phenomenon of competition in the ovules acts in favour of maintaining heterozygosity. It increases the frequency of vital seeds and minimizes the loss of zygotes by preventing the formation of female gametes that would give lethal embryos when combined with the pollen of the species. But in artificial hybrids with combinations of Renner complexes in which one lacks a zygotic lethal, such as *Oe. albicans·*^{h}*hookeri* where ^{h}*hookeri* proves to be stronger in providing the egg cells than *albicans*, which is originally an extreme egg cell complex, and in other cases where the Renner complexes break down in segregating hybrids, the competition with preference of the same genotype in both sexes reduces the chances for meeting of genetically unequal gametes and acts in favour of creating homozygosity and genetic homogeneity of the offspring. A discussion of this problem using the mathematical methods of population genetics is still missing.

4.1.9 Incompatibility

The question arises whether the loss of microspores or gametophytes can be reduced in other ways without endangering the main goal of maintaining heterozygosity. In the North American species the results of hybridization experiments indicate the existence of another mechanism which conforms to this purpose. Some species are constant heterozygotes with monotypic egg cells, whose pollen behaves as monotypic in controlled self-pollinations but as either ditypic or monotypic in crossing experiments with other species, depending on the partner. This behaviour has already been described for the pollen complex *excellens* of *Oe. chicaginensis* by Oehlkers (1943b), but was not explained. It was first detected as occurring regularly in the *biennis*-1 group of the North American *Oenothera* species and interpreted as an incompatibility reaction by Steiner (1956). The conclusion from the observations is that the species has two types of viable pollen, but that in these cases the female partner of the pollination decides which one of several pollen types will have the chance of fertilization. A similar observation was made by Renner (1956) who found an S-allele in the Renner complex *pingens* of *Oe. atrovirens* which inactivates the pollen grains of this genotype in self-pollinations and causes this Renner complex to behave as an egg cell complex. However, in hybridization with *Oe. atrovirens* as the male partner the pollen grains containing *pingens* are fully functioning.

To understand the mechanism the genetics of incompatibility, as revealed by investigations on *Oe. organensis* Munz., a homozygotic species from North America, ought to be discussed first. This species has a limited area where a small population of ca. 500 individual plants was discovered. Later a few smaller populations were added, together they totalled not more than 1000 plants. All the plants proved to be sterile after self-pollination, but gave abundant seed set after cross-pollination between plants. Investigation of the sterility in controlled progenies led to the discovery of a system with multiple incompatibility alleles (Emerson 1938, 1939, 1940). Extensive crossing experiments between the plants of the population revealed the existence of cross-sterility too. The mechanism does not work against self-fertilization, as was first assumed, but fertilization is inhibited completely when the same allele is present in the pollen and in the style. The basis is not sterility but incompatibility between certain genotypes.

In accordance with the first description of the phenomenon as self-sterility, the appropriate genes were called sterility genes or S-genes, in modern nomenclature incompatibility alleles, S_I. In the population of *Oe. organensis* Munz. first investigated, ca. 50 S_I-alleles were present. In *Oe. organensis* alleles are found that do not start the incompatibility reaction but allow fertilization following self-pollination in spite of identical alleles in pollen tube and style. These fertility alleles are designated as S_f-alleles.

This mechanism works as a very efficient prevention of self-fertilization, and that means a high degree of heterozygosity in the plants of the population, reducing the chance of homozygosity for rare alleles of any genes and preventing nearly completely homozygosity for new alleles of genes linked to the S-locus.

Self-incompatibility systems of the same type as in *Oe. organensis* Munz. are detected in several other species of *Oenothera* (summary in Cleland 1972; Stubbe and Raven 1979b) and also in the subgenus *Raimannia* (Linder 1952a,b, 1954; Crowe 1955; Linder et Brun 1957; Cleland 1960) and by Bali and Hecht (1965) in *Oe. rhombipetala*.

The great number of alleles raised the suspicion of a high mutability at the S-locus. In investigations on spontaneous and induced mutations, Lewis (1948, 1949, 1951) used the seed set after incompatible pollination as a test. The spontaneous mutation rate is ca. 1.87 per 10^6 cells. Induced mutations were found after X-ray treatment of pollen mother cells in different stages of meiosis. Treatment at metaphase I was most effective. Mutations could not be detected following treatment of the pollen. The interpretation of this result is that at later stages of pollen development the reaction of the pollen is already determined and cannot be changed by a mutation in the nucleus.

The analysis of the plants arising from fertilization with the mutated pollen revealed another peculiarity. The mutation first changed the pollen reaction of the S_I-allele so the pollen tube carrying the mutated allele could pass through the styles with the original allele. The plants which arose in this way were self-compatible and the pollen carrying the mutated allele could also fertilize other plants with the original allele, but their styles were incompatible for pollen with the original allele. The S-locus proved to be a complex gene consisting of at least two parts, one directing the reaction of the style and another the reaction of the pollen tube. Only the pollen part was mutated. The stylar part remained unchanged. However, this cannot lead to the conclusion that in the part of the S-locus controlling the stylar reaction mutations do not occur. Mutations affecting the stylar reaction could not be detected by the method used for screening the mutations. In appropriate crossing experiments crossing-over between the two parts was not observed.

The population genetics of incompatibility between pollen tube and style is a very interesting issue covering the relation between mutability of the locus, the size of the population and the number of S_I-alleles that can be present in a population. Other problems are the influence of incompatibility on the heterozygosity of loci in the linkage group carrying the S-locus and the population structure concerning genotype frequencies of homozygotes and heterozygotes of loci in other linkage groups, both in the generations immediately following the appearance of a new allele at any of the loci and in the long run until a stable population structure is reached. These questions have to be handled by methods of mathematical genetics. Their importance is not limited to *Oenothera* and the general problems cannot be discussed here in more detail.

The incompatibility reaction is mentioned for two reasons. First, *Oenothera* was involved at the start of investigations on the problem of self-incompatibility, contributing to the general clarification of the problem, and, second, the genus made good use of the system for its own purposes.

The importance of self-incompatibility in *Oenothera* reaches much farther than just one small species with limited distribution. As already mentioned, the species of the *biennis*-1 group consists of an *alpha* complex active in the egg cells and a *beta* complex transmitted by the pollen. The *alpha* complex contains an incompatibility allele S_I at the S-locus. In most populations, all plants have the same S_I-allele, while the *beta* complexes have an S_f-allele. Pollen tubes with this allele in the vegetative nucleus can grow through the style of the plant on which they originated, but those of the alternative pollen type carrying the S_I-allele are inhibited. The fertilizing pollen tubes are monotypic in spite of ditypic pollen grains. Following self-pollination, only the pollen tubes with the S_f-allele and consequently carrying the *beta* complex will reach the ovary for fertilization. The formation of *alpha·alpha* homozygotes is inhibited (Steiner 1956, 1957, 1961a,b, 1964; Steiner and Schultz 1958; Khafaji and Steiner 1970). When the *alpha* and *beta* complexes of the species differ in the alleles of many other loci, then heterozygosity is ensured by this system. A similar situation is found in other groups of North American species (Cleland 1972). The system can become more complicated by additional genes, but as these are not involved directly in the incompatibility reaction, they will be discussed in other chapters.

Self-incompatibility is supposed to have played a role in the evolution of complex-heterozygosity in the sect. *Eu-Oenothera*. Then there must exist self-incompatible homozygotic species in this group which, when involved in hybridizations, could have contributed the incompatibility alleles. A possible source of the S_I-alleles was found in *Oe. grandiflora* l'Her. The population structure of this species is rather variable, with complex-heterozygotic and homozygotic groups as well. In some of the latter, both self-fertile and self-sterile plants are found (Stubbe and Raven 1979b; Steiner and Stubbe 1984, 1986). The special problems of selection for S-Alleles are discussed by Charlesworth (1979).

When only one S_I-allele is present in a complex heterozygotic species, as is the case in most populations, then pollen with the *alpha* complex never has a chance for fertilization. Following crossing with a species containing the same S_I-allele, pollen with the *alpha* complex is inhibited. Fertilization is carried out only by pollen with the *beta* complex and the pollen is found to be monotypic. When, however, such a species is crossed with another one which differs in the S_I-allele of the *alpha* complex, then both types of pollen grains, those with the *beta* and the *alpha* complex as well, are able to perform fertilization. In such experiments, the pollen proves to be ditypic. In some populations crossing experiments revealed the existence of several S_I-alleles. In these cases, homozygotic *alpha·alpha* plants were expected but not observed. Inspection of the flowers disclosed the reason for the lack of homozygotes. In most species of this group the anthers are longer than the style, limiting the possibility for self-pollination. In contrast, in the populations with several S_I-alleles the anthers are of the same length as the style and dehiscence occurs before the flower bud opens. In this way self-pollination is ensured before pollen from other plants can be brought to the stigma by pollinators.

By cross-pollination of species with different S_I-alleles in their *alpha* complexes, viable *alpha·alpha* combinations can be obtained. Viability of the *alpha·alpha* homozygotes was demonstrated by inspection of the progeny from artifical pollination of emasculated flowers. These plants are completely self-sterile, as both types of pollen tubes will meet their S_I-allele in the style and become inhibited.

In some of the north American species the *alpha* complexes work with double security. They contain in their genome, in addition to the S_I-allele, pollen lethals which cause death of the microspores, pollen grains or germinating pollen tubes. This reaction is independent of the genetic constitution of the partner in the cross.

Crosses between 10 species from six different subgenera of *Oenothera* (two of them self-fertile, the other eight self-incompatible) revealed that they all have the same incompatibility system. The S-genes are homologous (Crowe 1955). Here too in some species the S-locus is closely linked to a pollen lethal whose lethal allele inhibits the transfer of one S-allele by the pollen.

The incompatibility system contributes efficiently to the maintenance of heterozygosity in the species but, at the same time, it enables the transmission of the complex with the S_I-allele to hybrids with other species, which is important in the evolution of *Oenothera* (Cleland 1972).

4.1.10 Selective Fertilization

In the subgenus *Raimannia* some of the heterozygotic species, e.g. *Oe. mollissima* (Schütz 1939), behave like *Oe. Lamarckiana* de Vries, being ditypic isogametic, with loss of homozygotic embryos by a system of balanced lethals. Some of the other cases discussed to date for maintaining constant heterozygosity are also found in species of the sect. *Raimannia*.

Some species of this section, however, have developed yet another system, leading to the same end of constant heterozygosity. Just as in the sect. *Euoenothera*, the basis is a system of gene complexes with balanced lethals. When homozygotic species of the sect. *Raimannia* are used as a parent in test crosses with these heterozygotic species both Renner complexes are transmitted by eggs and pollen in approximately equal frequency. Consequently, following self-pollination, a segregation for three types, one heterozygote resembling the parent and two homozygotes, which could eventually be lethal, is expected. However, homozygotic plants are not observed. Lethal embryos and empty seeds which could represent the homozygotes are extremely rare. The same holds for unfertilized ovules when superfluous pollen was applied in relation to the number of ovules present in the ovary.

The conclusion from these observations is that there must exist a special mechanism which ensures the efficient formation of heterozygotic offspring in the species without considerable loss of embryo sacs, inactive pollen grains or lethal embryos. This is only possible when fertilization is not at random, as is illustrated by the following scheme. Given is a species with the

Renner complexes A and B. In the progeny of self-pollination the following types are expected:

		Expected Complex in pollen		Observed Complex in pollen	
Complex in egg cells		A	B	A	B
	A	AA	AB	–	AB
	B	AB	BB	AB	–

When the homozygotes do not appear, in spite of two types of functioning pollen, this means that the A-egg cells are not fertilized by A-pollen, and the B-eggs are not fertilized by B-pollen.

The solution is found by experiments with homotypic pollen, either from homozygotic species C·C or from hybrids A·X and B·X with a pollen inactivator or a pollen lethal in the X-complex. A method used in addition to the usual scheme of hybridization is double pollination. Some time, usually 12–24 hours, after the first generous pollination, which should be sufficient to fertilize all the ovules, a second pollination is performed with genetically different pollen so that the hybrids resulting from both pollinations can be clearly recognized by their phenotype.

The design of the experiment is:

Cross ♀ × ♂	Progeny Expected	Observed
A·B × A·B	A·A, A·B, B·B	A·B
A·B × A	A·A and A·B	A·B and unfertilized ovules
A·B × B	A·B and B·B	A·B and unfertilized ovules
A·B × C	A·C and B·C	A·C and B·C
A·B × A, followed by × C	A·A, A·B, few A·C and B·C	A·B and A·C
A·B × B, followed by × C	A·B, B·B, few A·C and B·C	A·B and B·C

The phenomenon was observed in the laboratory of Schwemmle. As an example, experiments with *Oe. Berteriana*, a constant heterozygote with the Renner complexes B·l, in which the results were first observed, will be presented (Schütz 1939). The homozygote *Oe. argentinea* and two semi-heterogametic hybrids, l·II and B·II, with a pollen inactivatory in the II-complex, are used as pollen partners. In the design presented above, the names A and B have to be replaced by the Renner complexes B and l, and C by *ha*.

Following controlled self-pollination of *Oe. Berteriana*, the expected l·l-embryos cannot be found. Only a few non-germinating seeds are observed, each containing a very small embryo which dies in the early globular stage. These embryos can be identified as homozygotic B·B. From double pollination of *Oe. Berteriana* with the hybrid l·II and *ha* only seedlings with B·l and l·ha arose. Similar results were obtained using pollen of the B·II hybrid, again followed by *ha*. In the progeny only l·B from the first pollination and B·*ha* plants were found.

The results can be interpreted on the assumption of differences in attraction between gametes of different genotypes. All types are active in both sexes, but the chance for fertilization depends on the genotype of the partner.

In the progeny of crosses between two heterozygotic species of this type, the four possible combinations of the Renner complexes of the parents are expected to be realized at approximately equal frequency. However, in most cases only two or three types are found, or four types are observed but not in the expected numerical relations. Moreover, in reciprocal crosses, the observed genotypes and their numerical relations are not necessarily the same. Extensive reviews of the numerous papers on this topic have been written by Arnold (1958a), Haustein (1967) and Schwemmle (1968b).

The conclusion from the observation that not all possible combinations of gametes occur and, in spite of two types of viable gametes in each sex, only certain combinations are realized can only be that fertilization is not at random. The phenomenon is called *selective fertilization*. This system of non-random fertilization, combined with balanced lethals as an additional safeguard, is a very efficient mechanism invented by *Oenothera* species of the sect. *Raimannia* to guarantee the maintenance of constant heterozygosity and the favours of heterosis, at the same time minimizing the cost which has to be paid in the form of considerable loss of gametophytes, gametes and zygotes.

4.1.11 The Genetic Model

The consequent application of the methods of classical genetics, hybridization, raising of progenies for several generations and quantitative analysis of the types observed in the progeny, produced so much information on the genetics of *Oenothera* that a model of the genetic behaviour can be constructed.

In order to understand the extent to which the results found for *Oenothera* hybrids and their progenies caused confusion among geneticists, we will have to compare the genetic behaviour of "ordinary" objects with the observations on *Oenothera*. The laws of Mendelian inheritance, which were already known to hold true for many cases in plants and animals, seemed not to be applicable to the results observed in *Oenothera*.

The Mendelian rules of heredity are concerned first with the behaviour of hybrids and hybrid progenies from controlled crossings. The differences observed can be summarized in the following table:

	Mendelian genetics	*Oenothera*
F_1:	Uniformity	Multiple hybrid phenotypes
	Reciprocal identity	Difference between hybrids from reciprocal crosses
	Dominance or intermediary phenotype	New phenotypes
F_2:	Segregation	Constancy of F_1-phenotype
	Between two loci, either free recombination of alleles or linkage and crossing-over	Absolute linkage or variable linkage relations
	Random variability of crossover frequencies	Variable crossover frequencies

The third rule of Mendelian heredity states that the expression of a certain character in a hybrid is either equal to one of the parents, dominant, or intermediate between the parents. In *Oenothera* it is not extraordinary to find in a hybrid an expression of characters that is not seen in either of the parents.

The species can, irrespective of the taxonomic sections of the genus to which they belong, be allocated to two groups differing by their genetic make-up. There are homozygotic species which behave according to the Mendelian rules of inheritance, and constant hybrids which breed true in the field and also after experimental self-pollination. The latter contain two large groups of gene loci in absolute or nearly absolute linkage, the Renner complexes. Two different ones are to be found in every species. They are separated in the course of formation of the gametes, each haploid gamete containing one group. As the two groups differ in the alleles of many loci, the species are heterozygotic for many genes. In certain hybrids the absolute linkage can break down, releasing the alleles for recombination.

The heterozygosity of the species and some of their constant hybrids is maintained by several mechanisms, which all have a clearly defined genetic basis, but work at different costs for the species. This has to be considered for the population genetics and evolution of the genus.

The first genetic mechanism is a system of balanced lethals, one lethal allele at different loci in each Renner complex, by whose action the homozygotes are eliminated as lethal embryos. This means a sacrifice of 50% of the zygotes in each generation. Other mechanisms avoid this loss and do not work by eliminating the zygotes but by preventing the formation of lethal diploid genotypes. The mechanisms are in the female line competition in the ovules, resulting in monotypic egg cells, and in the male line competition between pollen of different genotypes, genes for pollen lethality, pollen inactivation and incompatibility between pollen and style with the same allele of a certain gene locus. Each of these genetic mechanisms acts towards maintaining heterozygosity of the species, all reducing the cost for achieving the goal by elimination of gametes of certain genotypes before fertilization, but having different advantages and disadvantages for the species, sacrificing genetic material in the process of sexual propagation and influencing the possibilities for evolution of the genus.

Finally the method of selective fertilization occurs in the sect. *Raimannia*, maintaining heterozygosity of the progeny by preventing fertilization of genetically identical gametes and, at the same time, making the best use of the material provided by the plant for sexual propagation and avoiding the loss of zygotes and gametes as well.

The structure of the species, consisting of two Renner complexes with special genetic properties for maintaining heterozygosity, creates difficulties for the taxonomy of the genus. "Pure species" in the classical sense of nearly homozygotic populations with a genetic variability limited to minor differences of unimportant characters are rare and the plants in a population of a "true breeding species" can be heterozygotes. This led to a discussion of the species concept which was not limited to *Oenothera*. The species concept in the classical sense is not applicable to *Oenothera*. The taxonomic units are not the species but the Renner complexes, as was pointed out long ago by Renner (1917) and Davis (1922) and confirmed by Cleland (1944, 1972) and many geneticists working with *Oenothera*.

As a conclusion from his first experiments with *Oenothera*, Renner made the statement that in *Oenothera* species hybrids do not exist, only hybrids between the genetically different gametes of two species which are hybrids themselves.

The Renner complexes can be combined to form constant hybrids which behave as "true breeding species" in isolated populations in the field and in experiments, and which are, at the same time, a source of continuous evolution of the genus by giving rise to new constant hybrids as soon as they meet in new habitats (Cleland 1972).

Knowing the solution, it is difficult to understand nowadays how revolutionary the ideas of Renner were. It took a long time before it was accepted not only by taxonomists but even by geneticists that a true breeding species, which behaves in the field according to all the rules of taxonomists like a "good species", could, at the same time, not be a homozygote as it should be, but that it could be a heterozygote of previously unknown, and therefore unexpected, complexity. Even in 1924 Broekels still rejected the idea of *Oe. Lamarckiana* de Vries not being a pure species.

4.1.12 Limitation of the Genetic Method

The restriction to the genetic methods applied in the preceding chapters led to the discovery of many interesting facts and to the solution of important problems. At the same time, it revealed that in spite of the analysis of genetic phenomena there are questions which remain unanswered. These require a change to other methods of investigation to find a solution. It is necessary to consider the chromosomes as the sites of the genes in order to clarify the occurrence of the giant linkage groups of the Renner complexes, and also to investigate developmental processes in sexual propagation to find the material basis for the observed elimination of certain genotypes from the fertilization process. This requires the application of methods of karyology and developmental biology to *Oenothera*.

4.2 Karyology of *Oenothera*

4.2.1 General Considerations

The chromosomes are the material basis of the genes. Therefore the question arises whether there exists a karyological equivalent for the genetic phenomenon of the Renner complexes. Because ultramicroscopy was not known, the old cytology was concerned only with the nucleus. Now the correct name is karyology.

Cytological investigations started soon after the first genetic results on *Oenothera* were published (Lutz 1907, 1908). The papers on chromosomes from the first 30 years of *Oenothera* research are of great historical interest because many questions of general importance were discussed, but the practical value is very limited due to the inadequate techniques of fixation and staining of chromosomes. Up to approximately 1935 only haematoxylin staining with different formulae was used. The first improvement was the introduction of crystal violet by Oehlkers (1940d), but both staining procedures needed paraffin sections, which had to be 7–10 µm thick for haematoxylin stains, and this left only a few complete nuclei for observation of meiotic stages from every anther. For crystal-violet staining the sections could be up to 25 µm thick, with the result that more nuclei were complete, but were partly covered by somatic cells, which made exact observation just as difficult. Furthermore, microphotography was unknown and later, using large cameras, glass-negatives and no light meter, microphotographic documentation was too complicated and too expensive for routine investigation on a large scale. Drawings, first freehand and later with the help of more or less complicated and reliable optical equipment, had to be the method of documentation of what the research worker had seen. The result was not an unbiased description of *Oenothera* chromosomes but a combination of bad slides and the unlimited phantasy of the observer, and the later suffered from prejudice based on theoretical expectations. All these factors had a rather large individual variability and there were endless discussions on problems which could only arise on this basis. I hope the reader will understand when I neglect most of them. If not, please look at the pictures yourself.

The chromosome number as determined from metaphase plates, 2n = 14, and the fact that no large differences in chromosome morphology could be recognized were observed correctly in the first cytological investigations (Cleland 1922; Boedijn 1924a). At the same time an important observation was made and considered to be a peculiarity that could be seen only in *Oenothera* chromosomes.

In some species like *Oe. Hookeri* de Vries a pairing of chromosomes in "normal" bivalents was seen in pollen mother cells (PMC) in the stages diakinesis and first metaphase of meiosis (Boedijn 1928). In most species, however, an end-to-end connection of chromosomes, which were arranged in large chains or circles, was observed (Cleland 1922, 1925, 1926; Sheffield 1927; Gates and Goodwin 1931). Some difficulties arose from the variability of the number of chromosomes in a chain and the number of short chains in a nucleus, but the problem was not discussed on a statistical basis, not only because of a lack of knowledge of appropriate statistical methods on the part of the investigators, but because it was just impossible to find in haematoxylin-stained slides a sufficient number of complete nuclei which could be analyzed properly.

The real problems arose in relation to the suspected arrangement of the chromosomes in meiotic prophase. As a result of the use of fixatives inappropriate for these stages, the greater part of the chromosomes stuck together and only some strands or loops were seen hanging from the chromatin mass into the nuclear cavity. In early leptotene the very thin structures were interpreted as unpaired, single chromosomes, and in slightly larger anthers the thicker chromosome threads seen in the nuclei of PMC were assumed to represent paired chromosomes. Up to this point everything was correct (Boedijn 1925a, 1928; Catcheside 1931a). The unsolved question was: How could the chromosome chains seen in the later meiotic stages arise?

The nuclei seen in the slides could not show the solution, so the discussion was determined by theoretical considerations. Some observers meant to see free, paired chromosome arms. Starting from the assumption that in any case homologous chromosomes pair in bivalents, they came to the conclusion that the end-to-end connections must arise later. Others, having slides of the same insufficient quality and seeing chromosome loops coming out of the mass of chromatin, decided that "normal" pairing must be impossible, considering the abnormal result in metaphase I, and that therefore *Oenothera* must have one extremely long chromosome. To explain the individual chromosomes which could be recognized in later stages, this chromosome should be subdivided into individual chromosomes somewhere between pachytene and diakinesis. Both groups were subject to the same error. They did not construct a theory based on reliable observations, but rather first developed a theory and gave to the observations an interpretation that was compatible with their theory.

What can happen is demonstrated in the publication of Kihara (1927). The problem is chromosome pairing. To begin with, early stages of meiosis are not investigated. On the basis of observations on the usual bad slides and hypothetical considerations metasyndesis is rejected. The theory on which the interpretation is based is completely wrong, but the final conclusions drawn are valid according to more modern interpretations.

Nowadays, looking at the pictures published in numerous publications, we can only wonder how the observers dared to "see" anything at all. The early cytologists just did not recognize how bad their fixation of chromosomes really was, but how could they have known? They believed in using the best possible technique, described what they thought to be facts and did not doubt the results. The solution of the unexpected behaviour of *Oenothera* chromosomes in meiosis came from quite another side, independent from any cytological techniques.

Therefore it seems better to look first for the observations on the general karyology of *Oenothera* before discussing the special problems. The breakthrough was made possible as soon as cytologists came to realize that they were not doing histology and that for their problems the arrangement of cells in the tissues was unimportant. The old methods, designed for observation of tissues, had to be abandoned in favour of methods which enabled observation on nuclei in cells, undamaged by cutting sections. The shift from sections to smears and squash preparations with the appropriate methods of fixation and staining, as described later by Darlington and La Cour (1962), was the important step in *Oenothera* karyology.

In spite of many cytological investigations, the first reliable description, which is still valid today, of the whole course of mitosis and meiosis in *Oenothera* was given by Marquardt (1937), who used an alcohol–acetic acid mixture as fixative and aceto-carmine squash preparations. His observations were in agreement with those of Wisniewska (1935) who had still used sections.

4.2.2 Mitosis

The haploid chromosome number is 7 in all species. In mitotic metaphase 14 chromosomes are seen. The chromosomes are metacentric, heterochromatic in the middle region on both sides of the centromere, and have long, euchromatic arms of nearly equal length. In mitosis, morphological differences between the chromosomes cannot be found, only two of them have rather large satellites. Notes made by different authors about differences in the morphology and size of the chromosomes are not reliable (Cleland 1957). By light microscopy 14 chromocenters and one or two nucleoli can be recognized in the resting nuclei of meristems but no further structures are visible. The first sign of entering mitosis in meristematic cells is an enlargement of the volume of the nucleus, accompanied by the disappearance of the chromocenters, called decondensation of the heterochromatin (Marquardt 1937). This stage is observed rather rarely, so it can last only for a very short time and was overlooked by earlier investigators. The following step is a reappearance of the chromocenters, recondensation of heterochromatin, accompanied by a change in the structure of the nucleus in which some very thin threads can be recognized. At the onset of mitotic prophase the centromeres lie close together and chromosomes can be distinguished as thin threads with a heterochromatic middle region. Later the chromosomes, apparently in correlation to their shortening and thickening as a consequence of spiralization, are separated and lie free in the nucleus.

Shortly before the maximum of chromosome contraction is reached, the nucleolus dissolves. Then the nuclear membrane disappears and the chromosomes arrange themselves as usual in the metaphase plate. These two processes must occur simultaneously and proceed very fast as cells without nuclear membrane and with contracted chromosomes outside the metaphase plate are not observed. Anaphase and restitution of the nucleus with chromocenters, nucleoli and the nuclear membrane are as usual. Compared with other plants the mitotic behaviour of *Oenothera* is as normal as possible.

4.2.3 Meiosis

Meiosis can be easily observed in pollen mother cells (PMC). Megaspore mother cells (MMC) are sometimes included in the investigations but are more difficult to obtain and are therefore seldom used (Hakansson 1928). The following description is based on the not very new but accurate and still valid observations of Marquardt (1937) on *Oe. Hookeri* de Vries, of Japha (1939) on the hybrid *Oe.* ^{h}*hookeri·flavens*, of Wisniewska (1935), and on the experience of the present author (Fig. 4).

In the microsporangia, the loculi in the anthers, the cells of the sporogenic tissue are arranged in a file parallel to the long axis of the anthers, and are surrounded by a monolayered tapetum. Following the last premeiotic division, the meiocytes enlarge considerably, contain a dense cytoplasm, and the nucleus increases in volume. The first visible steps of meiosis are morphologically the same as in mitosis: decondensation of the chromocenters, immediately followed by recondensation of the heterochromatic centers of the chromosomes. At the same time, the euchromatic chromosome arms can be recognized as very long, thin threads which cannot be followed individually. The chromocenters lie close together in a typical bouquet stage as described in early leptotene also for other species, and which is maintained during prophase until diplotene. Pairing of chromosomes starts at the heterochromatic part. During zygotene the structure of the chromosomes, with thick parts of the threads and loops of thin threads between them, makes clear that the middle heterochromatic parts can be completely paired while in the euchromatic ends unpaired regions are still present. Pairing does not occur along the whole length of all chromosomes at the same time, neither does it proceed from the middle part towards the end of the arms. In the euchromatic parts it can start at different places but is always finished in pachytene. Even after completion of pairing, with free chromosome arms visible, the structure of the nucleus is not suitable for detailed analysis. The long chromosome arms cannot be distinguished by any morphological criteria. A pachytene analysis, as performed in other objects, is not possible.

Sometimes, during meiotic prophase several asynaptic stages are recognized in which all the chromosomes collapse at one site in the nucleus. The first, occurring early between zygotene and pachytene, is called synizesis, and then, after unravelling of the chromosome mass, another, the second contraction, is observed in late pachytene. Both were recognized as artefacts by Hakansson (1928) but were taken for granted by others.

The pairs become separated by further contraction of the chromosomes. At the same time, the chromosomes within a pair start to separate and chiasmata can be recognized, but the chromosomes are still so long and twisted, even in cytologically "normal" species such as *Oe. Hookeri* de Vries, that an exact determination of the number of chiasmata in a nucleus is not possible. Only in some extraordinary cases can two or more chiasmata be recognized in a bivalent. In this stage, diplotene, investigation by light microscopy shows that the nucleolus starts to dissolve. Often, all or some chromosomes in a nucleus can again be seen sticking together. Then the chromosomes contract further and terminalization of the chiasmata is observed. In diakinesis the chromosomes are connected by terminal chiasmata, but these can be missing, leaving rod bivalents or two univalents instead of the ring bivalents in *Oe. Hookeri* de Vries. At this stage the nucleolus has disappeared completely. The nuclear membrane dissolves and the chromosomes are arranged immediately in the metaphase plate. Anaphase and telophase of the first meiotic division do not exhibit any peculiarities.

The terminalization of chiasmata is not completed by metaphase in all chromosomes. The "cross-arms" ("Querarme" in the German literature) were described by Wisniewska (1935) and discussed by Renner (1948a) as special structures which could be related to crossing-over. Hoffmann (1954) investigated them in more detail using the meanwhile improved staining techniques and sound statistics. The quantitative evaluation of the frequency of cross-arms in different meiotic stages from early diakinesis to metaphase led to the interpretation that they are the consequence of delayed terminalization of normal chiasmata. In anaphase the chromosome pairs with cross-arms separate just like those with end chiasmata do.

In other species the chromosome circles, already observed by the first cytologists working with *Oenothera*, can be clearly identified from diakinesis onwards. The chromosomes are connected end-to-end by terminal chiasmata. When these are missing, the circle breaks into one large or several smaller chains. In the chromosome rings, and also in the chains, the individual chromosomes are arranged mostly in a zigzag form during metaphase. From the equal length of the chromosome arms it follows that the centromeres are spaced at equal distances, such fascilitating the orientation of the chromosomes in the metaphase plate and the distribution of adjacent chromosomes to opposite poles of the spindle in anaphase I (Japha 1939; Cleland 1957). This is different from the behaviour of translocated chromosomes in maize which do not have a heterochromatic middle region and whose translocation points are located in the euchromatin. In these cases a regular distribution of homologous chromosome arms to opposite poles during anaphase I is not guaranteed (Figs. 5, 6).

At the end of telophase I two dyad nuclei are formed, each with seven chromomeres. In clear preparations the chromocenters are cross-shaped, indicating that the chromatids have already separated and are connected only at the centromere. At the end of the first meiotic division a cell wall is not formed. The meiocyte is binucleate. The second division follows immediately without a recognizable prophase. Following dissolution of the nuclear membrane, the chromosomes are arranged in two metaphase plates and the chromosomes are distributed to the poles in anaphase II. The two spindles lie perpendicular to one another. The four small haploid tetrad nuclei, each with seven chromocenters, lie in a tetrahedral arrangement in the cytoplasm of the meiocyte (Jean 1974). Cell division in the tetrad is simultaneous. The tetrad cells are immediately completely separated from one another as young microspores whose cell walls develop without any contact with the old wall of the meiocyte. The space between the meiospores and the cell wall becomes filled with callose. Apart from the formation of the chromosome circles in some species, the events during meiosis are, in principle, the same as in other objects.

The duration of meiosis in PMCs in the field is approximately 5 days. The method of investigation was to cut off from the inflorescences all the buds with anthers over 5 mm in length. At the same time, all buds from some other plants were used as a control to determine the correlation between the length of the anthers and the meiotic stage of the PMCs. From these data the meiotic stage of the PMCs in the buds which were left on the experimental plants could be estimated. From the next day on, measurement of anther length and fixation for microscopic observation were performed to determine the growth and progress in the meiotic stages in the PMCs of several plants from the experimental group, and this was repeated during the following days. The time necessary for the transition from a premeiotic nucleus to telophase I is 5 days and from late pachytene to tetrads up to 2 days. In buds of cut inflorescences in low-temperature rooms, meiosis is delayed and, probably, only PMCs which were in leptotene at the start of the experiment can finish meiosis under these conditions within a week (Linnert 1951).

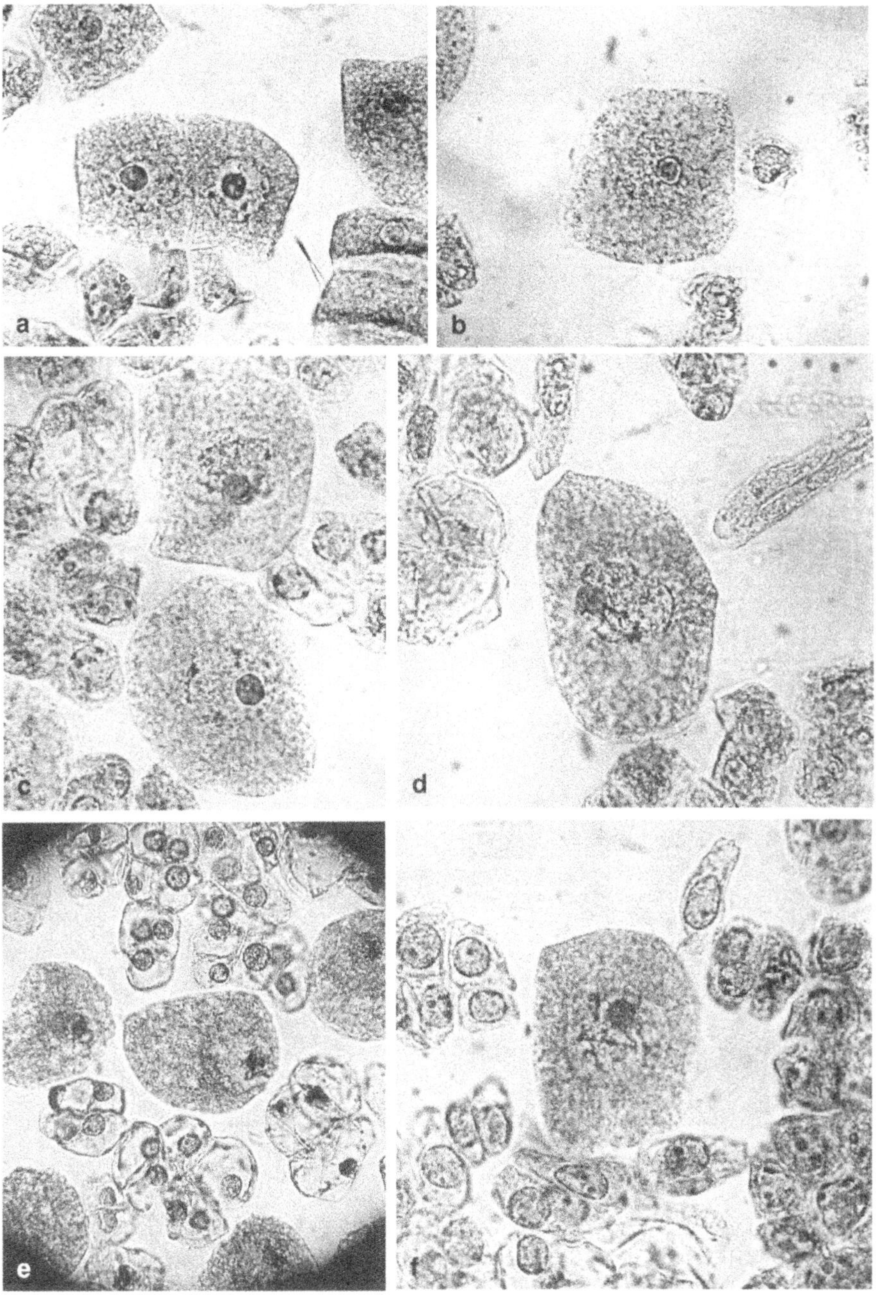

Fig. 4. Meiosis in pollen mother cells of *Oenothera Hookeri* de Vries, a homozygote with the 14 chromosomes arranged in seven bivalents during the first meiotic division. Microphotographs of aceto-carmine squash preparations of young anthers. **a** Premeiotic PMCs, nuclei with nuclear membrane, chromocenters and nucleolus.

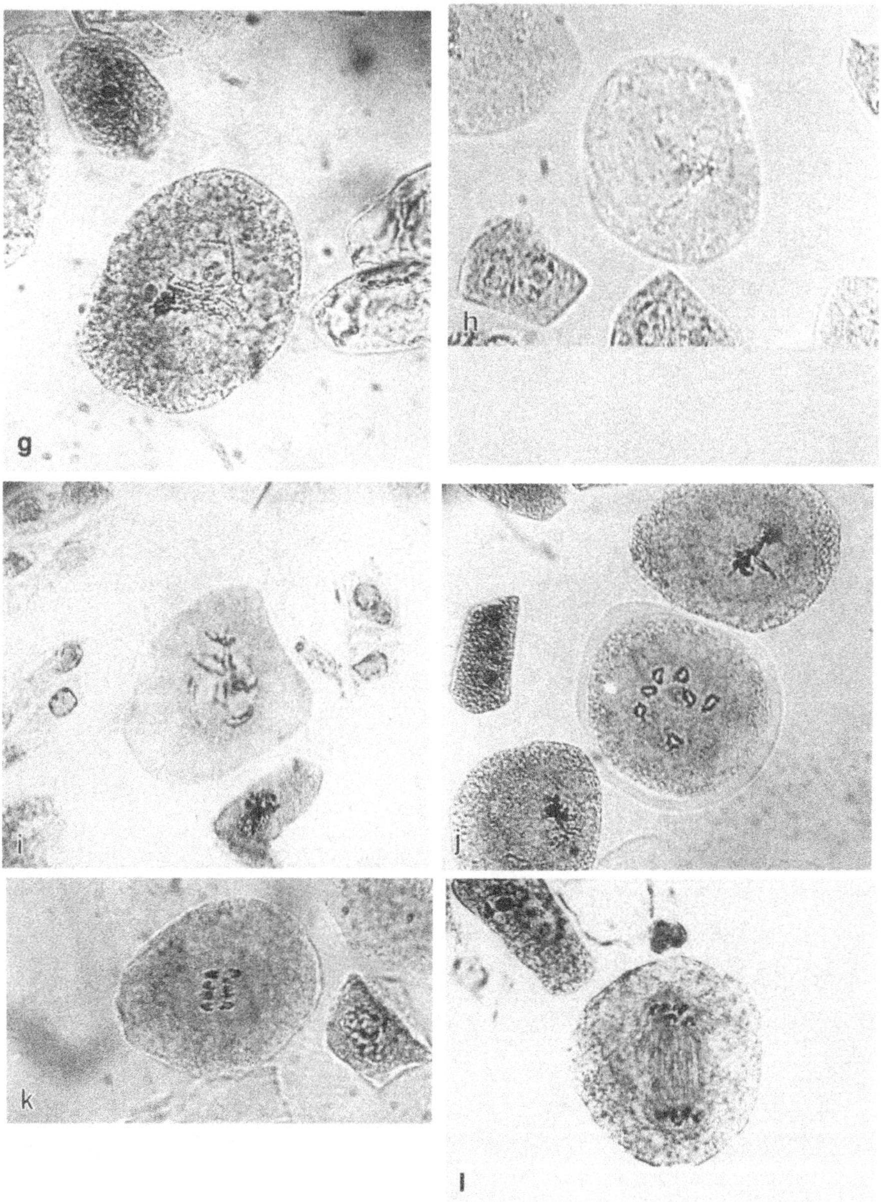

b Decondensation of chromocentres. **c** Two PMCs, recondensation of heterochromatin and early leptotene. **d** Leptotene, nucleolus still present. **e** "Synizesis", artefact at the end of leptotene. **f** Early pachytene. **g** Bukett stage at the end of pachytene, heterochromatic parts of the chromosomes lie close together, long euchromatic strands of paired chromosomes. **h** Pachytene, later stage. **i** Diplotene, seven bivalents, heterochromatic middle parts and euchromatic end clearly distinguished. **j** Diakinesis. **k** Metaphase I. **l** Anaphase I. **m** Metaphase II. **n** Anaphase II and tetrad nuclei. **o** Young tetrad cells still enclosed in the callose wall of the meiocyte. **p** Tetrad cells separated, each with its own callose wall within the callose wall of the PMC. **q** Tetrad of microspores released from the callose wall of the PMC. **r** Young pollen grain, mononucleate, still enclosed in the callose wall of the microspore

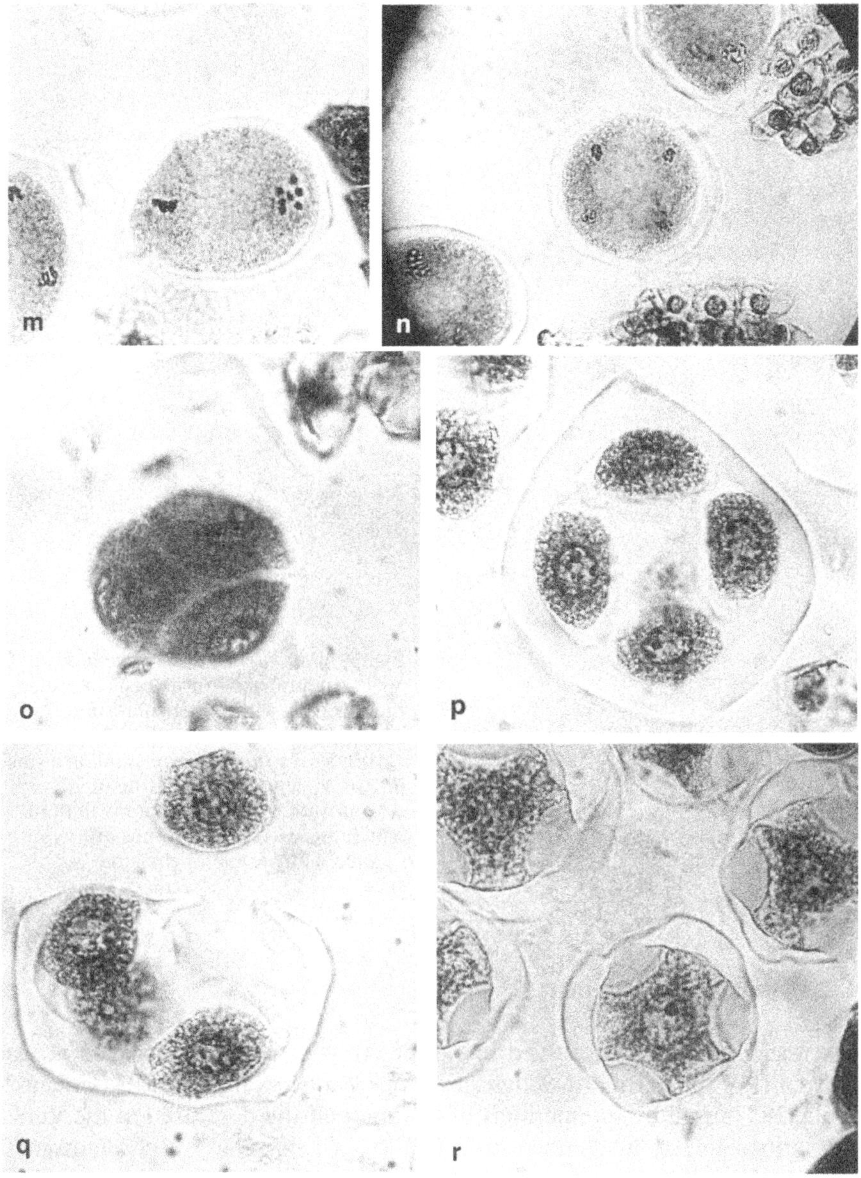

Fig. 4. *Continued*

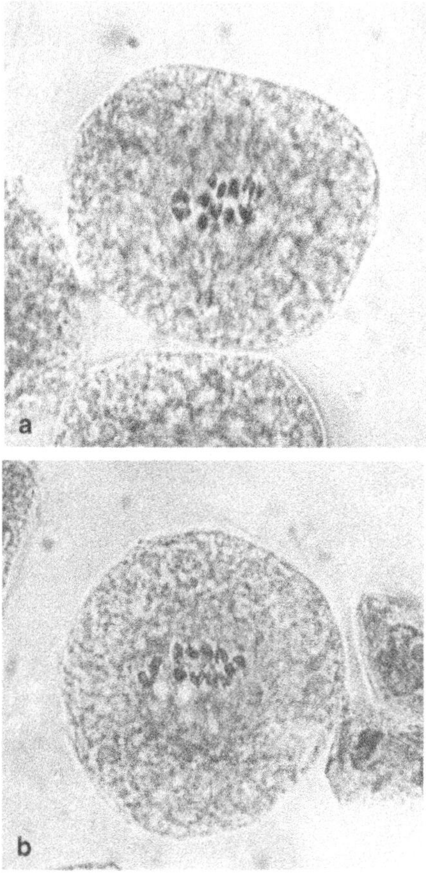

Fig. 5. *Oenothera Lamarckiana* de Vries, metaphase I of meiosis in pollen mother cells with the chromosome configuration 12 2. **a** Ring of 12 chromosomes and one bivalent as a ring of two chromosomes. **b** Ring of 12 chromosomes and an open bivalent in which the two chromosomes are connected by one end chiasma

4.2.4 Facts and Artefacts

Discussing the events observed in earlier investigations the suspicion arises that some of the peculiarities described for *Oenothera* are artefacts (Hakansson 1928) caused by the methods of fixation and the dyes used at the time. At various stages, as characteristic feature, a "clogging" of chromosomes into an unanalyzable mass is described by different names: contraction stage, synizesis, second contraction, stickiness of chromosomes.

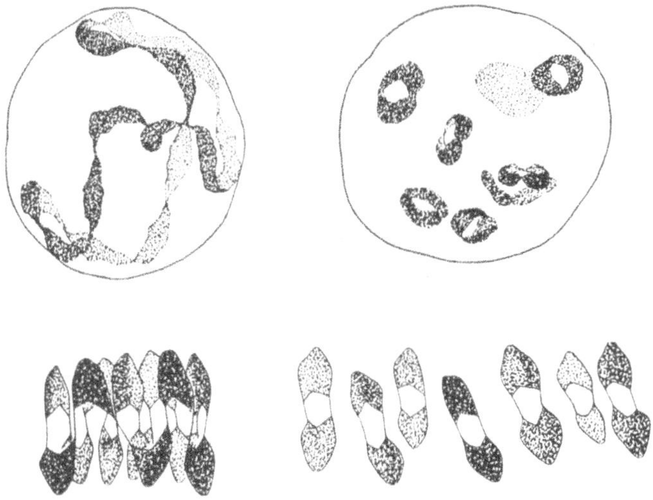

Fig. 6. Diakinesis and metaphase I in *Oenothera* from a type with seven bivalents and another with a ring of 14 chromosomes, all connected by end chiasmata (redrawn after Cleland and Oehlkers 1929 in Oehlkers 1956). This picture is presented as an example of good documentation in the classical literature on *Oenothera*

Comparing the drawings of different authors and the explanations of the figures, it seems that they certify nothing about chromosome behaviour but a lot about the influence of phantasy in scientific interpretation of microscopical observations (e.g. Cleland 1922, 1926; Boedijn 1924a; Shull 1928b; Kulkarny 1929a,b; Illick 1929, 1932; Lelieveld 1930, 1931; Emerson 1931b). The discussion centers on pairing behaviour and the interpretation offered is always that the clogging of the chromosomes indicates a change in the structure or the chemical composition of the chromosomes or their surface. This, however, is far from obvious.

In evaluating the pictures seen in cytological preparations it is often forgotten that the chromosomes are not lying in a vacuum but in what is called the nuclear sap. Their behaviour in a living cell and during the process of fixation depends not only on their own chemical composition but on the physical chemistry of that other nuclear component. In a living meiocyte the nuclear sap must be fluid to enable the movements of the chromosomes during prophase and up to metaphase. When, under the influence of the fixative, this fluid solidifies, the chromosomes are kept in position, but when it stays in the fluid state the dying chromosomes can collapse. When this is taken into consideration, the differences in behaviour of the chromosomes in the course of meiotic prophase, and especially the "synizesis", do not indicate important changes in the physicochemical structure of the chromosome surface but have their cause most probably in transient changes in the chemical composition or the physicochemical structure of the nuclear fluid occurring several times in pachytene and again in diplotene. If this change is such that the fluid is not solidified by the fixation procedures, it facilitates the collapse of the chromosomes. From late diplotene onwards the chromosomes are always seen separated. The nuclear sap must have again changed its state to become solid by fixation, keeping the chromosomes in place distributed in the nuclear cavity.

When the cell is killed and thereby the forces that keep the chromosomes floating apart in the nuclear fluid break down, the chromosomes must collapse under the influence of gravity. Sticking together after fixation can only be avoided when they have no chance to leave their position, and this will only be the case when the fluid or semi-fluid substance, called nuclear sap, is immediately transformed into a solid state by the fixative. When, after fixation, the chromosomes stay separated in one stage and collapse in another, this indicates a change not in the structure of the chromosomes but in the nuclear sap, causing a different reaction to the same treatment. When the nuclear sap stays fluid, the chromosomes collapse, undergo a contraction, become sticky, enter the stage of synizesis or do something else which you can give a nice name, but when the sap becomes solidified, we see fine, clear pictures of separated chromosomes. The different reactions of the chromosomes during the course of meiosis indicate primarily physicochemical changes in the nuclear sap. Changes in the surface structure of the chromosomes, as proposed by some investigators, cannot be excluded but the observations do not prove it.

When we do not stare only at the chromosomes but also consider the rest of the nucleus, the old observations, first given wrong interpretations and then discarded as artefacts, can provide valuable information about developmental processes within the nucleus of the meiocyte, processes that need further investigation by modern methods. What happens during meiosis in the nuclear sap when its reaction to the fixatives is changing and what is the importance of these changes for the physiology of meiosis, e.g. chiasma formation, contraction and movement of the individual or paired chromosomes? A first-class biochemist who can handle microchemical methods in cytochemistry, knows a lot of physical chemistry and something more about chromosomes has to become interested in the problem to answer the questions.

4.2.5 *Oenothera* Specialities

The meiocytes of *Oenothera*, however, would not have encouraged so many scientists to look for the chromosomes if the description of the stages of meiosis had been the only problem. The first investigations on the cytology of *Oenothera* revealed a problem which kept busy a whole generation of cytologists.

During the middle stages of meiosis, when the chromosome arrangement could be seen even with the available staining methods, in most species the chromosomes were not seen in pairs, bivalents, as in other objects, but were connected end-to-end in circles or chains. All the investigators of *Oenothera* chromosomes were in agreement as far as this phenomenon was concerned. The chromosome catenation, as it was later called by Gates (1931), was soon an established fact. The points of discussion and disagreement were:

How was this arrangement of the chromosomes achieved from the previous stages?
What was the right interpretation of the observations?
Are the rings of chromosomes just another peculiarity of a genus already accepted as exceptional or can any connections be established with the genetic phenomena observed in *Oenothera*?

A correlation was soon found. Those species which were genetic homozygotes had bivalent pairing of the chromosomes in the first meiotic division. On the other hand, in the heterozygotic species such as *Oe. Lamarckiana* de Vries and *Oe. biennis* L. large rings of chromosomes were found. This was in contradiction to every observation in other genera. The considerations centred on the problem of how the pairing of chromosomes in meiotic prophase must occur to give, in later stages, the chromosome rings. Did homologous chromosomes pair, as they are expected to do, and later form the end-to-end connections, or was the pairing already anomalous, with association of the chromosome ends from the beginning? Even the existence of a giant chromosome that was separated rather late in meiosis into smaller chromosomes was assumed.

Concerning the arrangement of the chromosomes in the circle, the question was whether there is a random assortment of chromosomes or a fixed place for each chromosome pair. Shull (1928b) as well as Cleland suggested a regular position for each chromosome, leading to an alternation of chromosomes from the two parents in the circle.

After many years of discussion on telesynapsis (metasyndesis, end-to-end pairing of the chromosomes in pachytene) or parasynapsis (parasyndesis, pairing of homologous chromosomes and formation of the connections between chromosome ends later at the end of prophase during the second contraction stage), in which many intelligent ideas were involved and more or less complicated hypotheses were presented and rejected, all on the empirical basis of the same bad slides, the solution turned out to be as simple as the famous egg of Columbus.

The cytologists had only to replace their beloved *idee fixe* of pairing of whole homologous chromosomes by the assumption of pairing of homologous *parts* of chromosomes. Then the two arms of one chromosome can pair in pachytene each with one arm of two other chromosomes to which they are homologous. The second arm of these chromosomes is free for pairing with homologous parts of other chromosomes. Following normal formation and terminalization of chiasmata in the paired arms, the end-to-end connections of the homologous chromosome arms appear to be the same as observed in any bivalent. If, in a diploid chromosome complement every chromosome arm is represented twice then one or more circles should come out at the end. The circles are nothing special but only the mechanical consequence of this type of pairing.

Even when with this assumption an excellent explanation of the chromosome circles observed in diakinesis and metaphase I could be presented, the observation of these stages did not give information on the pairing of chromosomes in the meiotic prophase. The assumption of pairing of half-chromosomes was still an unproved hypothesis. Considering the difficulties of pachytene analysis in *Oenothera*, it is not astonishing that it took more than 10 years to find the cytological confirmation of the new hypothesis. Japha (1939) analyzed the hybrid *Oe.* ^{h}*hookeri·flavens* which has as its chromosome configuration in meiosis a ring of four chromosomes and five bivalents. Using the same cytological techniques as Marquardt (1937), she could demonstrate the cross-pairing of four chromosomes in late pachytene as the basis of the ring-of-four observed from diakinesis onwards. However, the cytological problem could only be solved after there was clarification about the kind of phenomena in meiosis on which the cytologists had to concentrate their attention.

4.2.6 The Basis of Chromosome Catenation

Looking for the chromosomes of *Oenothera*, the problem could not be solved directly but shifted only step by step towards a solution. The problem was not only how the pairing of chromosomes had occurred, but how this homology of half-chromosomes had arisen. The solution was found with help from another genus.

A young *Oenothera* geneticist, Cleland, met at a genetics conference an older colleague, Blakeslee, who was working with *Datura* on genetic and cytological problems. Blakeslee had observed chromosomal abnormalities in *Datura* which could be explained by the assumption of translocation of parts of a chromosome to another chromosome. When the process of transfer involved not an arbitrary part of a chromosome but the end, and when the transfer was a reciprocal exchange of chromosome ends or even whole chromosome arms between two non-homologous chromosomes, then normal pairing of the homologous parts would result in a cross-like figuration in pachytene and a circle of four chromosomes in metaphase.

As soon as Cleland and Blakeslee combined in their discussion the results on *Datura* with the known observations on *Oenothera*, the answer was obvious: reciprocal translocations of chromosome arms were the solution to the *Oenothera* problem. The joint publication of these ideas became the breakthrough to a karyological model of chromosome behaviour in

Oenothera (Blakeslee and Cleland 1930; Cleland and Blakeslee 1930, this being the most extensive and most important of the two publications; Cleland and Blakeslee 1931) (Fig. 7).

Looking back at the discussions in the older papers on *Oenothera* chromosomes, it is interesting to find that several scientists were very near to the solution (e.g. Shull 1928b; Sheffield 1929), but all had built a barrier in their own mind by considering a chromosome only as an undividable unit, a chromosomal individual, and having a strong belief in pairing of homologous chromosomes instead of structures containing substructures and possible pairing of homologous parts. However, it would not be fair to expect that in 1930 those biologists, who had just become acquainted with the new fields of genetics and cytology and had been working with *Oenothera* themselves in some cases less then 10 years, and who had still to fight for acknowledgement against the established fields of anatomy and physiology, would be prepared to throw away the knowledge they had just acquired in favour of a new idea. Cleland and Blakeslee had really taken a great step forward. Remember that the question of the chemical nature of biological structures was not "in", macromolecules had still not been invented at the time (I mean by the chemists, in nature they already existed but nobody knew) and that just by progress in this field of organic chemistry the understanding of chromosome behaviour was made possible.

4.2.7 Chromosomes and Translocations, or: Chromosomes Unlimited

As soon as the basis was laid, *Oenothera* cytology was pushed forward by big steps. The concept of individuality of chromosomes was replaced by individuality of chromosome arms.

In the years following the publications of Blakeslee and Cleland, many geneticists and cytologists published papers confirming the idea of reciprocal chromosome translocations and declaring the final solution of the *Oenothera* problem (e.g. Darlington 1929, 1931, 1933; Catcheside 1930; Hakansson 1930b; Emerson 1932; Kuhn 1933; Renner 1933a). In some cases where the new interpretation was included in papers on other topics, the suspicion arises that some people feared to miss the connection to a new field in biology and were jumping on the train that had just started, without contributing anything new to the movement themselves. On the other hand, publications still appeared raising doubts about the new interpretation of old observations (e.g. Sheffield 1929; Gates 1931; Illick 1932; Schwemmle 1932) but by 1933 this discussion was finished.

Meanwhile Cleland had formed connections with the German geneticists Renner and Oehlkers, involved in *Oenothera* research for many years, and on the basis of this collaboration the next steps were prepared (Cleland and Oehlkers 1929, 1930; Renner and Cleland 1933).

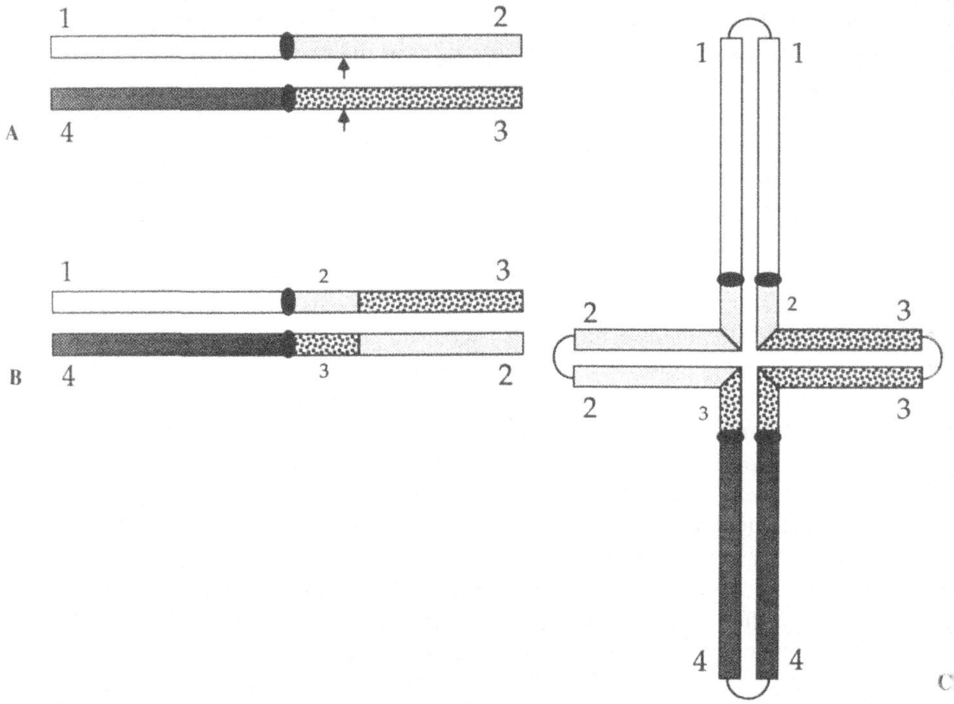

Fig. 7. Reciprocal translocation between two chromosomes. Schematic presentation of a reciprocal translocation between two chromosomes as an example for the chromosome configuration in meiosis of a heterozygote with one reciprocal translocation involving two chromosomes, combined with two chromosomes with the original arrangement of the same chromosome parts. The translocation points are supposed to lie in the heterochromatin which is not specifically indicated in the drawing (computer drawing by E. Rudloff). Movement of the chromosomes to the poles in anaphase will lead to a separation of the chromosome complexes of two chromosomes each in such a way that at the one end of the spindle the two translocated and at the other end the two original chromosomes will be included in the dyad cell. **A** Structure of the original chromosomes; breakage points for the reciprocal translocations indicated by *arrows*. **B** Structure of the translocated chromosomes with interstitial segments between the centromere and translocation point. **C** Cross figure in pachytene as result of pairing of homologous chromosome parts between two translocated chromosomes and two chromosomes with the original arrangement of the chromosome arms. **D** Ring of four chromosomes connected by end chiasmata in diakinesis, derived from the pachytene cross in which in all homologous parts chiasmata had been formed. **E** Chain of four chromosomes in zigzag orientation in metaphase I, connected by three end chiasmata. With four end chiasmata, an additional one in the chromosome arms No. 1, the result will be a zigzag ring of four chromosomes

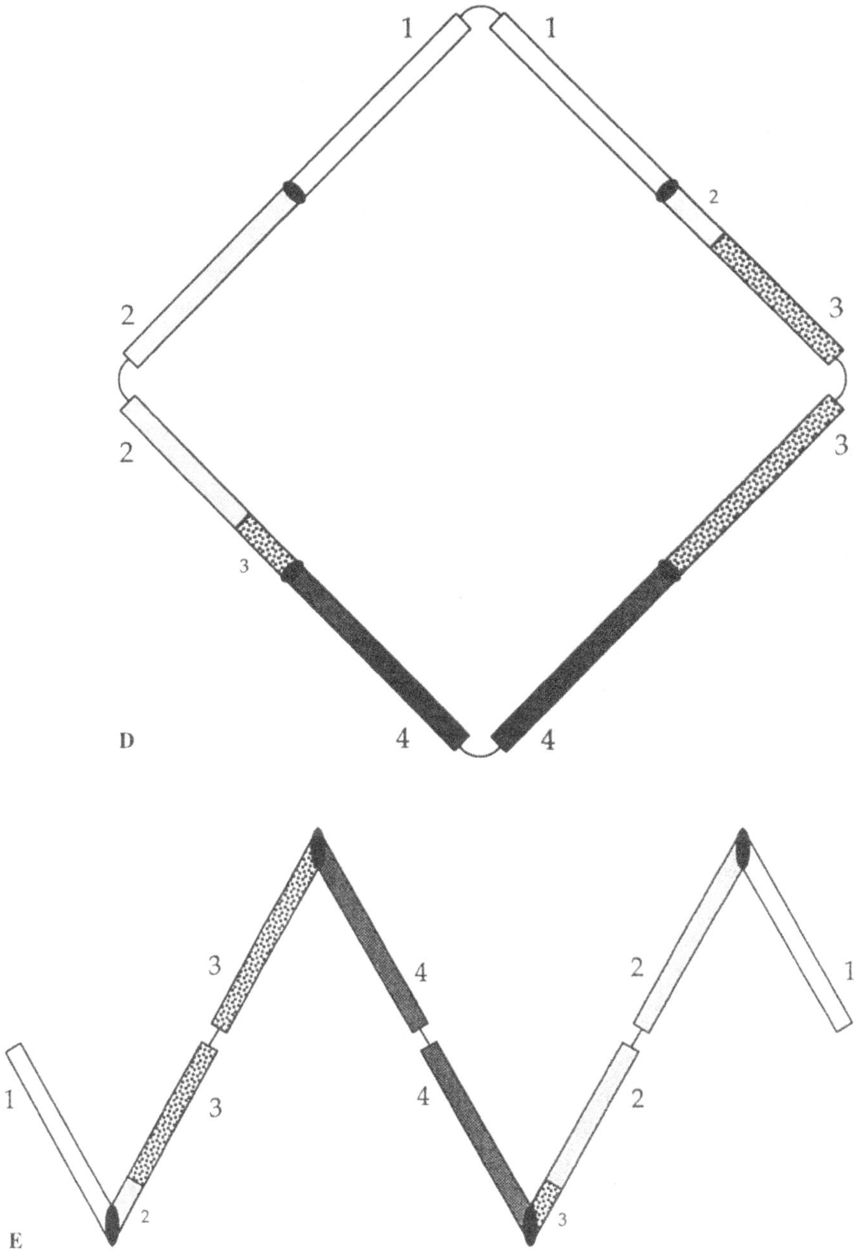

Fig. 7. *Continued*

Even before the solution was known, cytologists had recognized the formal problem and were discussing the possibilities of chromosome arrangement in larger groups. As a consequence of the new ideas the chromosome arms, and not the chromosomes as in other objects, were numbered for formal identification (Cleland 1933; Renner and Cleland 1933). The haploid set was considered to contain 14 individual chromosome arms, numbered from 1 to 14, which could be combined in different patterns to make seven chromosomes. The arrangement of these 14 chromosome arms in a given case is called the **chromosome formula**, describing a **chromosome complex**. The observations on *Datura* led to the conclusion that these patterns were not achieved by random combination but evolved by repeated reciprocal translocations, starting with a standard arrangement found in the cytological homozygous species with seven bivalents in meiosis. When two haploid sets of seven chromosomes are combined, they will arrange in meiosis in bivalents or chromosome circles, as expected from pairing of homologous arms. The figure formed by the chromosomes, as seen in diakinesis and metaphase I, is called the **chromosome configuration**. Each of these configurations can be made up of very different chromosomes, so the meiotic configuration considering the chromosome formula of the two haploid sets involved is called the **chromosome arrangement**.

On this basis the cytological chromosome configuration of a circle of four chromosomes, as observed in diakinesis and metaphase of the first meiotic division, can be taken as an indication of a difference of one reciprocal translocation between the two chromosome sets contributed by the parents. In this analysis some complications arose, but it was soon recognized that a translocation was not a safeguard against another translocation with a different chromosome (Darlington 1929). Two rings of four chromosomes each would mean a difference between the chromosome sets of the two parents made up by two translocations, involving two pairs of chromosomes. A ring of six chromosomes would also indicate a difference of two translocations but between three chromosomes. Three reciprocal translocations can lead to a meiotic chromosome configuration of either three rings-of-four, or two rings with six and four chromosomes respectively, or one ring-of-eight when combined with a set of the original arrangement of chromosome arms. Five translocation steps can lead to a ring-of-twelve plus a bivalent, but the same number of translocation steps can also produce two circles of either ten and four or eight and six chromosomes. Continuing this line, we find that six is the minimum number of reciprocal translocations necessary to connect 14 chromosomes in a circle.

There are exactly 15 different chromosome configurations which can arise from the combination of two sets of seven chromosomes each (Darlington 1931, Hedayitullah 1933). When a circle of chromosomes is designated by the number of chromosomes connected together in diakinesis and metaphase and a bivalent is considered to be a ring of two chromosomes, then the chromosome configuration seen in a meiocyte can be described by a simple formula. The possible configurations are then (for the moment look only at the first column):

Chromosome configuration	Number of different chromosome arrangements
2 2 2 2 2 2 2 (seven bivalents)	135 135
4 2 2 2 2 2	2 837 835
4 4 2 2 2	14 189 175
4 4 4 2	
6 2 2 2 2	
6 4 2 2	
6 4 4	
6 6 2	
8 2 2 2	
8 4 2	
8 6	
10 2 2	1 089 728 640
10 4	1 089 728 640
12 2	1 816 214 400
14 (circle of 14 chromosomes)	1 113 510 400

It has to be kept in mind that the chromosome configuration does not give any direct information about the chromosome formulas of the two sets of chromosomes. Each of these configurations can be made up of very different chromosomes.

The determination of the number of differences between two sets of chromosomes which could be interpreted as repeated reciprocal translocations, leading to the formation of chromosome circles in meiosis, gives only the minimum number that is necessary to make a circle of a given size. However, this simple consideration points towards an immense number of possible chromosome combinations.

To demonstrate this, some more theoretical considerations are necessary. With 14 chromosome arms 91 different chromosomes can be constructed (Cleland 1972). These can be combined to form haploid complexes of seven chromosomes. The combination cannot be arbitrary. A complete haploid chromosome complex should contain every chromosome arm, but only once. To calculate the number of possible different complete chromosome complexes is a mathematical problem of combination theory. Without going into details of the calculations, it is evident that the number of different complete sets will be limited but very large. There exist 135 135 different orders of 14 chromosome arms arranged in seven chromosomes. Each of these represents the chromosome formula for a hypothetical Renner complex. Without lethals this will make the same number of possible homozygotic species.

The chromosome complexes can be combined to make up diploids, the unimaginable total number of 9 130 801 680 different ones. Besides the already mentioned homozygotes there are 9 130 666 545 complex heterozygotes possible with one of the 14 different meiotic chromosome configurations with at least a ring-of-four.

Starting with a given haploid complex of seven chromosomes, each new reciprocal translocation will always lead to the arrangement of another one of these sets. Considering a given arrangement of chromosome arms, a chromosome complex, there are exactly 42 different reciprocal translocations between two chromosomes which will make a circle of four chromosomes together with the starter complex. This holds for any complex of seven chromosomes which is characterized by the arrangement of the 14 arms. Counting the number of possible different rings-of-four will still give a limited number. To find the exact number we have to start with the 14 chromosome arms. There exist 1001 different combinations of four chromosome arms out of 14, each with three possibilities for different arrangements of the four arms (e.g. 1·2 and 3·4, or 1·3 and 2·4, or 1·4 and 2·3), which can be combined to make up three different circles of four chromosomes.

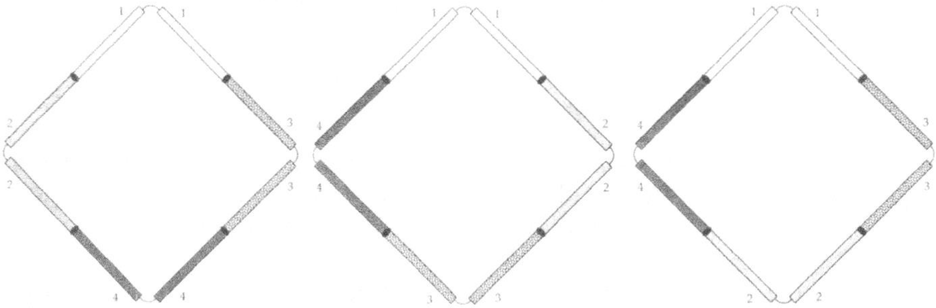

Considering 14 chromosome arms, that makes altogether 3003 different rings-of-four. To find the chromosome configuration 4 2 2 2 2 2, the remaining chromosomes of the two chromosome complexes must have the appropriate formulas to form five bivalents. For each of the 3003 rings-of-four there exist 945 arrangements of the 10 remaining chromosome arms in five chromosomes and the same number of possibilities to make five bivalents. That means there are 2 837 835 possibilities to achieve this simple chromosome configuration.

When there is one circle of four chromosomes, there can be a second, originating from another translocation between two chromosomes. From 10 chromosome arms 210 different groups of four chromosomes can be formed, each one making three different arrangements with the possibility of forming three different circles, as demonstrated above, and giving 630 different rings-of-four to combine with each of the first ones. The remaining six chromosome arms can make up 15 different groups of three chromosomes. Considering the rules of combination, that makes altogether 14 189 175 different cases of 4 4 2 2 2. Some more figures are presented in the list of chromosome configurations. Go on and see if you can work out how many possibilities exist for each of the remaining chromosome configurations.

On the other hand, starting with a given chromosome complex, the number of possible reciprocal translocations following one another is unlimited. There does not exist a limiting rule which prohibits new translocations from occuring, even when, after several steps, the original formula for the arrangement of chromosome arms is reached again. The inevitable consequence is a more complicated structure of the middle regions. This means that for a given chromosome formula of a chromosome complex it is possible to calculate how many steps are necessary for the shortest way from the standard arrangement of the chromosome ends, or from any other arrangement supposed to be the origin, to achieve this special structure. On the other hand, it can never be known how many steps were really taken and how many roundabouts were included on the way. Furthermore, new translocations cannot only occur between chromosomes of one complex but can also involve chromosomes of the other complex combined with it in a nucleus. The only restriction is that the resulting haploid chromosome complex must be complete, containing all the chromosome arms.

One point is clear: the number of Renner complexes identified in *Oenothera* is large, but it is still small compared with the possibilities of chromosome arrangements. In consequence the probability of a newly discovered Renner complex being identical with one whose chromosome formula is already known is extremely small.

This is enough to demonstrate the dimension of the problem. Let us stop calculating, otherwise the number of pages with possible chromosome formulas will reach astronomical dimensions without contributing much more to the solution of the problems.

Each reciprocal translocation has to be considered as an evolutionary event. In *Oenothera* there is a very large but limited number of possible chromosome formulas and chromosome arrangements in the chromosome configurations in meiosis, but an unlimited number of ways of reaching each of them. There is no limit to evolution.

4.2.8 Pairing Behaviour of Translocated Chromosomes

Besides abstract formulas for arrangements of chromosome arms, the behaviour of the chromosomes has to be considered. When two sets of two chromosomes each are taken, which differ by one reciprocal translocation, the pairing of the homologous parts will lead to a cross-shaped figure in late pachytene. But how this figure looks exactly will depend on the location of the translocation sites on the chromosome arms.

The construction of translocation configurations, based on the assumption of different locations of the translocation points on the chromosome arms and the consideration of the theoretical consequences in each case, led to the conclusion that formation and terminalization of chiasmata in the translocated parts will lead to end chiasmata in diakinesis and metaphase and thereby transform the pairing cross into a circle. Chiasma formation in the interstitial segment, between the translocation point and the centromere, however, will lead to disturbances during the process of terminalization and will give rise to complicated, unexpected chromosome configurations from diakinesis onwards, followed by disturbance of chromosome distribution to the poles in anaphase I. These are observed and discussed in investigations on mutagenesis but are not found in species or hybrids (Marquardt 1948; Harte 1973a). The conclusion is that the translocation points of the chromosomes must be located in those parts which are protected from chiasma formation, that means the heterochromatic parts of the chromosomes. The investigations of Japha (1939) demonstrated that in the rings-of-four included in her project the translocation point is located in the heterochromatic middle region of the chromosomes near to the centromere but not identical with it. The translocations are symmetrical so that the middle regions of the translocated chromosomes nearly all have the same size. When the length of all chromosome arms is nearly the same in all species of *Oenothera* and only reciprocal translocations of chromosome arms consisting of the euchromatic region and a part of the adjacent heterochromatin are considered, large differences of size between the translocated parts of the chromosomes can be excluded. This is very important for understanding the chromosome behaviour in *Oenothera* (Figs. 8, 9, 10).

The chromosome part between the centromere and the translocation point is called the **interstitial segment**. When a chromosome is involved in several translocations, which do not occur at exactly the same site, then the chromosome part between two translocation points will not be homologous to either of those chromosomes which are homologous to the two arms, but will be homologous to the third chromosome. This part is called the **differential segment**. When chiasmata are formed in this part, terminalization is not possible and very complicated configurations will result in diakinesis and metaphase I, making a regular distribution to the anaphase groups nearly impossible. In *Oenothera* the differential segments are located in the heterochromatic middle region and can therefore not be a source of meiotic complications caused by failing terminalization of chiasmata.

In types with large chromosome circles, evolved by a series of reciprocal translocations, this structure of the *Oenothera* chromosomes will cause the heterochromatic middle regions with the centromeres, the translocation points, interstitial and differential segments to lie so close together after completion of chromosome pairing at the end of the first meiotic prophase that an analysis seems impossible, not because of insufficient cytological techniques but from "technical reasons" on the part of the chromosomes.

Another conclusion is inevitable. In the pairing behaviour of large chromosome groups the middle part must have such a complicated structure of interstitial and differential segments of various lengths, including the centromeres, that for mechanical reasons unpaired regions, as assumed by Catcheside (1933b), must occur. This would be an additional safeguard against unwanted complications by chiasmata situated between translocation points.

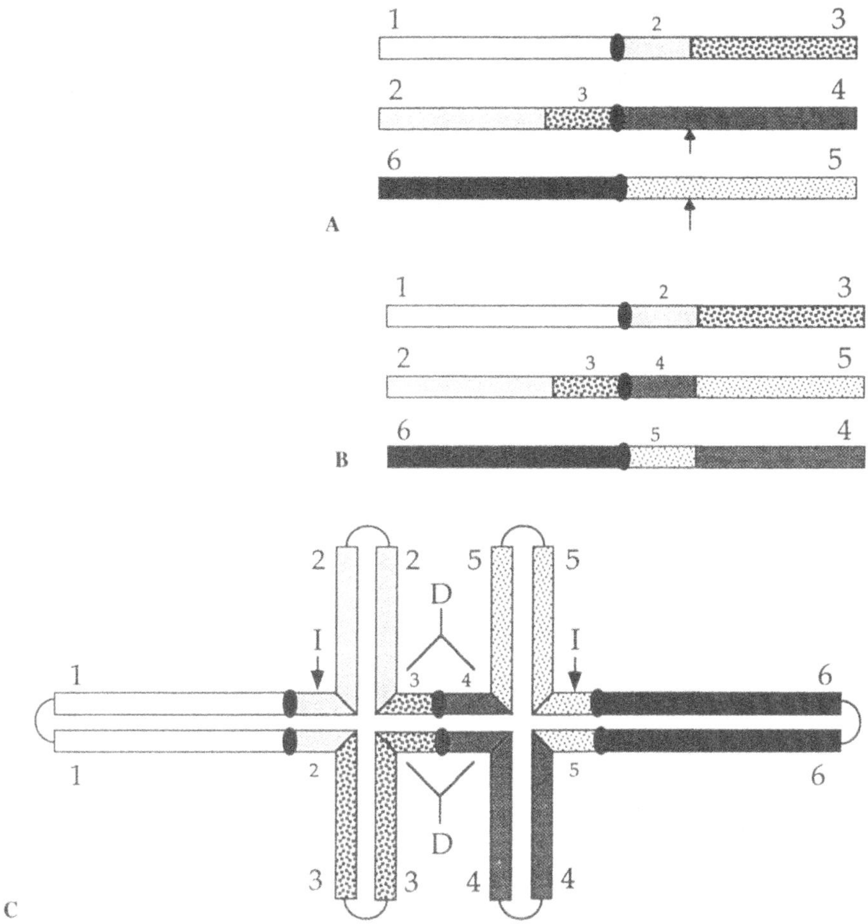

Fig. 8. Reciprocal translocations involving three chromosomes. Schematic representation of some possible results of two successive reciprocal translocations between three chromosomes which illustrate the different results of translocation events, even when the formal description, based on the exchange of chromosome ends, is the same. The examples demonstrate chromosome configurations in complex heterozygotes with three translocated chromosomes, combined with three chromosomes which have the original arrangement of the chromosome parts. The translocation points are supposed to be in the heterochromatic middle region of the chromosomes which is not specifically indicated in the drawings. When the anaphase movement of the chromosomes is finished, the result will be a separation of the two chromosome complexes, the one with the three translocated chromosomes, the other with the original arrangement. In Figs. 8–10: *I* interstitial segments, *D* differential segment (computer drawings by E. Rudloff). First example: **A** Arrangement of chromosome parts before the second reciprocal translocation, sites of breakage indicated by *arrows*. **B** Translocated chromosomes with interstitial segments between centromere and translocation point and a differential segment between two translocation points in the same chromosome. **C** Pachytene pairing in a complex heterozygote with three translocated chromosomes and three chromosomes with the original arrangement of chromosome arms. **D** Ring of six chromosomes in diakinesis, following formation of chiasmata in all homologous chromosome ends. **E** Chain of six chromosomes in metaphase I in zigzag arrangement and with five end chiasmata. When an additional end chiasma connects the two homologous free chromosome arms, a ring of six chromosomes in zigzag arrangement will be present

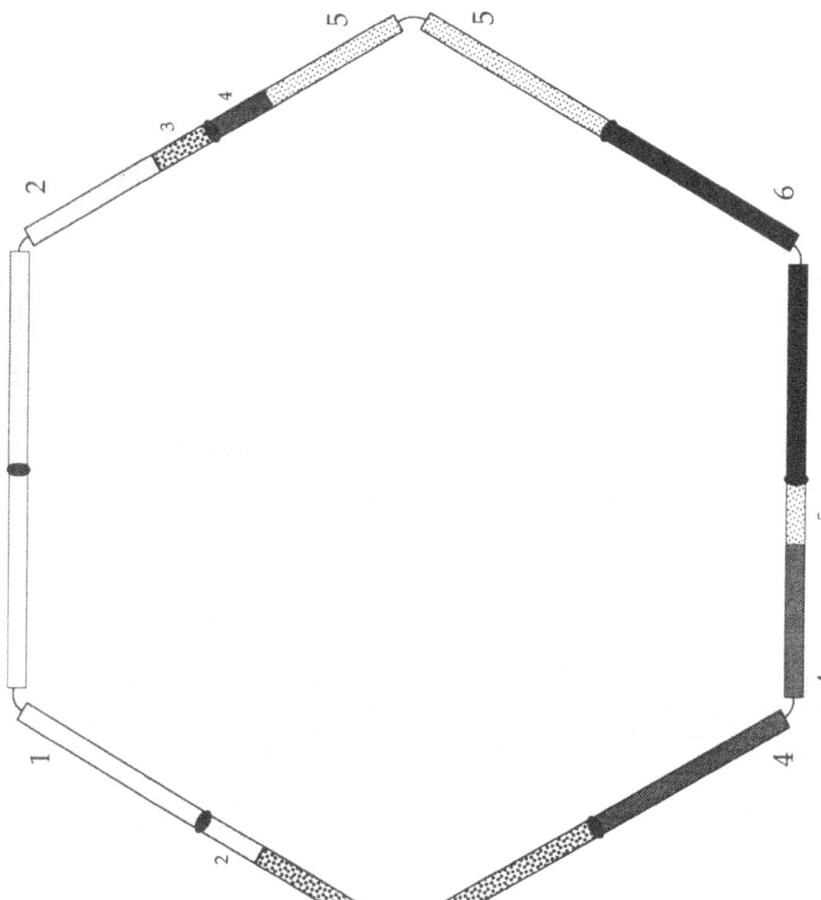

D

Fig. 8. *Continued*

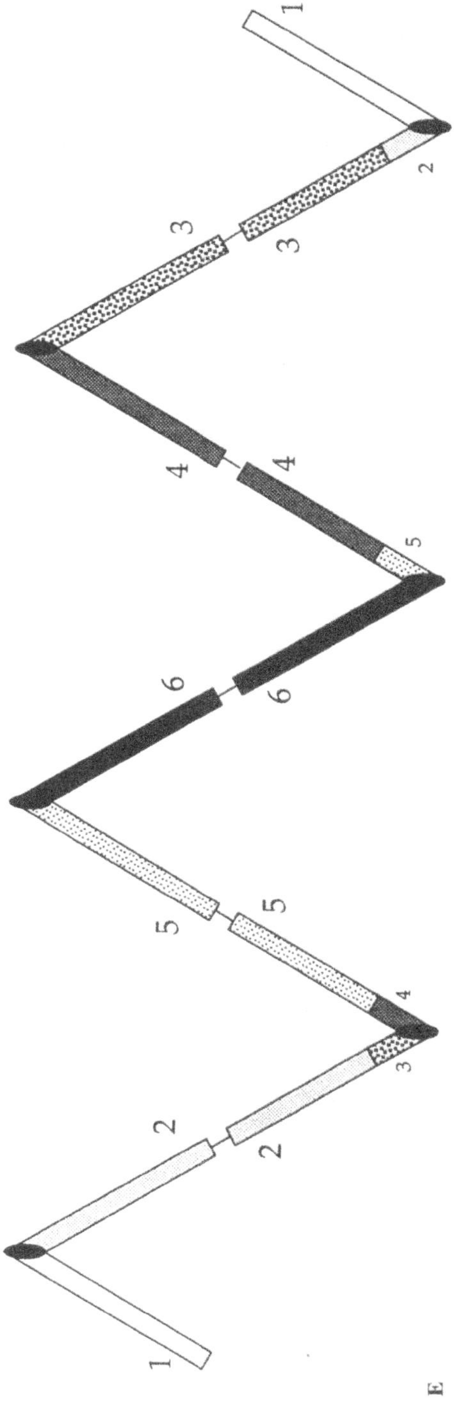

E

Fig. 8. *Continued*

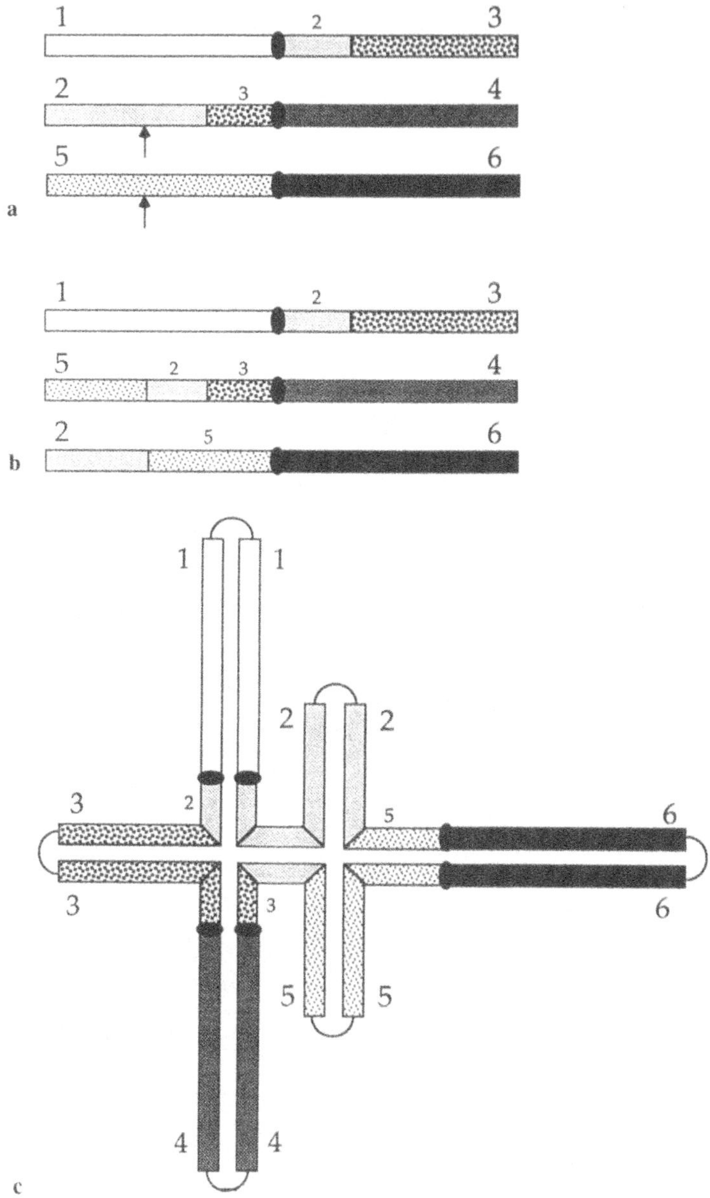

Fig. 9. Second example for two reciprocal translocations involving three chromosomes. From diakinesis onwards to anaphase I the arrangement of the chromosomes will be in analogy to the first example presented in Fig. 8. **a** Same initial situation, but with different location of the sites of breakage for the second reciprocal translocation as indicated by *arrows*. **b** Arrangement of the chromosome parts after the second reciprocal translocation according to this modus. **c** Pachytene pairing between the homologous chromosome parts in the three translocated chromosomes and in the chromosomes with the original arrangement of chromosome arms in a complex heterozygote

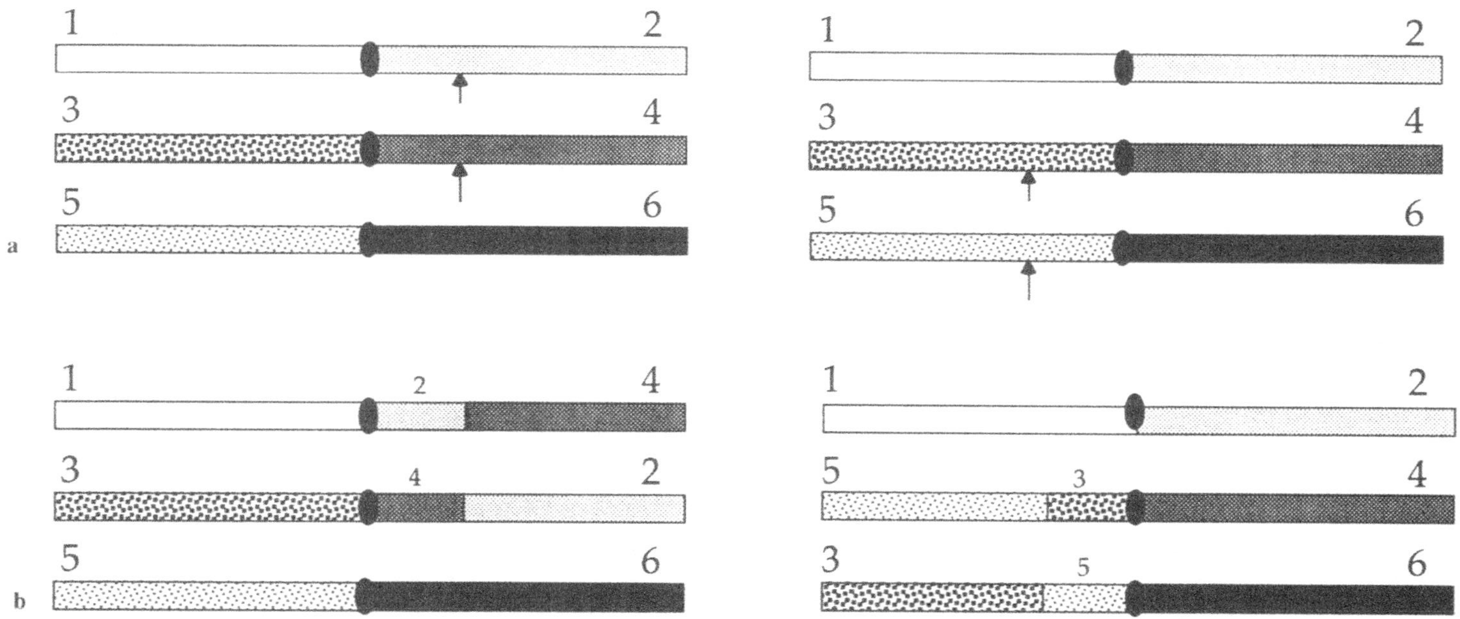

Fig. 10. Third example for a meiotic chromosome configuration of six chromosomes. This is formed in a heterozygote by combination of two chromosome complexes with one reciprocal translocation each, involving together three chromosomes. By the chromosome movement in anaphase I the two chromosome complexes will be separated with the same arrangement of chromosome parts as introduced into the heterozygote by the gametes, each with one chromosome in the original arrangement of chromosome arms and two translocated chromosomes. **a** Initial arrangement of the chromosome parts in the two chromosome complexes involved. **b** Structure of the translocated chromosome of the two chromosome complexes. **c** Pachytene pairing of homologous chromosome parts in a heterozygote with two chromosome complexes, each with a different reciprocal translocation. **d** Chain of the six chromosomes, connected in metaphase by five end chiasmata

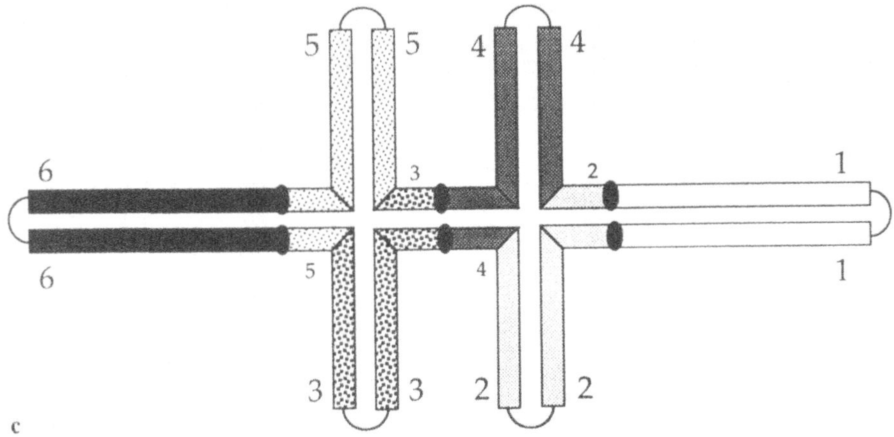

c

Fig. 10. *Continued*

The structure of translocated chromosomes and their pairing behaviour lead to an alternation of parental chromosomes in the circles. In metaphase the circle is orientated in the spindle of most meiocytes in such a way that the adjacent chromosomes must pass to opposite poles in anaphase I. This behaviour is not seen in newly arisen chromosome translocations in other genera like maize. It must have evolved in *Oenothera* as a regular event. Most probably it is due to the position of the translocation points near the centromeres in the large heterochromatic middle parts of chromosomes with arms of approximately equal length. In this way the regular mechanics of chromosome movement will cause the chromosomes of each set to assemble at the same pole and be included together in one dyad nucleus. A recombination of parental chromosomes, as can occur after bivalent pairing, is impossible. As a result the chromosomes of a complex will manoeuvre together in meiosis. The haploid set of chromosomes that was brought into the zygote by a gamete will stay together and be passed to the next generation as a unit.

4.2.9 Physiology of Meiosis and Chiasmata

With the introduction of better staining techniques for chromosomes it became possible to ask questions other than those about chromosome structure and the description of stages of nuclear division. As soon as the improved slides made it possible to investigate not only single cells but populations of PMCs in the anthers, an experimental and quantitative approach to meiotic phenomena could begin.

This was seen by Oehlkers who started research on the physiology of meiosis. One point is whether the chromosomal events during meiosis can be experimentally changed. An important step in meiosis is the formation of

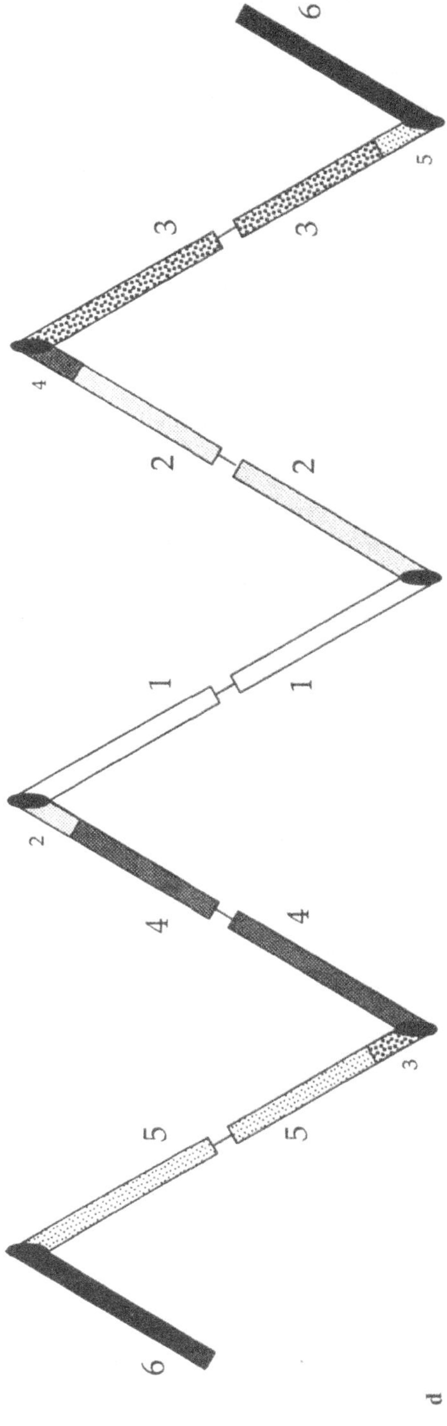

d

Fig. 10. *Continued*

chiasmata in the paired chromosomes. The problem which Oehlkers chose to investigate was the influence of external factors on chiasma formation. Considerations about the cytological basis of the genetic phenomenon of crossing-over were the starting point. If crossing-over was caused by exchange of parts between homologous chromosomes during prophase pairing, then a change in chiasma frequency must be correlated to similar changes in crossing-over frequency (Oehlkers 1935b, 1936, 1937). The problem was to find an object which had sufficient analyzed genes and also had chromosomes suitable for quantitative estimation of chiasma frequencies under experimental conditions. As an *Oenothera* geneticist he chose his favorite object and consequently started with the critical point of finding out whether chiasma frequency was variable and whether correlations with experimental influences could be established.

In *Oenothera* the chiasmata cannot be counted directly. Instead, the number of end chiasmata, or end connections as they are sometimes called, in diakinesis was chosen as a test. Thus, not the chiasmata but the number of chromosome arms with at least one chiasma are considered. These can be counted reliably. Between 14 chromosomes there are, irrespective of the chromosome configuration, 14 end-to-end connections, each indicating chiasma formation in a special chromosome arm. When, in one arm of paired chromosomes which should appear as a ring bivalent, no chiasma is formed, then from diakinesis onwards there is a rod bivalent with two free ends and only 13 end chiasmata in the whole nucleus. A bivalent without chiasmata leads to two univalents. When, in paired arms of a larger group of catenated chromosomes no chiasma is formed, then from diakinesis onwards the end-to-end connection will be missing and the circle is transformed into a chain with two open ends. Several chromosome arms without chiasma formation in a translocation figure will cause a disruption of the circle into two or more chains or univalents. The frequency of end chiasmata per nucleus varies between 14 and 0. From populations of PMCs from flower buds of inflorescences kept under experimental conditions, the frequency of nuclei with different numbers of end chiasmata can be determined and a statistical evaluation of the results is made possible. Besides the crystal-violet staining, and later aceto-carmine squash preparations, the introduction of statistics was the second step which made progress in this field possible.

In the evaluation of the results of experiments on influencing chiasma frequency it must be reminded that the number of chiasmata in a sample can considerably increase or decrease without influencing the frequency of end chiasmata, as long as in every chromosome arm at least one chiasma is formed. Only when the chiasma frequency becomes so low that paired chromosome arms without any chiasmata occur more frequently, can the difference between the experimentally treated sample and the contol be detected with this method by statistical analysis of the frequency distributions.

In a series of publications on the physiology of meiosis Oehlkers and his students, using both field material and cut inflorescences in temperature-controlled rooms, reached the conclusion that the chiasma frequency is not a constant but is variable under experimental conditions. The frequency of chiasmata is decreased by low temperature (Zürn 1936). In experiments in which changes of temperature during the treatment are the important variable, the lowest temperature decides the reaction (Haselwarter 1937). Photosynthesis is an important factor, as is proved by a correlation between chlorophyll content and chiasma frequency (Zürn 1936) and as is established further by using chlorophyll-defective types (Oehlkers 1940b,c). Simulating an increase in photosynthesis by feeding cut inflorescenses with glucose in a controlled environment increases the chiasma frequency too (Zürn 1939; Wiebalck 1940). Application of salts, e.g. KCl, in a series of concentrations led to a decrease in chiasma frequency. Organic substances like Urethan or Saponin had no detectable influence, or caused an increase in chiasma frequency (Oehlkers 1943a). Hydration of the plant has an influence too. In extreme conditions of drought or superfluous watering of the plants, fewer chiasmata are observed than in plants which had an optimal water content (Kisch 1937). After an experimental drought period it takes several days with normal water conditions for the normal chiasma frequency to be regained (Harte 1942). At the end of the experiment the plant will need some time to regain the normal hydration of the cells. The chiasmata observed in diakinesis are the result of events that occurred some time before. This means that the time lapse between the stage of chiasma formation and observation cannot be less then 2 or 3 days. Because pachytene lasts for 3 or 4 days (Linnert 1951), the stage of chiasma formation cannot be later than mid-pachytene.

With the improved method of squash preparations it became possible to analyze not only a selection of undamaged PMCs, as on slides, but nearly all PMCs in the appropriate stages of meiosis in an anther and include in the analysis the variability between anthers of a flower bud, between buds of a plant and between different plants grown under the same conditions and prepared for investigation at the same time. The differences between these groups, which are expected to have the same chiasma frequency, are significant in several cases. The assumption of equal chiasma frequency under similar conditions cannot generally be confirmed. However, the differences are smaller than between the experimental groups in earlier investigations (Harte 1953c, 1954a,b).

The chiasma frequency in different chromosome configurations in the nucleus demonstrated that not all chromosomes react in the same way to experimental treatment (Catcheside 1933b; Harte 1954a,b). The chromosomes of a ring-of-four generally have fewer end chiasmata than the bivalents of the same sample. The relative chiasma frequency in circles and bivalents is not a constant but depends on the genotype investigated. Because the circles in different hybrids contain different chromosomes, this observation proves that differences between chromosome arms in chiasma formation exist. Differences between chromosomes in circles and bivalents and between circles of the same size but with different chromosome formulas are also found in the reaction of increasing or decreasing chiasma frequency (Harte 1954a). In the large circle of *Oe. albicans·stringens* with 12 chromosomes two chromosome arms are present with a chiasma frequency considerably less than in the others, while the chromosome arm connected to these two, lying between them in the ring, has a very regular end chiasma (Harte 1954b). Catcheside (1933b) observed differences in chiasma frequency between a ring-of-four and bivalents and assumed that unpaired regions in pachytene in translocation chromosomes could be the cause of reduced chiasma frequency. According to the observations of Japha (1939) and Marquardt (1948) disturbance of pairing can happen mainly in the heterochromatic interstitial segment where no chiasmata are expected. Combining these considerations with the observations on different chromosome circles it must therefore be concluded that a disturbance of pairing cannot be the cause for the observed difference in chiasma frequency, but that real differences in the probability of chiasma formation between chromosome arms do exist.

4.2.10 Meiosis in Polyploids

From the first investigations of *Oenothera* chromosomes onwards, observations on aberrant chromosome numbers are recorded, mostly for *Oe. Lamarckiana* de Vries (van Overeem 1921; Boedijn 1925a; Hakansson 1926). These include differences of ploidy as well as aneuploid chromosome numbers. In both cases, in addition to the determination of the chromosome number, the pairing behaviour, as far as it can be derived from the chromosome configurations in metaphase, was a main point of interest.

In haploids seven chromosomes are mostly seen as univalents, but occasionally pairing and chiasma formation is observed (Emerson 1929c; Kulkarny 1929a). Catcheside (1932) presented a theoretical discussion on chromosome pairing in haploids, based on the assumption that only complete homologous parts, not even alleles of a locus, can pair. Because in *Oenothera* haploid pairing and chiasma formation are observed, the haploid set in *Oenothera* must contain several identical parts in different chromosomes to enable the chromosome configurations observed in diakinesis of the haploid PMCs. The possibility of irregular pairing, in case the ideal homologous partner is not available, is not considered. Because pachytene analysis is not possible, the question of pairing of non-homologous parts in haploids has to remain open until the nucleotide sequences of all *Oenothera* chromosomes are available for comparisons. At the moment it is only possible to conclude that in diploids homologous pairing is preferred.

Triploid plants with 21 chromosomes, called hemigigas, are described by Boedijn (1925b) and Capinpin (1930, 1933). Hakansson (1931) and Catcheside (1931b) investigated trivalent configurations. Descriptions of meiosis in tetraploids, the gigas plants, are given repeatedly (e.g. Boedijn 1925a; Hakansson 1926; Davis 1933; Renner 1933b; Seitz 1935). Darlington

(1929) compared the ring formation in tetraploids with translocation figures. An extensive discussion on multivalent pairing became possible when, by new techniques, quantitative investigations on the frequency of chromosome configurations in polyploids were presented.

Linnert (1948) found in tetraploids of *Oe. Hookeri* de Vries and the related species *Oe. franciscana* and *Oe. Johannsen*, that the four homologous chromosomes can join as a group, but in any part only two of them are paired. Chiasma formation in the paired regions and a change of partner in the group occur regularly. As a consequence of chiasma formation and a change of the pairing partner, in diakinesis three or all four chromosomes of a quadrivalent can be connected by end chiasmata to form a range of different configurations, but combinations of univalents, bivalents and a trivalent are observed too. When a change of the partners occurs in a quadrivalent, then the probability of another one occurring in the same group is drastically reduced. There is a double correlation observed: when the chiasma frequency increases, the possibility of detecting changes of partners increases too but, at the same time, the frequency of chiasma formation is reduced by frequent exchange of the partners. This can be explained by the occurrence of small unpaired regions in the euchromatic parts when a chromosome changes its pairing partner in the quadrivalent.

When a form with a circle of four chromosomes becomes polyploid, as is the case in the hybrid *Oe. Hooker* 4n × *Johannsen* 4n, then a double cross, or a figure of eight chromosomes with change of partners should be possible in pachytene. However, the expected configurations of eight chromosomes are much less frequently observed than expected. The frequency of chiasma formation must be reduced in the pairing configurations which become very complex by the combination of exchange of partner in the homologous euchromatic parts and at the translocation points in the heterochromatin. On the basis of the empirical observations on the frequency of different quadrivalent formations, the possibilities of quadrivalent pairing are discussed (Linnert 1948). Similar observations on pairing in tetraploids which are heterozygotic for a translocation have been made by Hecht (1942). The double bivalents are expected to appear as quadrivalents and the double ring-of-four as a figure-of-eight in diakinesis, but both chromosome configurations are not observed in the meiocytes. Only circles of four or eight chromosomes are discussed, and not the more interesting complicated figures observed by Linnert.

4.2.11 Meiosis in Aneuploids

Especially in the progeny of *Oe. Lamarckiana* de Vries aberrant forms with 15 chromosomes were observed (van Overeem 1921; Boedijn 1925a,b; Herzog 1940). Later the chromosome configuration in diakinesis became interesting (Catcheside 1933a, 1936; Renner 1949). The results of trivalent pairing, either in triploids or trisomics, are essentially the same as for quadrivalents. As the consequence of a change of partners within the group of three chromosomes unpaired chromosomes and chromosome regions are observed. Chiasma formation occurs in the paired regions and leads to configurations of three chromosomes from diplotene onwards, connected by end chiasmata in diakinesis. A polarization of the trivalents during the first metaphase and anaphase in the meiocytes cannot be observed. The supernumerary chromosome is distributed at random to the poles of the meiotic spindle, irrespective of whether it is a univalent or a member of a trivalent configuration (Herzog 1940; Renner 1949). The same holds true when in poly-trisomics more then one supernumerary chromosome is present. Similar observations were made on *Oenothera* species of the sect. *Raimannia* (Haustein 1956).

Capinpin (1933) investigated the progeny of triploid *Oe. Lamarckiana* de Vries obtained by self-pollination and hybridization and found plants with chromosome numbers ranging from 14 to 21. In PMCs, not only the expected trivalent configurations were observed, but also bivalents, univalents, and broken and open long chains of chromosomes.

4.2.12 Some Additional Observations

Extra diminutive chromosomes were observed by Cleland (1951, 1967) in several races of *Oe. Hookeri* de Vries. They do not pair with the normal chromosomes. The investigation of hybrids revealed that in two of the races the small chromosomes are homologous. They can pair and form chiasmata, even when the size of the extra chromosomes is different (Cleland and Hyde 1963). The origin of these peculiar chromosomes is probably found in an inter-subgeneric hybridization between *Oe. Hookeri* de Vries and an undefined species from another subgenus, and subsequent backcrossing with *Oe. Hookeri* de Vries.

Chromosome pairing in a hybrid *Oe. Berteriana* × *muricata* can occur. The origin of the chromosomes can be recognized by the differences in size. At least three pairs of a large and a small chromosome each and chains with alternating large and small chromosomes are observed, as well as many univalents and rod bivalents; however, viable progeny could not be obtained (Schwemmle 1927).

4.2.13 The Karyological Model

In the evolution of the genus *Oenothera* reciprocal translocations of whole chromosome arms must have occurred repeatedly. The consequence is the formation of cross-shaped figures or more complicated star-like arrangements of chromosomes during the pairing of homologous chromosome parts in the prophase of the first meiotic division. Following regular chiasma formation in the euchromatic chromosome arms, the process of terminalization from diplotene onwards will lead to chromosome circles, as observed in diakinesis and metaphase I.

The location of the translocation points in the heterochromatic middle parts of the chromosomes protects the interstitial and differential segments from crossing-over. The approximately equal length of the arms is the cause for equal distances between the centromeres in the chromosome configuration, and both factors facilitate the formation of a zigzag orientation of the chromosomes at metaphase I, enforcing the movement of homologous arms to different poles in anaphase I. By this mechanism complete sets of chromosomes, containing all 14 chromosome arms in the same order as in the parents, are kept together during the first meiotic division. There is no recombination of parental chromosomes.

The result is the formation of chromosome complexes which are transferred to the meiospores together and are consequently inherited as a unit, even in the case of complicated translocation patterns.

4.2.14 Limitation of the Karyological Method

The application of karyological methods to investigate *Oenothera* chromosomes led to a description of chromosome behaviour and to an interpretation of the specialities in chromosome pairing observed in meiocytes. However, a particular method can give information only in the field for which it is designed. Karyological methods can tell us about chromosomes but nothing else.

The question remains whether the most intriguing observation of chromosome catenation reveals just another peculiarity of *Oenothera*, as suggested by some investigators, or has a genetic meaning. The next step must therefore be to bring together the genetic conclusions about Renner complexes and the karyological observations on chromosome complexes and look for a causal relationship between the genetic and karyological models. Cytogenetics is the next area to be investigated.

4.3 Cytogenetics of *Oenothera*

4.3.1 Combination of Two Methods

The result of research on *Oenothera* by the genetic method is the concept of Renner complexes, each characterized by the combination of alleles of many genes and inherited as a unit. The result of research with the methods of classical karyology is the discovery of chromosome complexes, characterized by the arrangement of chromosome arms and inherited as a unit. The logical consequence is to combine the two sets of results, achieved by the application of different methods to the same object. The result of this approach to the problems will be a cytogenetic model of inheritance in *Oenothera*.

4.3.2 Renner Complexes and Chromosome Circles

Before the concept of reciprocal translocations between the chromosomes was introduced, the first cytogenetic investigations had revealed an important correlation. Those species which proved to be genetically homozygotic have the meiotic chromosome configuration of seven bivalents. On the other hand, large chromosome circles are observed in the heterozygotic species with two different Renner complexes. It was soon recognized that when one large or several short chains in a nucleus are considered to be parts of a circle, the chromosome configuration in meiosis is a constant for each combination of Renner complexes, irrespective of whether it is in a species or a hybrid. The logical conclusion from the coincidence of genetic and karyological observations is the assumption that the chromosome complexes could be the material basis for the Renner complexes.

If this is true, then it should be possible to allocate a certain arrangement of chromosome arms to every genetic Renner complex and, following this idea, to locate identified genes or small linkage groups in numbered chromosome arms, irrespective of the fact that these cannot be identified by cytological criteria. Furthermore, there should be predictable interdependence of other karyological and genetic phenomena.

To test the hypothesis, a formal analysis is performed. The chromosome arms are given arbitrary numbers. Two chromosome arms making up a whole chromosome are connected by a dot, e.g. 1·2. The standard set of chromosomes contains, by definition, the 14 chromosome arms in numerical order. A reciprocal translocation is indicated by an exchange of numbers in the chromosome pair involved in the process. Because standardized rules would be of great help in identifying the pattern of chromosome arms in the Renner complexes, Renner suggested such a set of rules and applied them in many papers (e.g. Renner 1933a, 1938b).

The identification should start with the chromosome 1·2 and any translocation discovered should always involve the chromosome with the lowest remaining numbers. As the numbers are arbitrary, the rule was introduced that the first translocation discovered between two chromosomes should be described by an exchange of the two last numbers, i.e. a translocation between the chromosomes 1·2 and 3·4 will result in the chromosomes 1·4 and 2·3. By this procedure the chromosome arms involved are identified and the numbers given to them cannot be further changed at random.

The first step was the identification of the chromosome configurations of as many species and hybrids as were available. Many were already known (e.g. Cleland and Oehlkers 1929, 1930; Gerhard 1929; Hoeppener and Renner 1929; Langendorff 1930; an extensive list by Emerson and Sturtevant 1931). The homozygotic *Oe. Hookeri* de Vries, already widely used in genetics, was elected as a standard and the Renner complex ^{h}*hookeri* was given the chromosome formula:

^{h}*hookeri* 1·2 3·4 5·6 7·8 9·10 11·12 13·14.

Next, hybrids with a circle of four chromosomes were sought. The first was *Oe.* ʰ*hookeri·flavens*. The ring-of-four indicates a difference of one translocation between the chromosome complexes and the chromosomes involved were designated as 1·2 and 3·4. Because these are introduced into the hybrid by definition from *Oe. Hookeri* de Vries, the translocated chromosomes are 1·4 and 2·3 and were introduced into the hybrid by the Renner complex *flavens*. The chromosome formula connected to *flavens* is therefore 1·4 2·3 and the remaining chromosomes, pairing in bivalents with those from *Oe. Hookeri* de Vries, have to be identical with them:

flavens 1·4 2·3 5·6 7·8 9·10 11·12 13·14.

The hybrid *Oe.* ʰ*hookeri·velans* also has a ring-of-four, again indicating a difference of one translocation. To find out if any of the already identified chromosome arms are involved, the hybrid *Oe. flavens·velans* has to be investigated. One of four possibilities must be realized:

Possible translocation in *velans* compared to ʰ*hookeri*	Expected chromosome configuration in *flavens·velans*
Same translocation in *flavens* and *velans*	Seven bivalents
Translocation between the same chromosomes as involved in *flavens*, but in another order	Circle of 4 chromosomes
Translocation between two chromosomes not involved in *flavens*	Two circles of four chromosomes each
Translocation between one of the chromosomes involved in *flavens* and another one	Circle of six chromosomes

In all cases the remaining chromosomes should be arranged in bivalents.

The chromosome configuration that is actually found is 4 4 2 2 2 (two circles of four chromosomes each and three bivalents), so *flavens* and *velans* differ by two translocations. The Renner complex *velans* cannot have the translocated chromosomes 1·4 and 2·3 found in *flavens* but must have these chromosome arms in the same arrangement as ʰ*hookeri*, making up the first of the two circles with *flavens*. Applying the given rules, the translocation in which *velans* differs from ʰ*hookeri* is 5·6 7·8 to 5·8 7·6. The remaining chromosome arms, pairing in bivalents with ʰ*hookeri* and *flavens*, must also have the same arrangement as in these two Renner complexes.

velans 1·2 3·4 5·8 7·6 9·10 11·12 13·14.

Oe. flavens·stringens also has a ring-of-four and *Oe.* ʰ*hookeri·stringens* has two circles of four chromosomes each, so *stringens* could have the same chromosome formula as *velans* but, instead of seven bivalents, the combination *Oe. velans·stringens* has the chromosome configuration 6 4 2 2 (two rings, one with six and one with four chromosomes). Applying to *stringens* the same procedure as demonstrated for *velans*, the combination of the data leads to the conclusion that *stringens* must have in common with *flavens* the two chromosomes which are different from ^{h}*hookeri* in this Renner complex and an additional translocation involving one of the chromosomes in which *velans* differs from ^{h}*hookeri* plus another one. According to the rules, this had to be the chromosomes 5·6 and 9·10 from ^{h}*hookeri*, giving 5·10 and 9·6 for *stringens*. The complete chromosome formula is:

stringens 1·4 2·3 5·10 6·9 7·8 11·12 13·14.

This method can be continued unlimited for every Renner complex introduced into genetic and cytological studies. It is only necessary to determine the meiotic chromosome configuration of the hybrids of the new Renner complex with a set of Renner complexes of known chromosome formulas.

When the assumptions on which the analysis is based are correct then the meiotic chromosome configuration of a new hybrid can be predicted, provided that the chromosome formulas of the two Renner complexes are known (Cleland 1933). This is, at the same time, an additional test for controlling the procedure and it was used as such in several cases.

An analysis with a description of the method was performed on *Oe. chicaginensis* by Renner and Cleland (1933). Cleland determined many chromosome formulas (summary in Cleland 1949, 1972) and Renner contributed, along with his students, the chromosome formulas for the Renner complexes of his extensive collection of species (summary in Renner 1938b, 1942a, 1956) using, as far as was possible, known genes to identify the chromosome arms properly.

Soon other geneticists applied the method (Emerson and Sturtevant 1931; Gates and Catcheside 1931, 1932; Baerecke 1944). Some confusion arose when Catcheside (1940) neglected most of the rules and the formulas already published by Renner and gave new chromosome formulas to many of the Renner complexes. Renner (1942a) replied to this offence, but as he could suppose that meanwhile the formulas of Catcheside were being used by other British and American authors, he himself suggested in the same paper use of the new formulas in order to avoid hopeless confusions about mere numbers instead of discussing the real problems. This attitude impressed me as that of a really great scientist who put his personal feelings aside to make way for further progress. The result of these efforts is that it is possible to allocate to every Renner complex the appropriate chromosome formula.

The same method of chromosomal analysis was applied to the Renner complexes of the sect. *Raimannia* (Haustein 1939a, 1952). Investigations on

Oe. mollissima and *Oe. affinis* revealed that each of these species can be subdivided to form several races which differ considerably in the chromosome formulas of their Renner complexes. The races of *Oe. mollissima* differ in more reciprocal translocations from one another than from *Oe. affinis*. The results of the cytogenetic analysis lead to the conclusion that the two species cannot be distinguished properly and have to be combined as one polymorphic species *Oe. mollissima* L. (Hecht and Tandon 1953; Tandon and Hecht 1955, 1956).

Investigations to clarify the relations between the chromosome formulas of the sections *Euoenothera* and *Raimannia* were not successful. Either the crosses were not fertile, or disturbances of pairing of the chromosomes or of chiasma formation in the meiocytes of the hybrids occurred to such an extent that many univalents were seen from diakinesis onwards. The occurrence of large chromosome circles can be suspected in some cases, but the meiotic chromosome configuration of the hybrids could not be determined with the certainty necessary to identify the chromosomes (Arnold 1963a).

4.3.3 Genetic Segregation and Chromosome Catenation

The final proof that Renner complexes and chromosome complexes are only two sides of the same phenomenon is presented when predictions about the behaviour of genes, which are derived from the behaviour of chromosomes or vice versa, can be verified. If the hypothesis of identity of chromosome complexes and Renner complexes is right, then in the progeny of hybrids with large chromosome circles the genetic Renner complexes should be maintained. This is observed in many hybrids. Small chromosome circles and bivalents in hybrids, which enable a recombination of the parental chromosomes during meiosis, should be accompanied by genetic segregation in the progeny. Renner (1928) and also Cleland and Oehlkers (1929) had already assumed a relationship between chromosome configuration in a hybrid and segregation of certain genes, but they could not present an explanation of this. Soon it became obvious that in all cases of breakdown of the Renner complexes, described as a change of linkage relations, the chromosome configuration of the F_1-hybrid is characterized by small circles and bivalents (Cleland 1932; Harte 1948) (see table in 4.1.6, p. 39).

The combination of cytological and genetic data of hybrid progenies with known chromosome formulas of the two Renner complexes made it possible to allocate genes to ring or bivalent chromosomes and even to numbered chromosome arms (Renner 1942a; Harte 1948). By this method, the lethal factors can be located too. The lethal of *velans* cannot be in the chromosomes 3·4, 9·10 or 11·12 but is probably located in one of the chromosome arms 5 or 6. The lethal from *gaudens* cannot be in the chromosome 5·6, but any other location is possible (Renner 1933a). In this way it could be proved that the lethals of the Renner complexes *velans* and *gaudens* cannot be alleles of one locus. The lethal of *flavens* can be located in one of the chromosomes of the ring-of-four in *Oe. flavens·stringens* (Harte 1948). By using the translocation point in the ring-of-four of the hybrid *Oe. flavens·^{h}hookeri* as a locus, genes which are responsible for competition between pollen tubes and for competition between megaspores in the ovule could be localized together with the loci *s*, *p* and some others in the linkage group of the chromosome arms 1-2-3-4 (Harte 1961, 1969a,b).

4.3.4 Chromosome Catenation and Crossing-Over

From cytological observations it is obvious that the euchromatic parts of homologous chromosome arms are paired with chiasma formation in meiotic prophase. End chiasmata are frequent in bivalents and chromosome circles. When pairing occurs, then alleles can be transferred from one Renner complex to another by crossing over (Cleland and Brittingham 1943). In species and most hybrids with high catenation of the chromosomes, however, segregation of genes by crossing-over is extremely rare, usually less then 1%. From the observation of constancy of the Renner complexes while chiasmata occur regularly, the conclusion is drawn that those genes which are responsible for the genetic differences between Renner complexes and which do not segregate in the species and hybrids with large chromosome circles must be located in parts of the chromosomes which are protected from exchange by chiasma formation. It is supposed that these are the heterochromatic middle regions or the adjacent parts of the euchromatin (Renner 1941). This, however, would create difficulties in expression of these genes, which are undoubtedly present. An interesting, but hypothetical, possibility would be to assume that while pairing is regular in the euchromatic parts of the chromosome arms, chiasma formation would be restricted to the distal parts. For as long as there is no method to locate chiasma formation in pachytene chromosomes, the problem will remain unsolved.

An interesting example is the gene *sulfurea*. The locus $s/+^{s}$ is located proximally in chromosome arm number 3. In *Oe. biennis* L., the recessive allele *s* can be transmitted occasionally from *albicans* to *rubens*. The *Oe. biennis* L. *sulfurea*, the result of fertilization of a normal *albicans-s* egg cell by a male gamete with *rubens-s*, originating from crossing-over, is observed repeatedly in the field and in experimental gardens (Oehlkers 1924; Kappus 1958). In the hybrid *Oe. albicans·^{h}hookeri* crossing-over of *s* from *albicans* to *^{h}hookeri* occurs more often, but is still rare.

Oehlkers (1933) investigated the recombination of the closely linked loci *s* and *p* in several combinations of Renner complexes which contain arm number 3 in circles of different size. The frequency of crossing-over in this region is variable and depends on the genotype of the hybrid, but it is extremely rare when the linkage group is located in a bivalent.

The chromosome region containing the loci P^s-S can be duplicated intrachromosomally, most probably by unequal crossing-over, as was observed by Catcheside (1947a). In *Oe. blandina* with a translocated chromosome 3·11 this was observed repeatedly and occurred independently 24 times in the experiments. The corresponding deficiency in arm number 3 of the chromosomes 3·4 and 3·11 was also observed. Within the duplicated region, crossing-over occurs frequently and leads to several anomalies. The deficiency is not transferred by the pollen. In another translocation chromosome 3·12, which has arisen from 3·4 and 11·12 in *Oe. blandina*, a position effect is observed. In plants which are heterozygous for the translocation and the locus $s/+^s$, the dominance of $+^s$ (yellow petals) is partially missing and the petals are variegated sulphur and yellow. When the allele P^r (red sepals) from *Oe. rubricalyx* is transferred by crossing-over to the translocated chromosome, the sepals of the heterozygotes p/P^r are variegated too. The explanation presented by Catcheside (1939, 1947b) is that the *p-s* region, located proximal in the chromosome arm, is very near to the heterochromatin. In exceptional cases, e.g. a translocation near to this point, the borderline between euchromatin and heterochromatin could be variable, causing inactivity of the region and leading to the expression of the recessive phenotype in some cell lineages. This is the only case of variegation as a position effect in *Oenothera*.

Those genes which segregate from hybrids with large chromosome circles are supposed to be located near to the end of chromosome arms to enable crossing-over at a frequency of nearly 50%. This conclusion, however, can only be drawn when recombination for an allele is frequent. Such exceptions are the loci *Co* (size of flower) and *br* (*brevistylis*). Both segregate from every hybrid, irrespective of the chromosome configuration, in nearly Mendelian ratios (Oehlkers 1940c; Renner 1942a).

Theoretical considerations about crossing-over contributed to the solution of other problems. Darlington (1936) discussed the limitation of crossing-over in *Oenothera*.

The concept of Oehlkers on the relation between chiasmata and crossing-over, which formed the basis for his experiments on the physiology of meiosis, was tested and verified by parallel cytological and genetic observations. When the variability of chiasma frequency can be influenced experimentally, then the same should hold for crossing-over. Because the reactions of chiasma frequency and crossing-over to experimental influences are not the same in all objects, conclusions about their interdependence can only be drawn from observations on the same experimental material. This was done in *Oenothera* (Oehlkers 1937, 1940c). Plants of the same genotype received the same experimental treatment at the same time. Then, from some inflorescences young flower buds, which had passed through the early meiotic stages during the experimental treatment, were taken for cytological observations. Buds of the same size were marked on other plants and their pollen was later used for crossing with appropriate genotypes for determination of crossing-over frequency. In the hybrid *Oe. albicans*·^{h}*hookeri* with a circle of 14 chromosomes segregation for flower size, depending on the gene *Co*/*co*, can occur only by crossing-over. The frequency of this event changes parallel to chiasma frequency (Oehlkers 1940b,c). For other loci the same was observed by Harte (1942). The observations on *Oenothera* were the first to give experimental proof of this relationship.

4.3.5 Trisomics

Plants with supernumerary chromosomes were observed and described by de Vries in the first years of his experiments with *Oenothera*. All the investigators found that pollen with eight chromosomes is not viable. The extra chromosome is only transmitted to the next generation by the egg cells. The same is observed in species of the sect. *Raimannia* (Arnold 1955, 1958b).

Gametes with supernumerary chromosomes can arise by random distribution of univalents in anaphase I or by non-disjunction in a chromosome circle, when three adjacent chromosomes move to the same pole. The frequency with which trisomics are observed depends on the genotype of the mother plant. In progenies of *Oe. Lamarckiana* de Vries, trisomics are observed frequently, but they are only rarely observed in *Oe. biennis* L. (de Vries 1925; Renner 1943b). The occurrence of similar types in two species was described as "parallel mutations" by Gates (1912).

When trisomics are needed for further research, an approved method is to start with the *gigas* type 4n, obtain triploids by crossing with pollen from normal diploids, and then crossing 3n ♀ × 2n ♂. In the progeny many aberrant types with one or more supernumerary chromosomes will appear (e.g. Zenke 1950). In the progeny of the hybrid *Oe. albicans*·^{h}*hookeri* the frequency of trisomics can be increased 10-fold by treatment of the mother plants with extreme temperature and drought (Oehlkers 1940c).

In homozygous types, seven different primary trisomics are possible. Catcheside (1954) described the trisomics of the homozygous *Oe. blandina*. The supernumerary chromosomes of the aberrant types were identified by crossing with pollen containing Renner complexes of known chromosome formulas and determination of the trisomic chromosome configuration in the hybrids. In *Oenothera*, however, seven different trisomics can arise from every Renner complex. From *Oe. biennis* L. (two circles, one with six and one with eight chromosomes) 14 different trisomic types are expected. *Oe. Lamarckiana* de Vries, from which most of the observed trisomics arose, has a circle of 12 chromosomes and a bivalent and consequently 13 different trisomics should be possible. Following the first observations of de Vries and the cytological investigations of van Overeem (1921), Boedijn (1924a), Cleland (1929), Hakansson (1928) and Emerson (1936), the chromosome configurations observed in trisomics are interpreted on the basis of pairing of homologous arms, including the extra chromosome (Hakansson 1930a).

In trisomics the pairing of three homologous chromosomes with a change of partner in a chromosome arm and chiasma formation will lead to complicated configurations in diakinesis with end chiasmata connecting three chromosome arms. These can only be analyzed with modern techniques and were not mentioned in earlier publications. In bivalents the resulting figures are relatively simple, but in chromosome circles the configurations can be different in every nucleus and can be so complicated that analysis is often not possible. Cleland (1967) describes two special cases of circles of five chromosomes, resulting from supernumerary chromosomes in hybrids between unrelated races of North American *Oenothera* species. They originated by non-disjunction, as is proved by the chromosome formula. Earlier, Herzog (1940) described circles of three and of nine chromosomes in trisomics in which the supernumerary chromosome is supposed to have arisen by translocation.

The trisomic types were described again by Catcheside (1933a, 1936), Emerson (1935, 1936) and, including trisomics from *Oe. biennis* L., by Herzog (1940). A thorough analysis of the trisomics derived from *Oe. Lamarckiana* de Vries with determination of the chromosome arms involved in the different types, was carried out by Renner (1940b, 1943b, 1949). Interdependence between the phenotype and the supernumerary chromosome is obvious, as was confirmed by Renner and again by Catcheside (1954). In the trisomic *lata*, for example, which is male sterile, the extra chromosome is 5·6 from *gaudens* and the trisomic *broad* from *Oe. blandina*, which also has the extra 5·6 chromosome from *gaudens*, is male sterile too (Catcheside 1937b).

Catcheside used the trisomics for further genetic experiments. The allele *br* was first transferred from *Oe. Lamarckiana* de Vries *brevistylis* to *Oe. blandina* and then into some of the trisomics. Normal segregation of *br* is observed from *blunt* and *glossy*, but *golden*, with the extra chromosome 11·12, gave trisomic segregation ratios for *br*. The same segregation is observed from the tertiary trisomic *subwhitish*, which has chromosome arm number 12 in common with *golden*. Because *br* segregates in nearly Mendelian ratios, even from hybrids with large chromosome circles, it was possible to locate the locus *br* in a numbered chromosome arm only by this method. The very high frequency of crossing-over suggests a location near the end of the chromosome arm.

The observation of Hance (1919) on the mitotic chromosomes of the trisomic *scintillans* of *Oe. Lamarckiana* de Vries is a very special case. In this type, which was identified by the morphology of the plants, the supernumerary chromosome should cause chromosome breakage of other chromosomes, especially the SAT chromosome, in mitosis. This was discussed by many cytologists as very important. Re-investigation of *scintillans* and other aneuploids of *Oe. Lamarckiana* de Vries with one or more extra chromosomes, with the special purpose of obtaining quantitative data on this type of chromosome breakage, could not verify the older observations

(Zenke 1950). Careful reading of the publication by Hance and a thorough study of his drawings, however, reveals that he himself was fully aware that he presented not firm facts but an interpretation of his observations. In slides stained with haematoxylin thin chromosome threads can often not be recognized. This poor staining occurs rather often between the proximal part of the chromosome arm and the distal satellite or between chromosome arms at the centromere. Hance stated that he would evaluate as a chromosome break every case in which the connecting chromosome thread could not be seen. However, his drawings prove that in the nuclei the parts of the "broken" chromosomes lie so close together that there can be no doubt about the effect of insufficient staining. This was confirmed by the aceto-carmine-stained preparations of PMCs investigated by Zenke in which the thin connection between the chromosome parts could be clearly identified. This is again an error by interpretation of bad slides, but in this case the fault is not on the part of the author but of the readers (did they read only the summary without looking for the proof in the paper? The text leaves no doubt about the intention of the author).

4.3.6 Evolution of Translocations

The evolution of the Renner complexes and translocations in *Oenothera* is described perfectly in the book by Cleland (1972); therefore this section will be very short.

When the right interpretation of the cytogenetic observations on Renner complexes and chromosome catenation was evident, it was clear that the differences between the arrangements of chromosome arms must have arisen during evolution by repeated reciprocal translocations.

As soon as the chromosome formulas of many Renner complexes were determined, the question of how many reciprocal translocations are necessary to proceed from a given formula to another could be answered. Such considerations include the evolution of the chromosome formulas.

An important conclusion drawn by Cleland from a comparison of the chromosome formulas of Renner complexes from many species was that the shortest route from the standard arrangement with the minimum number of translocations could not always have been the actual course. As an example, it is evident from the genetic records that the Renner complex ^{h}*decipiens* arose from *velans* in such a way that the chromosome formulas differ by two reciprocal translocations, but at the same time the formula of ^{h}*decipiens* differs from ^{h}*hookeri* by only one reciprocal translocation. The evolutionary step from *velans* to ^{h}*decipiens* must include translocations by which the original arrangement of ^{h}*hookeri* is reconstituted in some chromosomes (Cleland 1942). The same holds for some Renner complexes with more complicated translocation patterns.

In every case a theoretical translocation history, leading from the standard arrangement to a certain chromosome formula with the required minimum of events, can be constructed but, considering the relations with the chromosome formulas of other Renner complexes which could be part of the evolutionary pathway, it becomes evident that in most cases more steps must have been included.

Furthermore, when genetic information about alleles of known genes is added, it is confirmed that new Renner complexes can arise by translocation between chromosomes belonging to two different haploid sets. The only condition is that, at the end, the new set must contain all the different chromosome arms, irrespective of their origin from the two original Renner complexes. Such a mixing of chromosome arms from different Renner complexes must be the case, e.g. for the Renner complex ^{h}*deserens* (*subvelans*) which contains chromosomes from *velans* and *gaudens* and has new translocations connecting them (Renner 1943d). It is suspected that in this way some other Renner complexes arose which are genetically very similar to a well-known one, but differ mainly in the loss of the lethal allele. This can be achieved by a reciprocal translocation such that the chromosome arm with the normal allele from the partner is present in the new set which consists mainly of genetic material from the other Renner complex. An example is ^{h}*blandina* which arose in experiments from *velans* and is nearly identical with this Renner complex, except that it contains the chromosome 5·6 from *gaudens* and does not have the lethal allele from *velans*. Therefore it can be homozygotic in *Oe. blandina*.

The differences in the alleles between Renner complexes are not correlated to the differences in the chromosome formulas, as is revealed by the investigations on some species of the sect. *Raimannia* (Hecht and Tandon 1953; Tandon and Hecht 1956).

From all the observations and considerations about the role of translocations in the evolution of the Renner complexes it is evident that the number of differences in the chromosome formulas does not give complete information about the relation between two chromosome complexes (Cleland 1942). The same was stated for the Renner complexes in the sect. *Raimannia* (Haustein 1952). A large number of differences points to many steps in evolution, but similarity in the arrangement of chromosome arms, as far as can be concluded from the formulas, does not mean a close relationship in the evolutionary pathway. To clarify this relation, the interstitial and differential segments must be included in the analysis, but this is not possible with the classical methods of either karyology or genetics.

4.3.7 The Cytogenetic Model of *Oenothera*

In the course of evolution the chromosome catenation of *Oenothera* arose by repeated reciprocal translocations between non-homologous chromosomes.

The mechanics of meiosis enables the translocated chromosomes to stay together in the meiospores as chromosome complexes. The balanced lethals keep the chromosome complexes with their alleles heterozygotic in sexual propagation. Even alleles which would otherwise be deleterious can be stored until they become profitable by combination with new alleles of other genes. By this system of gene complexes on the basis of high chromosome catenation the species are constant hybrids. When two Renner complexes, which differ only by small differences in their chromosome formulas, are combined in a hybrid, then the meiotic configuration is characterized by bivalents and small chromosome circles in meiosis and Mendelian segregation is made possible by recombination of the parental chromosomes. As a consequence the original Renner complexes break down. When, by hybridization, new combinations of Renner complexes with high chromosome catenation arise, they can become new constant species. The constant heterozygosity which is enforced by the formation of the Renner complexes makes the advantages of heterosis available for the species.

Some peculiarities of crossing-over, especially the low frequency of crossing-over in species with a high frequency of end chiasmata in meiosis and differences in crossing-over observed in hybrids, could not be explained until now.

4.3.8 The Limitation of the Cytogenetic Methods

With the combination of karyological and genetic methods applied to the same experimental material, valuable information about the *Oenothera* sys-tem, especially on how it is constructed and how it works, can be obtained, but it can only lead to a description of the present state of affairs. The remaining question is: How did the differences arise which have been investigated by this research method?

4.4 Mutations and Mutation Theory

4.4.1 The Start of Mutation Research

Genetics and cytogenetics of *Oenothera* had become most interesting, but they were not the first speciality of this genus to be discovered. That was actually another most unusual observation. Taxonomists had been busy describing "good species" as correctly as possible since Linnaeus created a system of classification. Their work was based on the assumption that the characters of species are constant and reproduced in every generation. Variability was most disturbing and therefore unwanted. The problem of evolution had to remain unsolved in spite of many theories, because it involved the assumption of changes in species characters. Then Darwin put forward his theory of selection and the survival of the fittest. But the crucial question remained: How can good species evolve at all? Either their characters are constant, then Darwinian selection has no point from which to start, or they are variable to an extent that selection can work. Then the "good species" with its nice constant characters, giving rise only to identical progeny, disappears, along with the taxonomists who are bound to their beloved "good species". This cannot be! Would You voluntarily destroy your raison d'etre just for a funny idea of somebody else? You must fight **for** it, and consequently **against** every idea that touches the basis of your scientific credo.

The crucial point is the variability of the characters of plants and animals. As soon as this was recognized, the behaviour of the variable characters in the progeny became the interesting point. In the decennia following Darwin this problem was so clearly recognized that it was made the basis of research only by two scientists: Mendel and de Vries. In both cases evolution was the problem and genetics the tool.

Mendel started work with the variable material he had at his disposal and with the working hypothesis: Can new types arise by hybridization? The answer was an undoubtable NO. By crossing plants which differ in one or a few characters nothing new could be made because of the segregation of the parental characters in the following generations. They are based on hereditary units which remain unchanged by hybridization. At first de Vries did the same with variable characters he collected from different sources, testing their behaviour in controlled progenies, just as Mendel did. But then he found not only the segregation of the parental characters with the gene *br*. He observed also in an *Oenothera* population the appearance of new characters, and not just once as an accident but regularly. Some types appeared every year, others only occasionally. Together they gave proof of a new phenomenon. The supposed good taxonomic species *Oe. Lamarckiana* de Vries reproduced itself in every generation, as it was expected to do, but it also gave rise to new types. It was not constant at all!

The new types were called mutants and the process by which they arose was termed mutation. In most cases the mutants differed from the parental species in several characters, just as species in a genus do. This led de Vries to the interpretation of his observations that he had found the source of genetic variability and the key to evolution. Stepwise changes in the genetic material created the basis for selection and origin of new species. The observations and the theory built upon them were presented in a number of publications and two thick books *Die Mutationstheorie* (de Vries 1901, 1903) and *Gruppenweise Artbildung* (1913). The principal ideas were elaborated by de Vries in a series of lectures delivered at the University of California in the summer of 1904 and published as *Species and Their Varieties, Their Origin by Mutation* in 1905. A shorter summary, stressing the importance of the mutants from *Oe. Lamarckiana* for the mutation theory, was presented by Renner (1919a).

For their impact on the development of genetic theories and the influence they had on the start of mutation research, the mutants of *Oenothera* as observed by de Vries require special consideration. The importance is independent of the facts that in modern cytogenetics the first observations have been given new interpretations and that their role in evolution is very limited, restricted to one type of mutation in some special cases in *Oenothera* and a few other plant genera.

The genetic variations observed in *Oenothera* and called mutations were of a very different kind. The use of the term mutation for every genetic change was proposed by Gates (1931), irrespective of whether the change involves the multiplication of the whole genome, of a single chromosome or part of it, or structural changes of the chromosomes or point mutations.

First the types of mutants observed by de Vries will be described and classified according to the modern interpretation. Then their influence in the evolution of the genus can be evaluated, and the role of *Oenothera* in mutation research will be discussed.

4.4.2 The Mutants of *Oenothera*

The genetic behaviour of the first mutants was very strange. Only a few behaved like Mendelian genes. Other types, usually differing from the parent species in several characters, proved to be constant after self-pollination from the beginning onwards. In the progeny of some other mutants a segregation was observed, either of the normal *Oe. Lamarckiana* de Vries and the mutant type, or the mutant and a new type that was constant in the following generations. The results from hybridization with *Oe. Lamarckiana* de Vries were just as difficult to describe. Some mutant characters were inherited only by the egg cells, or the backcross segregated like a progeny from self-pollination. Some mutants were sterile in one sex, *brevistylis* (female sterile) or *lata* (male sterile). Moreover, some mutant types, similar to those of *Oe. Lamarckiana* de Vries, were observed in *Oe. biennis* L. and other species too. This process was called parallel mutation by Gates (1912).

There were lengthy discussions about the mutants and their origin, but these can be neglected because meanwhile the problems were solved. The mutants can be classified and the process of mutation understood according to the present knowledge about the cytogenetics of *Oenothera*.

4.4.2.1 Polyploids

The *gigas* mutants were observed in many species. They proved to be tetraploids, originating by polyploidization during early embryogenesis. The first identification of a mutant as tetraploid was made by Gates (according to Dart 1966). Tetraploids, or amphidiploids in the case of constant heterozygotes, are observed repeatedly in the progeny of the hybrid *Oe. franciscana* × *biennis* as 4n *Oe.* ^{h}*franciscana·rubens* (Davis 1943). The mutants *semi-gigas* are triploids arising from fertilization of an unreduced gamete, either pollen or egg cell, by a normal partner. The opposite process, the appearance of a mutant *diplo-gigantea* with 14 somatic chromosomes from the tetraploid *Oe. gigantea* (syn. *Oe. Lamarckiana* de Vries *gigas*) was observed by Heribert-Nilsson (1939). It was not a pure *Oe. Lamarckiana* de Vries but had a mixture of characters from the mutants *rubrinervis* and *lata*. This was explained by the mixing of chromosomes of the original Renner complexes, or parts of them, in the tetraploid stage.

4.4.2.2 Trisomics

The next group of mutants which was repeatedly observed and found to be inherited only through the egg cells are the trisomics. They can arise from meiotic disturbance during anaphase, when, either as a consequence of non-disjunction in a chromosome circle or from random distribution of univalents, tetrad cells with supernumerary chromosomes are produced. The first trisomics were observed as mutants by de Vries who did not recognize them as chromosomal abnormalities. A description of the trisomics from *Oe. Lamarckiana* de Vries was later presented by Gates (1923a,b). In the progeny from the egg cells of the triploid mutant *semi-gigas* a large number of aberrant types, many of them already known as "mutants", were identified as trisomics (Dulfer 1926). De Vries (1925) described as "latent mutability" the fact that in the progeny of the triploid hybrid *Oe. biennis gigas* × *Oe. biennis* many mutant types appeared which are really trisomics, although they were not recognized as such at the time.

In microspores the supernumerary chromosomes are the cause of non-viable pollen grains, but egg cells with an unbalanced chromosome complement can function normally, when all the chromosome arms are represented. In each species as many phenotypically different primary trisomics can arise as there are different chromosomes. Meanwhile for many of the trisomics from *Oe. Lamarckiana* de Vries the chromosome formula of the supernumerary chromosome has been determined (Renner 1940b, 1949). A special case is the male sterile *lata* with the supernumerary 5·6 from *gaudens* (Catcheside 1937b) and described as a parallel mutation in *Oe. biennis* L. with the same additional chromosome from *rubens* (Renner 1943b). Trisomics are also observed to arise from the homozygote *Oe. blandina* (Catcheside 1933a).

Some special cases of trisomics containing eight-chromosome Renner complexes, I_A, I_B and I_C, are described for the sect. *Raimannia* by Arnold (1955, 1958b) and Arnold and Bina (1957). The egg cells are partially sterile. Pollen development is normal, but at anthesis the pollen grains with eight chromosomes are either dead or inactive. When occasionally a pollen grain containing I_B survives and germinates, the pollen tube can grow normally and reach the ovules. In *Oe. odorata*, I·v, the appearance of these trisomics is explained by double non-disjunction during metaphase I. In I_A one chromosome of the Renner complex I is replaced by two chromosomes of the Renner complex v. The trisomic I_B arises by double non-disjunction. In this process two I-chromosomes are replaced by three from the partner complex v. The eight-chromosome Renner complex I_C can arise from *Oe. odorata* by double non-disjunction with a distance of five chromosomes, causing the replacement of three I-chromosomes by four from v (Buchner 1956). A schematic example of non-disjunction in a heterozygote with the Renner complexes A·**B**, in which the chromosome arms are represented by letters and the two complexes are distinguished in the text by normal and **bold** type and by shading in the drawing, is described in the following scheme:

The normal metaphase configuration is:

Parent A·**B**.

Renner complex A:

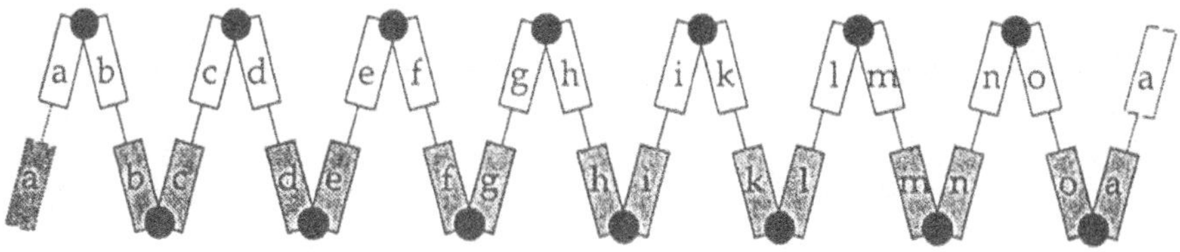

Renner complex **B**:

In a scheme for meiotic non-disjunction in metaphase I it is illustrated how in Renner complex A the chromosome l·m is replaced by **k·l m·n**:

Altered Renner complex A:

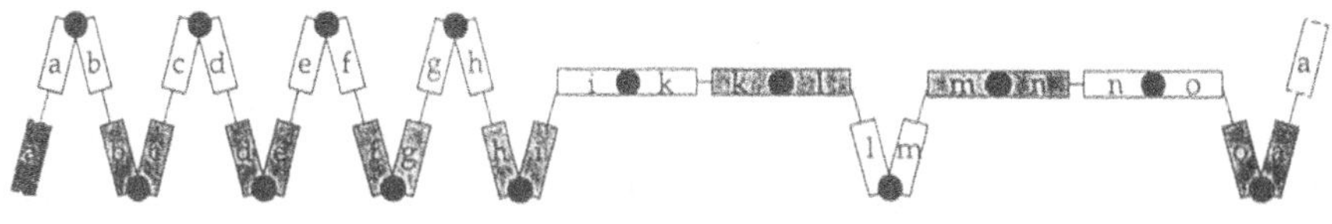

Altered Renner complex B:

Gametes with altered Renner complex A (duplication of chromosome arms k and n and replacement of chromosome arms l and m by **l** and **m** from the partner complex) are only viable as egg cells. Gametes with altered Renner complex **B**, which is missing chromosome arms **k** and **n**, are not viable.

A trisomic zygote with the altered A and the original A is only viable when the lethal of A is located in chromosome l·m or, when located in one of the chromosome arms k or n, the normal allele in either **k** or **n** will be dominant over the two recessive lethal alleles in either k or n.

The chromosome configuration in such a trisomic is four bivalents, formed by homozygotic A-chromosomes, and a chain of three chromosomes containing the two **B**-chromosomes with bivalents made up of A-chromosomes attached with trivalent connections at both ends.

A scheme illustrates the pairing in trisomics:

Altered A:

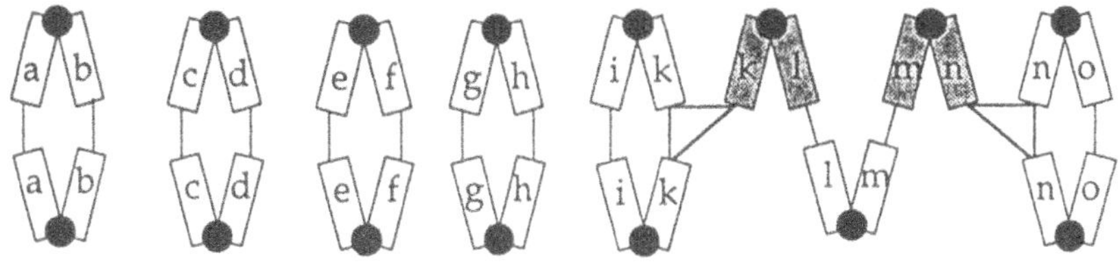

Original A:

Altered A with **B** is viable when the lethal of **B** is not in chromosome arms l or m, or when the lethal is located in n or k and the normal allele in the chromosome arms n or k is dominant over the two recessive lethal alleles of the two Renner complexes A and **B**.

The chromosome configuration of this trisomic is an open chain of 11 chromosomes, six from A and five from **B**, and two bivalents of **B**-chromosomes attached by trivalent connections at the ends.

Altered A:

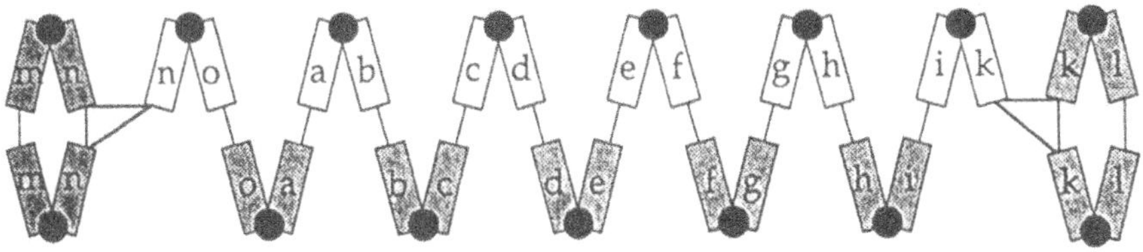

Original **B**:

In these cases there is not a single supernumerary chromosome that can be lost without further damage, but the duplicated chromosome arms are located in two different chromosomes. The trisomics of this type are characterized not only by the chromosome number but also by the mixture of chromosomes from the two Renner complexes of the parent. An identification of the chromosome formulas of these types was attempted by Dittmann (1956).

The analysis of the trisomic mutants in the subgenus *Raimannia* reveals a special problem of *Oenothera* trisomics. The change in phenotype, which is important for the characterization of the mutants, cannot only be brought on by genes in the supernumerary chromosome, as is known for other species. The change can just as well be the consequence of the combination of chromosomes from the two parental Renner complexes creating new combinations of alleles. Such a really new eight-chromosome Renner complex with a new chromosome formula behaves like an egg cell complex with a pollen lethal. It is inherited only through the egg cells, is inactive in the pollen and will be genetically and cytologically different from both Renner complexes present in the parent. This happens in spite of the fact that neither point mutations nor new reciprocal chromosomal translocations are involved, but only duplication of two chromosome arms. Such things can be found only in *Oenothera*.

4.4.2.3 Chromosomal Translocations and Lethals

A most important and interesting group of mutants arises by new reciprocal chromosomal translocations. Most of them are observed first as "half-mutants" in which an unchanged Renner complex is combined with another one with a new chromosome formula. The term "half-mutant" was used by de Vries for the new Renner complex which was first observed in combination with an original one, but Hakansson introduced the same term for the aberrant plant containing such a Renner complex in combination with an unchanged one. Hakansson's definition was used by all subsequent authors.

When the reciprocal translocation occurs between chromosomes of an established Renner complex, an influence on the phenotype cannot be expected. The mutation is only detected by accident when, by cytological investigations of the species or a hybrid, an unexpected chromosome configuration is found. When, however, the new translocation occurs in a structural hybrid and involves chromosomes from both Renner complexes, then recombination of alleles becomes possible. Suppose the homologous chromosome arms in the two Renner complexes have different alleles of one gene locus. Together with the exchange of chromosome arms these alleles, which are responsible for the genetic diversity of the original Renner complexes, can become homozygotic and cause a change in the phenotype which is recognized as mutant.

Furthermore, when the new Renner complex contains a lethal from the original complex, a balanced lethal situation is present and the mutant is constant in its progeny, recognized as a permanent half-mutant. However, when in the process of chromosomal translocation a lethal is lost because the chromosome arm containing it is replaced by the same from the partner complex, the new Renner complex can be realized as a homozygote after self-pollination. The aberrant type observed first is classified as a half-mutant with a segregating progeny consisting of the half-mutant and the new homozygote. The phenomenon has been known for a long time but the process was investigated in detail by Renner (1941).

Earlier Emerson (1935) discussed the origin of translocation mutants in *Oenothera*. In real gene mutations the chromosome formula of the Renner complexes cannot be altered. He explains the origin of new translocations by crossing-over in the interstitial segments of the paired translocated chromosomes between the centromere and the translocation point in a constant hybrid. This could give rise to mixed Renner complexes, containing chromosome arms from both partners. The cytological process is theoretically possible. Considering that the translocation points of the Renner complexes are located in the heterochromatic middle parts of the chromosomes, very complicated chromosome arrangements must result from crossing-over in these regions. However, they are not described in the cytological literature on *Oenothera* species. The occurrence of this mechanism for creating new translocations must therefore remain an open question.

Renner (1943d) suggests another possiblity for the origin of new translocations. By interlocking of chromosome rings, as is observed rather often in diakinesis, chromosome breaks and fusion of the broken ends could be induced. As there is no evidence that interlocking is followed by chromosome breaks or that at such late stages in meiosis broken chromosomes can re-unite, this proposition is even more hypothetical than the one just mentioned.

In trisomics of *Oe. odorata* and *Oe. Berteriana* a high rate of new translocations is observed, often involving the supernumerary chromosome (Arnold and Kressel 1965b). The authors discuss the possibility of crossing-over following non-homologous pairing or exceptional pairing of extremely small homologous regions in otherwise non-homologous chromosomes as a source of new translocations (Arnold et al. 1963; Arnold 1967a). This explanation is again hypothetical.

In all cases, instead of these complicated hypotheses, the occurrence of real reciprocal translocations by a process of break-and-fusion, as observed in mutations in many objects, can be assumed. It seems only that this process occurs at a higher frequency in *Oenothera* than in most other plants.

When the supernumerary chromosome of a primary trisomic is involved in a new reciprocal translocation, secondary trisomics can arise in which the extra chromosome has a chromosome formula that is not represented in the parent.

A possible origin of homozygotic mutants was observed in several cases and was discussed in detail by Renner (1941). When, in a balanced lethal situation, the lethal allele is lost, irrespective of the event leading to this result, either crossing-over, a point mutation or something else, the alteration makes homozygotes of the Renner complex possible. This can happen in constant hybrid species, such as *Oe. suaveolens* Desf. and *Oe. grandiflora* Solander, and in permanent half-mutants. Several cases of half-mutants and their segregating homozygotic translocation mutants have been described. A summary is presented by Renner (1943d). The translocation mutants *erythrina*, *rubricalyx* (ring of six chromosomes) and *Afterglow* (ring of eight chromosomes) arose first from *Oe. Lamarckiana* as permanent half-mutants containing *velans* and the modified Renner complex, both with a lethal. By loss of the lethal from the new translocation complex they became segregating half-mutants (Shull 1942). *Oe. dulcis* as a constant half-mutant was described by de Vries (1935) and gave rise to *Oe. blandina* after loss of the lethal in the Renner complex *velans*; ^{h}*blandina* is nearly identical with *velans* but without the lethal. The way in which new translocation mutants arise can be very complicated. As an example consider the primary half-mutant *Oe. rubrinervis*, derived from *Oe. Lamarckiana* de Vries. The new Renner complex differs from *velans* by one reciprocal translocation. Then the type *Afterglow* (syn. *rubricalyx*) arose as a secondary half-mutant. The new Renner complex ^{h}*latifrons* differs from *velans* by two translocations and in the process the lethal from *velans* is lost. *Oe. latifrons* is a homozygote that behaves as a good species, according to the criteria of both genetics and taxonomy (Cleland 1931).

Some cases of species and segregating half-mutants and their homozygotic mutants are:

Original type	Renner complexes	Mutant	Homozygotic Renner complex	Reference
Species				
Oe. suaveolens	*albicans·flavens*	*lutescens*	*flavens*	(Renner 1941)
Oe. grandiflora	*acuens·truncans*	*ochracea*	*acuens*	(Gerhard 1929)
Half-mutants				
Oe. erythrina		*decipiens*	^{h}*decipiens*	(Shull 1942)
Oe. rubrinervis		*deserens*	^{h}*deserens*	(Shull 1942)
Oe. rubricalyx		*latifrons*	^{h}*latifrons*	(Cleland 1931)
Oe. dulcis		*blandina*	^{h}*blandina*	(de Vries 1935)
Oe. rubrisepala	*convelans·subvelans*	*subvelutina*	*subvelans*	(Renner and Hirmer 1956)

The opposite process, the creation of a balanced lethal situation by new lethals, which arise as mutations, is observed repeatedly. Such a case was reported by Davis (1947). The cross *Oe. Lamarckiana* × *franciscana* gives rise to the hybrid *Oe. gaudens*·^{h}*franciscana* which is similar to *Oe. Lamarckiana* because of the genetic and structural similarities between the Renner complexes *velans* and ^{h}*franciscana*. The homozygotes *Oe.*

^{h}*franciscana*·^{h}*franciscana* are not realized as empty seeds but as yellow lethal seedlings. In one line heat-shock treatment produced a new lethal mutation in the Renner complex ^{h}*franciscana*, causing death of the homozygotic embryos. This made the hybrid behave genetically like *Oe. Lamarckiana* de Vries with 50% empty seeds. The new lethal is not an allele of any known gene locus.

4.4.2.4 Gene Mutations

Spontaneous mutations of genes with Mendelian segregation in the progenies were detected in *Oenothera* as soon as geneticists started looking for them. Relatively few mutants can be described as point mutations (e.g. *brevistylis*, *vetaurea*). Several mutants have already been mentioned in the chapter on genetics of *Oenothera* and some more will be discussed in relation to developmental biology. Here, it is sufficient to state that *Oenothera* behaves in this respect just as any other plant under close observation. The great difficulty is that the mutational event can only be observed directly in homozygotes. In the constant hybrids the appearance of a new phenotype, showing by Mendelian segregation that a new genotype has arisen, only provides proof that at some time in the past a mutation must have occurred. But a recessive allele can be hidden in the Renner complexes for an unlimited period of time, until, after crossing-over to the partner complex, it is expressed in the phenotype. To this type belong *nanella* in *Oe. Lamarckiana* de Vries and *sulfurea* in species with the Renner complex *albicans* (Oehlkers 1924). The great genetic diversity of the Renner complexes which is observed by segregation following hybridization demonstrates that gene mutations cannot be rare events in *Oenothera*. In *Oe. biennis* heterozygosity for biochemical characters is observed, which could only arise by mutations (Levy and Levin 1975; Levy and Winternheimer 1977).

If mutations occur as evolutionary events then this must certainly be correlated to changes in DNA. This aspect has been introduced recently into *Oenothera* research. First results indicate that this can be proved by comparison of the DNA pattern in two species of the sect. *Raimannia*, *Oe. villaricae* Dietrich (syn. *Oe. Berteriana*) and *Oe. picensis* Dietrich (syn. *Oe. odorata*). Six plants of each species have been investigated and the similarity of the specimens of each species as well as the differences between the two species are obvious. Similar results, but with different patterns, are obtained using other oligonucleotide primers (Chapman pers. comm.) (Fig. 11).

4.4.2.5 Summary: The Role of Spontaneous Mutations in Oenothera

The Renner complexes of *Oenothera* are characterized by two specialities, the reciprocal chromosomal translocations and the balanced lethals. Once the Renner complexes are created, recombination of other mutated alleles is

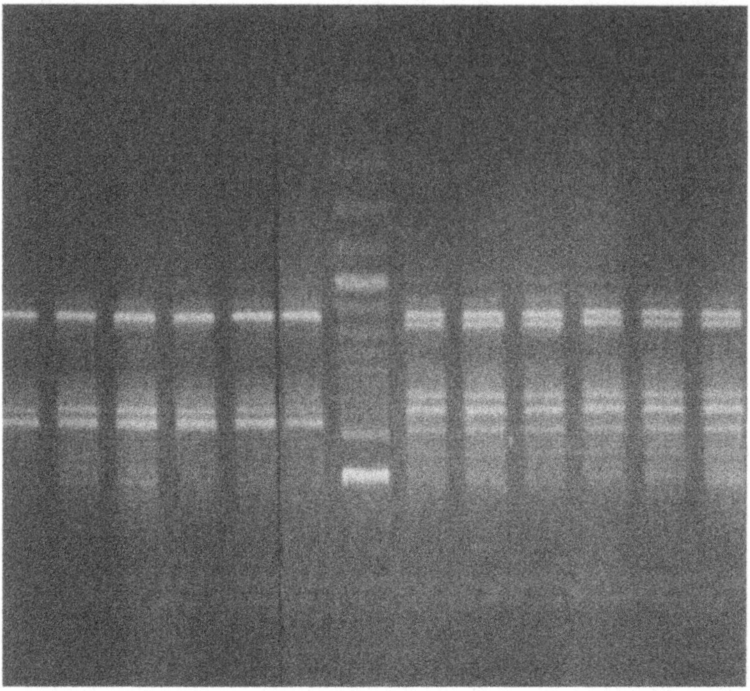

1 2 3 4 5 6 7 8 9 10 11 12 13

Fig. 11. Differences in DNA between two species of subgenus *Raimannia*. DNA fingerprints by PCR (polymerase chain reaction) of six randomly chosen plants of *Oe. villaricae* Dietrich (syn. *Oe. Berteriana*) (*lanes 1–6*), 100-bp standard DNA ladder (*lane 7*) and six plants of *Oe. picensis* Dietrich (syn. *Oe. odorata*) (*lanes 8–13*). (Courtesy of M. Chapman)

inhibited. Consequently, selection, irrespective of whether it favours positive changes or acts against unfavourable alleles, cannot work. Only mutations promoting heterosis will develop a positive selection value. On this basis, the normal mutability of gene loci leads to genetic diversification of the Renner complexes, as described in the preceding section. *Oenothera* does nothing special any more. It developed its genetic variability by a combination of processes which occur also in other plants.

4.4.3 Experimental Mutagenesis

4.4.3.1 Gene Conversion or Multiple Mutable Alleles?

In *Oe. biennis* L. a mutant *cruciata*, also described as *Oe. biennis* L. *apetala*, from the Botanical Garden at Leiden (Netherlands), differs from *Oe. biennis* L. only by the extremely narrow petals. It was observed in wild populations by de Vries (1903) and described by Stomps (1912, 1913). The *cruciata* type arose three times independently of one another from families of *Oe. biennis* with normal petals. The *cruciata* allele is also present in both Renner complexes of *Oe. atrovirens*, *pingens·flectens*. *Oe. Lamarckiana* de Vries *cruciata* has been known since 1923 and was described by Oehlkers (1930).

This *cruciata* gene is a special case. At first it looked like an ordinary gene mutation, with dominance of the normal allele *Cr* over the recessive *cruciata cr*, but soon unexpected genetic behaviour was observed. In the hybrids the character proved to be variable. Sometimes the F_1-hybrids had flowers of intermediate form, varying from the nearly complete *cruciata* type to the nearly normal with only small defects at the margins of the petals. Moreover, in progenies from crossing the hybrids with the parents or other types, the supposed hybrids behaved like homozygotes in some cases, producing only normal or only *cruciata* progenies. These phenomena were investigated by two German scientists, both very reliable, well-known *Oenothera* geneticists, Oehlkers and Renner and their students. These two were good scientific friends who agreed on all the other points of genetics in general and of *Oenothera* in particular. They performed nearly the same experiments, with the same stocks, and came to very different conclusions. The question was whether the observed mutability at the *cr* locus is due to instability (in German: Labilität) of some alleles or to conversion, the induction of mutation by the partner in a heterozygote. Thus, the disagreement was not on the observations but on the interpretation of the observed facts.

Oehlkers (1930, 1935a, 1938) started from the observation that the first *cr*-alleles arose from homozygotic *Cr*/*Cr* parents. They were therefore in any case real mutations from the dominant to the recessive state of the gene. This could not be called conversion as heterozygosity was not involved in initiation of the process. The phenotype of *Oe. biennis* L. *cruciata* is variable. In the progenies a selection for strong or weak expression of the character is possible, and this process can be reversed in every generation. His explanation was that at the *cruciata* locus there exist mutable alleles which could change their state in either direction to normal or to *cruciata* in varying degree, irrespective of the partner allele in the other Renner complex of the hybrid. The results of Siebs (1939) support the interpretation of Oehlkers. At the automutable locus *cruciata* a series of multiple alleles was found, mutating easily in both directions, irrespective of the partner allele. Only *Cr* from *Oe. Hookeri* de Vries and *cr* from *Oe. atrovirens* are in a stable condition. Her results, based on a large series of experiments with large progenies, are, however, not compatible with the theory of gene conversion supported by Renner. Instead, for many of the results obtained by Renner and his students, crossing-over of the *cr* alleles could be presented as a plausible explanation, when this can reach nearly 50% even in hybrids with high catenation of the chromosomes. That this is possible in *Oenothera* is proved by the loci *br* and *Co*.

Renner, on the other hand, found that Renner complexes which entered a hybrid with *cr* could reappear in the next generation bearing the dominant allele, and vice versa (Renner 1937b, 1939, 1942c, 1943c, 1957, 1958b,c,d,e, 1959b; Renner and Sensenhauer 1942; Haustein 1951). His interpretation was that alleles of the *cr* gene, which were stable in the homozygotic condition, either *Cr*/*Cr* or *cr*/*cr*, became mutable when combined in a hybrid *Cr*/*cr* and changed their status in the direction to that of the other allele with which they were combined. This mutability directed towards the status of the partner allele and induced by heterozygosity was interpreted as gene conversion in the sense of Winkler, whose ideas were discussed (and mostly rejected) in several publications by well-known geneticists such as Stern (1930, 1932) and, with special reference to *Oenothera*, by Goldschmidt (1958). Immediately following the supposed conversion the alleles remained in an unstable state, again mutating easily, but after several generations a stable state could be achieved. The alleles of the extreme types, *Oe. Hookeri* de Vries with broad petals and *Oe. atrovirens* with extremely reduced petals, seemed to be stable themselves but were able to induce conversion of the allele in a partner complex in a hybrid. In the F_1-hybrids, *Oe. biennis* L. *cruciata* × *Oe. Hookeri* de Vries, variability of the phenotype is observed, ranging from *cruciata* to normal, that is due to variability in the dominance relations of *Cr* and *cr* without influencing the state of the alleles. The Renner complexes ^{h}*franciscana* and ^{h}*purpurata*, containing the allele *Cr*, behaved like ^{h}*hookeri* (Renner 1958b). Summaries of his experiments are presented by Renner (1958d,e, 1959b).

The gene *sepaloid* (*sep*), when homozygotic, produces a phenotype with extremely reduced petals, similar to *cruciata*, but conversion at this locus is not observed (Rossmann 1961, 1963a,b).

At the end there were two common points: (1) the alleles *Cr* from *Oe. Hookeri* de Vries and *cr* from *Oe. atrovirens* were stable and (2) there existed other alleles which could mutate. The difference of views was on the cause of the mutability, gene conversion in direction to the partner, or general instability. When, regarding the number of experiments performed with this gene, no new facts could be produced by making more hybrids and raising more progenies, both rivals gave up the discussion without an agreement in sight and remained scientific friends as before. The solution came from quite a different source. Cleland (1970) concluded from his experiments that the *cruciata* case is not a problem of genetics but of development. The alleles at the *cr*-locus do not mutate constantly, either by instability or by conversion. The variability is not in the gene locus but in the circumstances under which the gene has to express itself. This will be discussed further in another Section (6.2.7.2).

4.4.3.2 Radiation and Temperature as Mutagens

The first contribution of *Oenothera* to mutation research was to give the starting shot for an investigation of the problem. However, *Oenothera* was not in the leading group in experimental mutagenesis. In plants this was first tried on *Antirrhinum* by Erwin Baur. When, through the experiments by Muller on *Drosophila*, irradiation became a tool for experimental mutagenesis, *Oenothera* was soon made a target (Brittingham 1931a). There are two types of experiments: (1) pollen is treated with the suspected mutagenic influence and the progeny is tested for mutations, and (2) inflorescences with young flower buds are treated to observe chromosomal mutations in the pollen mother cells. From the beginning both chromosomal and gene mutations were considered.

X-ray treatment of *Oe. blandina* gave rise to plants with one or two chromosomal translocations and 50% pollen sterility (Catcheside 1935). In the X_2 generation of this experiment only one segregating mutant was observed, a narrow-leaved form with a ring of four chromosomes, indicating a reciprocal translocation. In further experiments more recessive mutants were found (Catcheside 1937a). In the progeny of X-irradiated pollen of *Oe. Hookeri* de Vries a dominant mutant with polyphenic expression and associated with a reciprocal translocation was observed (Marquardt 1943).

A similar case was observed by Harte (1972c, 1973a). In an experiment on *Oe. Hookeri* de Vries irradiated by microwaves, two dominant mutants (named Sm-1 and Sm-2) with narrow leaves were observed and found to be lethal when homozygotic. Each was associated with a reciprocal translocation with one translocation point in the euchromatin and the other in the heterochromatic part. This are several cases, observed by different authors, in which a morphological effect on the phenotype is connected with a chromosomal rearrangement (Figs. 12, 13).

Trisomics were observed in the progeny of X-irradiated seeds of *Oe. odorata* (Arnold and Kressel 1965a). The two Renner complexes of *Oe. Berteriana*, 1·B, differ in their susceptibility to mutagenic influences. After X-radiation (Krautblatter and Arnold 1967) and after treatment with ethyl-methyl-sulphonate (Kressel and Arnold 1967) mutations are observed more frequently in l than in B.

The influence of temperature and, more importantly, the interaction of temperature treatment and radiation effects has been observed in several experiments. Mutants with abnormal chromosome numbers, monosomics, trisomics and double trisomics, appeared in *Oe. Hookeri* de Vries and *Oe. Lamarckiana* de Vries after heat-shock treatment (Michaelis 1930).

In the experiments on mutagenesis in PMCs carried out in the laboratory of Oehlkers the inflorescences were cut and kept in Erlenmeyer flasks filled with water or the test solutions in temperature-controlled rooms for 6 days. The investigations of Linnert (1951) on duration of meiosis proved that under these conditions PMCs observed at the end of the experiment in diakinesis, the stage most favourable for detection of chromosome mutations, must have been treated in early pachytene. The conclusion is that in this meiotic stage breakage and fusion of chromosomes is possible.

In the hybrid *Oe. flavens*·^{h}*hookeri* the treatment of inflorescences with different temperatures (10, 15 or 20 °C) before, during or for 12 hours

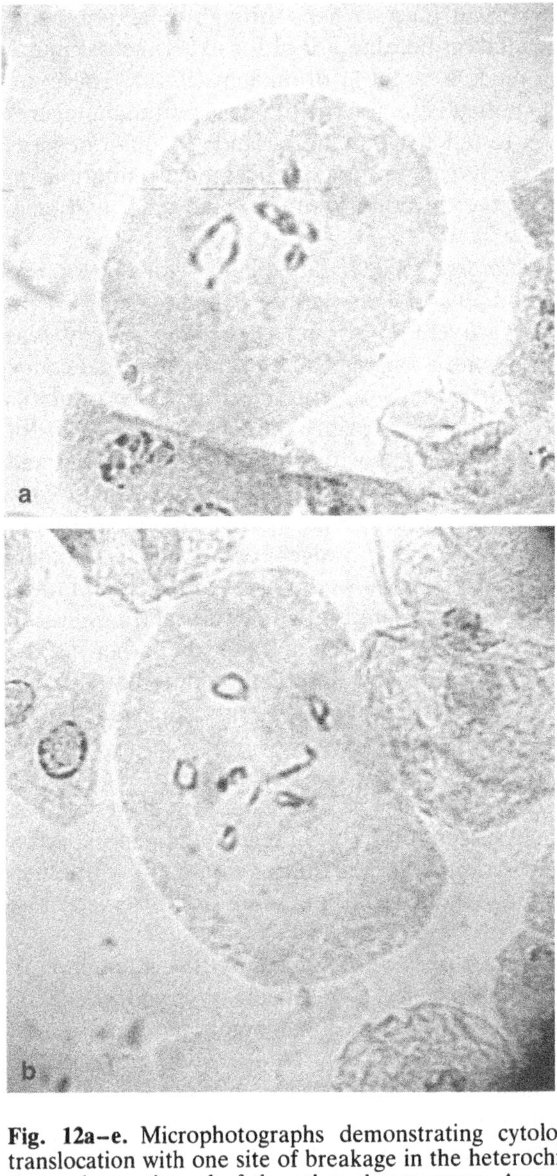

Fig. 12a–e. Microphotographs demonstrating cytological effects of a reciprocal translocation with one site of breakage in the heterochromatin and the second site in the euchromatic end of the other chromosome. As an example observations on the mutant *Sm-1* (dominant, narrow leaves, not viable as homozygote) are presented. The microphotographs show some of the chromosome configurations observed in diakinesis and metaphase which are explained by the schemata in Fig. 13. **a** Ring of four chromosomes in diakinesis, demonstrating the five heterochromatic regions (compare H in Fig. 13). **b** Small ring with attached chain of two chromosomes (compare G in Fig. 13). **c** Small ring with two attached chromosomes (compare F in Fig. 13). **d** Metaphase I with chain of four chromosomes (compare J in Fig. 13). **e** Anaphase I with chromosome bridge as a result of chiasma formation in the translocated chromosome part between the two heterochromatic regions

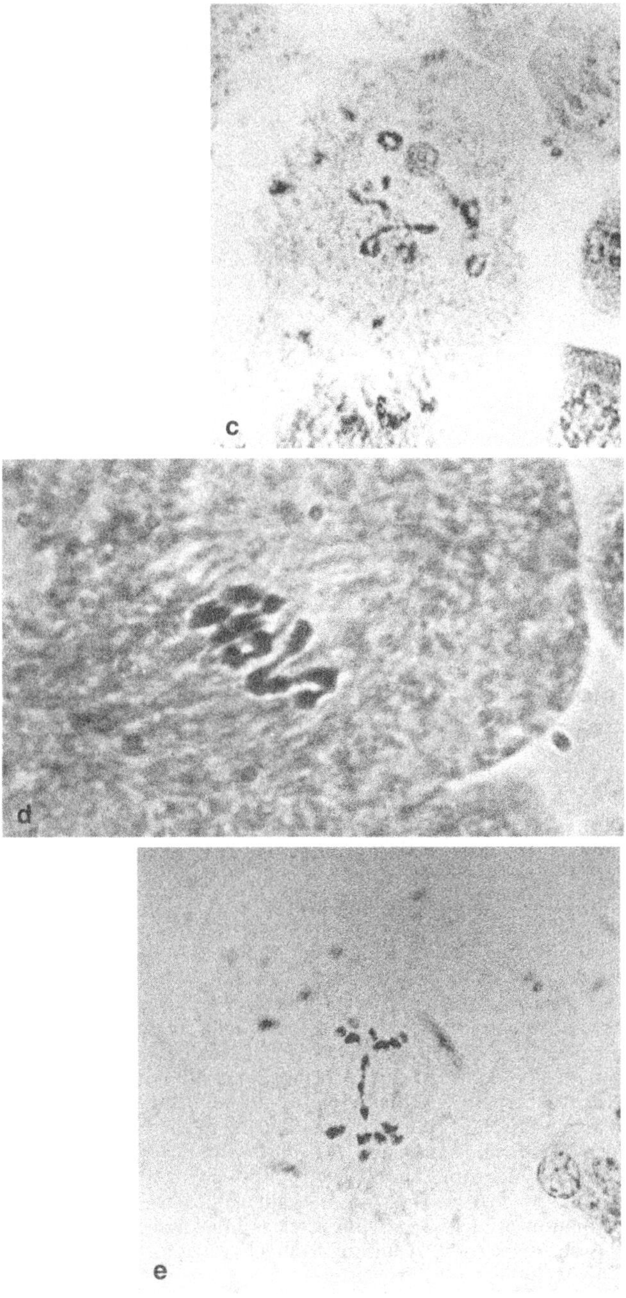

Fig. 12. *Continued*

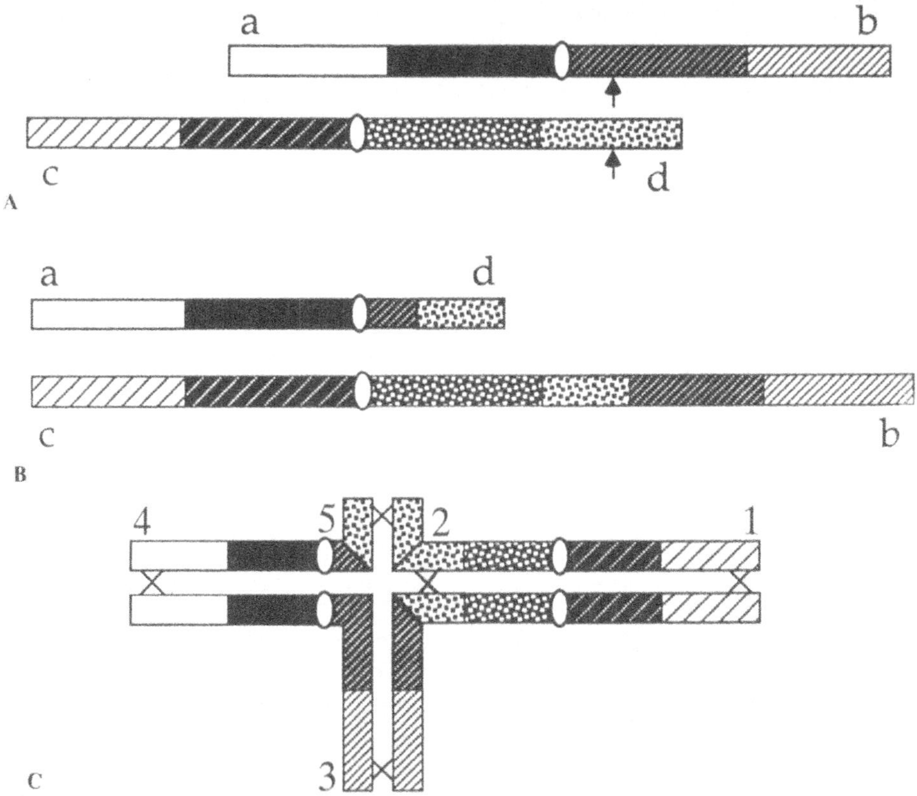

Fig. 13. Schematic representation of the chromosome arrangement in the mutant Sm-1. The chromosome configurations are expected on the basis of varying chiasma formation in the five euchromatic regions which are available for this process. The original drawings of PMCs in diakinesis were made either directly from microscope slides (aceto-carmine squash preparations) or from microphotographs. They are transformed into schemata to demonstrate the different possibilities of cytological realization of one reciprocal translocation in the meiocytes of a heterozygote. They give proof of the difficulties in interpretation of cytological observations. In this special case many pollen mother cells had to be inspected before the identification of the translocation event was possible. **A** Scheme of the original non-homologous chromosomes, the breakage points for the reciprocal translocations are indicated by *arrows*. The chromosomes are not identified. In order to avoid confusion with genetically identified chromosomes, they are not indicated by numbers but as *a*, *b*, *c*, and *d*. **B** Structure of the chromosomes after the reciprocal translocation. **C** Schematic representation of the pachytene pairing in the heterozygotic mutant, with two translocated chromosomes and two chromosomes with the original arrangement of the chromosome arms. The five sites of possible chiasma formation are *numbered* and indicated by *crosses*. **D** Figure of eight, appearing in diakinesis as the result of chiasma formation at all five possible sites as indicated in **C**. **E–I** Different arrangements of the chromosomes, resulting from chiasma formation on four of the five possible sites as indicated. **J–L** Different arrangements of the chromosomes, resulting from chiasma formation on three of the possible sites as indicated. **M** Example of a chain of four chromosomes as a result from three chiasmata. According to the four possibilities for lack of chiasma formation at the sites 1, 3, 4, and 5, four different arrangements of the four chromosomes in a chain are possible, the one shown here with the smallest chromosome at one end as result of a missing chiasma at position No. 5 being the most frequently observed. The arrangements resulting from chiasma formation in the other combinations of three or two or only one of the five possible chiasmata were all observed, but are not presented here. They can be imagined easily by eliminating one, two or three of the possible chiasmata in **E** to **J**

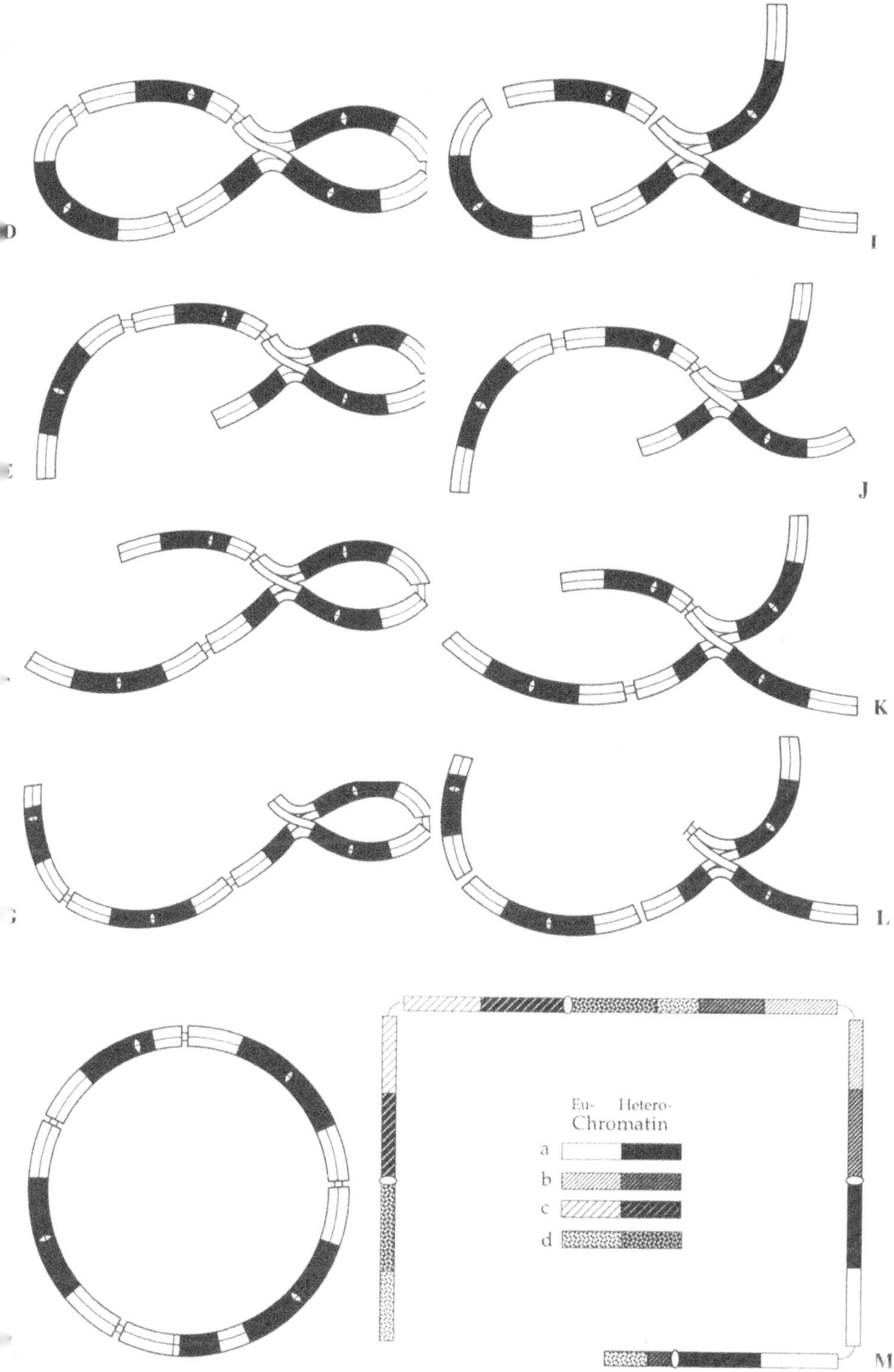

ig. 13. *Continued*

immediately following irradiation did not influence the frequency of chromosomal mutations in the PMCs observed 6 days later. However, treatment at 20 °C for 36 hours immediately after irradiation, compared with treatment at 10 °C, raised the frequency of fragmentations considerably while the frequency of translocations was reduced. An influence on the process of fusion of the broken chromosome ends must be assumed (Stange 1951a). Treatment of plants and inflorescences of *Oe. Hookeri* de Vries at 10 °C during 6 days following irradiation raised the frequency of chromosomal mutations in the PMCs by three- to five-fold. Chromosomal and chromatidal breaks were involved, leading to fragmentations and reciprocal translocations of chromosomes (Marquardt 1952). In relation to the dose of X-rays, the frequency of fragmentations follows a one-hit curve, that of translocations a two-hit curve (Stange 1951a). The results of treatment of the inflorescences with several salts following irradiation revealed an increase of the mutagenic action by LiCl. A similar influence of KCl seems possible. In the controls, treatment with LiCl or KCl without irradiation, an increase in the spontaneous mutation rate was not observed (Stange 1951b). These substances were chosen for the treatment because of the influence of KCl in other experiments (see Sect. 4.4.3.3).

In experiments with microwave irradiation of pollen, inflorescences and intact plants in the field, the induction of chromosomal mutations in PMCs (Harte 1949b) and of chromosomal mutations and point mutations in the progeny of treated pollen was observed (Harte 1972a, 1972c, 1973a). A special effect in these experiments was the observation of mosaic plants, consisting of two sectors, one normal and the other bearing a dominant mutation, in the X_1 generation (Harte 1973b).

Radiation treatment by the application of solutions of ^{32}P and ^{35}S to inflorescences of plants in the field, by the method of split stems as described below, induces chromosome mutations in PMCs as expected (Oehlkers and Bergfeld-Gärtner 1962). The frequency depends on the dose. The distribution of chromosome breaks in the genome is the same as observed by Oehlkers and Linnert (1949) following treatment with chemical mutagens.

4.4.3.3 Chemical Mutagenesis

Up to this point, experimental mutagenesis in *Oenothera* does not differ considerably from the results obtained on other objects. But in *Oenothera* you can always expect a surprise. In the experiments by Oehlkers on the influence of different treatments on chiasma frequency the question arose, what would be the effect when the inflorescences of *Oe. flavens*·^{h}*hookeri* were treated with a combination of two substances, KCl and urethane, one raising chiasma frequency in diakinesis and the other reducing it. I remember the day in the institute at Freiburg and the excitement which followed when the technician Mrs. Vanselow, a very reliable person and specially trained in handling *Oenothera* chromosomes, who carried out the chiasma counting for the professor, came and told him that there was something wrong in this experiment. There were many cells in which the chromosomes looked strange and she did not know what to do with them, and she asked him: "Please come and have a look at these chromosomes yourself". He went to the microscope, had the required look at the chromosomes and did know what to do with them. What he saw during the following hours and days, inspecting many slides himself and demonstrating them to his assistants, were chromosome fragmentations and many reciprocal translocations of chro-

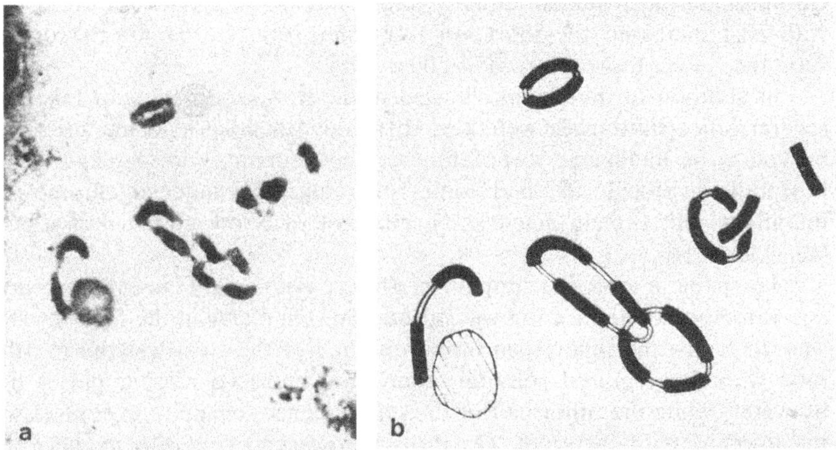

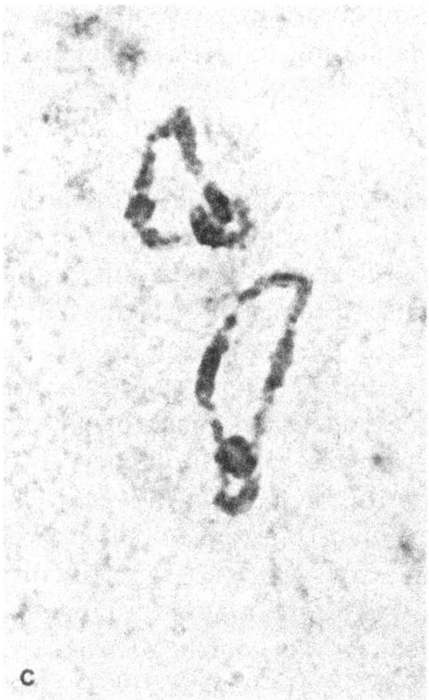

Fig. 14. Chemical mutagenesis. Pollen mother cells in diakinesis with chromosome breaks and translocations from experiments by Oehlkers and Linnert on chemical mutagenesis with urethane/KCl (microphotographs and drawings contributed by G. Linnert). **A** Microphotograph of a cell with two rings of three chromosomes each, two ring bivalents, one open bivalent and two univalents. **B** Schematic interpretation of the chromosome configuration in this PMC. **C** Microphotograph of another cell with several translocations involving eight chromosomes (the missing bivalent out of focus)

mosomes and chromatids. The conclusion was the detection of chemical mutagenesis. The results were confirmed in further experiments and published in a series of papers (Oehlkers 1943a, 1946, 1949, 1952; Oehlkers and Linnert 1949, 1951) (Fig. 14).

The different substances tested in experiments during subsequent years did increase the frequency of chromosome mutations only to a small extent, mostly at the limit of statistical significance, when applied alone, but in combinations of organic compounds with salts, especially KCl, a dramatic increase in mutagenicity was observed. In one experiment the addition of KCl to urethane raised the mutation frequency to 38% of the PMCs investigated. The average effect of this combination was 16% of the cells with chromosomal aberrations. Equally, the combination of glucose or saponin with KCl increased the mutation frequency from 1–3% for the organic substance alone to over 10% (Oehlkers 1952).

In addition to the chemicals used in the first experiments of this kind, several other substances were tested, mainly alkaloids and narcotics, and proved to be mutagenic to different degrees (summary in Oehlkers 1952). Morphine hydrochloride and some other alkaloids induced chromosome mutations, but Codein acted as an inhibitor of X-ray mutagenesis (Tiele-Winckler 1956).

To enable a longer treatment than was possible with cut inflorescences, a new method of application was introduced. On plants in the field the stem was slit below the inflorescence and one half of the stem was put in a test tube with the required solution, which was replaced after a preset time by water, while the other half of the inflorescence remained in contact with the lower part of the stem. On these inflorescences meiosis in the PMCs in the young buds could be observed after different times (Oehlkers 1946). The method of application of the chemicals did not have any influence on the results. Following the treatment of inflorescences of *Oe. Hookeri* de Vries with Zefirol (an invertsoap) chromosome mutations of the same kind as occurred after X-ray treatment were observed (Marquardt 1951).

In diakinesis, the chromosomes of *Oenothera* cannot be distinguished, but the hybrid *Oe. flavens*·^{h}*hookeri* has the chromosome configuration 4 2 2 2 2 2. This allows one to analyze new translocations and to differentiate between the reaction of the two groups of chromosomes in bivalents and the chromosome circle. The statistical analysis of the translocations in these experiments and the use of other hybrids with different rings-of-four, containing other chromosome arms, led to the conclusion that not all chromosomes are involved in rearrangements with the same frequency. There are two chromosome arms in the ring-of-four of *Oe. flavens*·^{h}*hookeri* which are very susceptible to breakage, leading to many rearrangements between the ring chromosomes, and other chromosome arms which are rarely involved (Oehlkers and Linnert 1949). This is in contrast to the random arrangement of chromosome breaks after X-ray treatment as was concluded from comparisons of the two treatments (Oehlkers et al. 1951; Oehlkers 1952). There is no correlation between the influence of chemicals on chiasma formation and mutagenic activity (Linnert 1953).

Treatment of seeds of *Oe. Hookeri* de Vries with solutions of N-methyl-N′-nitro-N-nitrosoguanidine resulted in a high proportion of sublethal plants with disturbances in the morphogenesis of leaves and chlorophyll deficiencies. In the progeny a high proportion of segregating mutants was detected (Kutzelnigg 1972).

With the detection of chemical mutagenesis a new chapter was opened in mutation research and *Oenothera* was there at the start. When it was discovered, chemical mutagenesis was a sensation. Now, 50 years later, it is discussed in every newspaper and sometimes we can only wonder if there are any chemicals at all which can be called bona-fide non-mutagenic. However, that is not the fault of *Oenothera*.

4.5 Genes Without Chromosomes: Extrachromosomal Inheritance

4.5.1 The Plastome

Together, the genes of the Renner complexes and chromosomal translocations can explain most of the cytogenetic peculiarities of *Oenothera*, but some problems still remain. As soon as more extensive crossing of species of *Oenothera* started it became evident that the Mendelian rule of reciprocal identity of hybrids did not hold true in all cases, even when both Renner complexes involved in hybridization were active in both sexes and the genetic identity of hybrids from crosses of the type A ♀ × B ♂ and B ♀ × A ♂ was beyond doubt (Renner 1922, 1924). The main difference observed was that green, vital plants were obtained from crosses in one direction and weak, white or yellow seedlings, supposedly of the same genotype, were obtained from the reciprocal cross. These first observations clearly showed that chlorophyll synthesis in the plastids had been disturbed.

Two examples will be presented. When *Oe. Lamarckiana* de Vries and *Oe. Hookeri* de Vries are crossed, the hybrid *Oe. gaudens·^{h}hookeri* is always green. The twin, *Oe. velans·^{h}hookeri*, is green only with *Oe. Hookeri* de Vries as mother but appears as yellow seedlings, which die as small rosettes, when *Oe. Lamarckiana* de Vries is the female parent (Renner 1934). The hybrid *Oe. flavens·curvans* is generally white, but when the same genome is obtained from the cross *Oe. muricata* × *suaveolens* as hybrid from the rare egg cells with the pollen complex *curvans*, fertilized by *flavens* pollen, the seedlings are green and fully viable (Renner 1936a). These results were confirmed by repeated observations, so the assumption of a genetic influence not located in the chromosomes was the logical consequence. Similar reactions were found in the subgenus *Raimannia*. When the combination of Renner complexes l·v is obtained from a cross with *Oe. Berteriana* as the mother, the plants are green. However, when *Oe. odorata* is the mother, the plastids in the seedlings of the same genotype are without chlorophyll. The same holds for other genome combinations (Renner 1937c). All these results seemed to be clear cases of maternal inheritance.

At first a diffuse "plasmon" was declared to be responsible for the unexpected observations. But soon it was recognized that the genetic difference must be located in the plastids (Renner 1919b, 1922, 1924) and later called "plastome" (Renner 1934). In some publications the expression "plastidome" is used instead but the meaning is the same and "plastome" is now generally accepted.

To make things complicated (what else can you expect from *Oenothera*!), the character "lack of chlorophyll in the plastids" is not constant but can change in a plant according to certain time patterns. Furthermore, in such a pair both hybrids can be chimaeras, either green with white or yellow sectors, or generally without chlorophyll but with green parts (Renner 1936a). This makes three problems: non-chromosomal inheritance, formation of chimaeras and variability of chlorophyll synthesis, which could not be solved together in one step. As is often the case in *Oenothera*, the first problem belongs to genetics, the last one to development, and the second one to both.

When more details were known, a survey of the results was presented by Renner (1937c). The observations on numerous pairs of reciprocal crosses between species can only be explained when widespread genetic extrachromosomal differences, located in the plastids, are assumed. Their detection was possible because of incompatibilities between genomes and plastomes which led to disturbances of chlorophyll synthesis in certain combinations and which were the cause of yellow or white plastids incapable of normal photosynthesis.

The idea of genetic differences located outside the nucleus was neglected by other geneticists at first. Remember: DNA was still unknown, as was the fine structure of cytoplasm and plastids. Biologists had just become acquainted with the idea that inheritance is performed by genetic units which need a material basis. They had acknowledged the chromosomes in the nucleus as responsible for this task, and then they were told that this was not enough, there were other possibilities for inheritance. If this was true, then there must be a material basis for these phenomena. And that idea was more than most biologists could stand.

The next step was official doubt, presented by Goldschmidt, who declared the term "plasmon" to be not only superfluous but dangerous. He said just what nearly everyone in genetics was thinking, and followed the argument of "it cannot be true because it is not allowed to be" (. . . denn, so schließt er messerscharf, nicht sein kann, was nicht sein darf! citation from a poem by Eugen Roth).

I remember the story that Oehlkers told his young collaborators when, in 1939, he returned from a visit to the USA where he had lectured at all the great eastern universities before audiences of important geneticists. After a lecture on plasmatic inheritance, as it was called at the time, he was asked by one of these famous scientists (not in the formal discussion but afterwards in the lobby): "Do you *Really* think that there are *Bodies* in the cytoplasm?". In the private talks in the laboratories several colleagues wanted to discuss results which they had not published because they felt uncomfortable about them but which pointed clearly in the direction of "plasmatic inheritance". The phenomena were known but not acknowledged. We young people really could not decide whether the story was funny or not. It seemed ridiculous yet too serious to laugh at. The problem is discussed from the point of historical differences in scientific thinking between Germany and the USA by Harwood (1987). Meanwhile the existence of these at that time hypothetical "plasmatic bodies" bearing genetic information is verified by the detection of DNA in the plastids and mitchondria.

By that time Renner (1934) had already summarized the results and discussions and made the clear statement that the only important question is whether or not the cytoplasm has independent qualities, constant over generations, which have a meaning for the organism. The idioplasm, as a genetic substance, can be present as the genome in the nucleus, as the plasmon in the cytoplasm and as plastome in the plastids in those organisms which contain plastids. Since the "idioplasm" has been identified as DNA and the presence of this substance in plastids and mitochondria is familiar to young biologists, everything is clear. But this knowledge would never have been achieved if observations had not been made and taken seriously, the formal description developed and the questions asked at a time when the scientific background did not allow the solution to be found at once. In a later review Renner (1952) summarized his newer experiments.

The reaction pattern of the chlorophyll-deficient hybrids was investigated thoroughly by Stubbe (1953, 1955, 1959, 1960, summary 1989c). It turned out that there exist five different plastomes in *Oenothera*. They can be identified by differences in plastid development with certain genomes. As a result of the same experiments, in a similar way the genomes can be allocated to six groups, formed by the combination of three types of Renner complexes, A, B and C, characterized by the type of chlorophyll synthesis with the five plastomes.

The types of the five plasmones, designated by Latin numbers according to Stubbe (1989c), are represented by:

- I in *Oe. elata* HBK (incl. *Oe. hookeri* Torr. & Gray) and *Oe. villosa* Thunb.
- II in *Oe. suaveolens* Desf. and *Oe. biennis* L.
- III in *Oe. grandiflora* and special cases of *Oe. biennis* L.
- IV in *Oe. parviflora* L. and *Oe. oakesiana* (A. Gray) Robbins
- V in *Oe. argillicola* Mackenzie.

Plasmon III is also present in the strain of *Oe. Lamarckiana* de Vries, collected in Sweden by Heribert-Nilsson and used by Renner and Stubbe. At several locations in Germany strains of *Oe. Lamarckiana* de Vries are observed which are identical with Renner's reference stock as far as the Renner complexes are concerned but which contain different plastomes (Kappus 1957). The type *Oe. Hookeri* de Vries has plastome I and behaves like *Oe. elata* HBK.

The three basic types of the genomes are designated by capitals and represented as homozygotes:

A in *Oe. elata* HBK.
B in *Oe. grandiflora* l'Her.
C in *Oe. argillicola* Mackenzie.

The species used as reference types are, according to Stubbe (1989c):

Species	Genome/plastome combinations
Oe. elata HBK (incl. *Oe. Hookeri* Torr. & Gray)	AA-I
Oe. grandiflora l'Her.	BB-III
Oe. argillicola Mackenzie	CC-V
Oe. villosa Thunb.	A_1A_2-I
Oe. nutans Atkinson	B_1B_2-III
Oe. biennis L.	AB-II, BA-III
Oe. parviflora L.	BC-IV
Oe. oakesiana (A. Gray) Robbins	AC-IV

A scheme constructed by Stubbe (first version 1959, last 1989c), summarizing the results of numerous experiments, clarifies the relations (Fig. 15).

The research on plastomes suffers from one severe bias. The parents of the hybrids differ in both the genome and plastome. For exact comparisons, however, it is necessary to have material where every difference observed eventually, by comparing the results of different experimental groups, can be brought back to only one difference in the experimental material. This means that the different plastomes must express themselves on a homogeneous genetic background. With an intelligent scheme for a sequence of crosses and backcrosses using the transmission of plastids by the pollen, Stubbe constructed such a series of reference types in which the plastomes are combined with the homozygotic genome of *Oe. elata* HBK strain *Johannsen* (Stubbe 1989c) and another series consisting of the plastomes I to IV with the genome *Oe. albicans·undans* (Stubbe, pers. comm.) which can be used for the same purpose.

The plastome is always inherited through the egg cells which contain proplastids. In *Oenothera*, however, the generative cells in the pollen tube carry, in addition to a nucleus, proplastids whose number varies only slightly

Stubbe

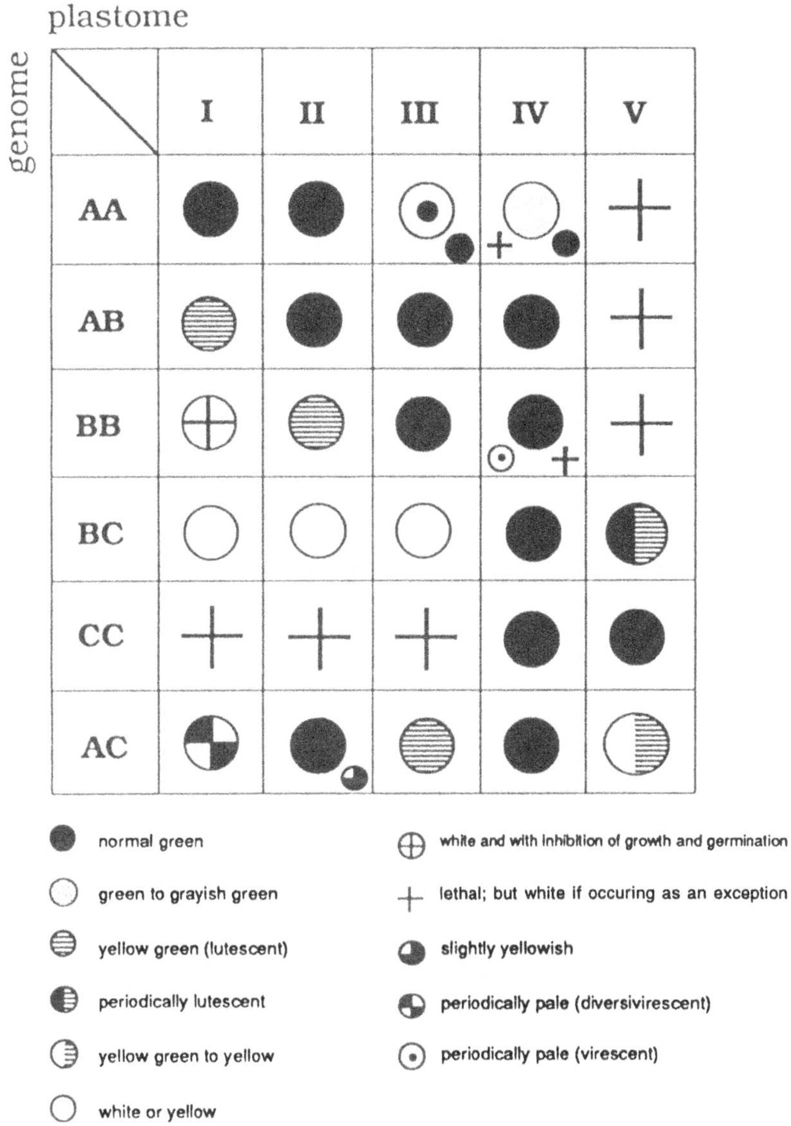

Fig. 15. Compatibility relations between different diploid genotypes and plastome types. There are three main genome types *A*, *B* and *C*, which can form six different diploid types. These can be combined with five different plastome types *I*, *II*, *III*, *IV* and *V*. Each of the 30 combinations is characterized by the development of the plastids, including synthesis and degradation of plastid pigments, as indicated by the *symbols*. Plastome IV is compatible with all genotypes and is therefore considered to represent the evolutionary oldest type. Type V could only evolve in genotypes of type CC (Stubbe 1964)

(Renner 1952; Stubbe 1963a). They are transferred into the zygote and are incorporated in the embryo. Inheritance of the plastome in the male line is normal, as has been observed by Renner and investigated in several cases (Stubbe 1963a; Stubbe et al. 1978). The result is the appearance of embryos which combine two different plastomes with one genome. The maternal and paternal plastids in a chimaera, even within a hybrid cell, maintain their genetic identity and do not influence each other.

In all the experiments with the plastomes of *Oenothera* the basis of the interpretation is the assumption of morphological continuity of plastids in the gametes. Mühlethaler and Bell (1962) interpreted their EM observations with the hypothesis that plastids should arise de novo from the nuclear membrane and therefore could not be considered to bear and transmit constant genetic information independent from the nucleus. Stubbe (1962), Schötz (1962a) and Schötz and Stubbe (1962) pointed out that this hypothesis is not in agreement with the observations on the genetic continuity of the plastomes in *Oenothera* and therefore could not be true, at least for this genus. This discussion was only ephemeral because the discovery of plastid DNA has solved the problem in favour of *Oenothera*.

Chlorophyll synthesis is not the only character influenced by the genes in the plastome in *Oenothera*. Very important is the inactivation of pollen, depending on the incompatibility of a haploid genome with a certain plastome (Renner 1919b; Göpel 1970). In all cases of pollen inactivation on the basis of plastome–genome incompatibility, the round starch grains are a characteristic feature (see Sect. 6.4.3 on pollen development).

The Renner complex *albicans* from *Oe. biennis* L. and *Oe. suaveolens* Desf. is incompatible in the pollen with every known plastome. The Renner complex *flavens* from *Oe. suaveolens* Desf., on the other hand, is compatible with the plastome of both species but incompatible with plastome IV from *Oe. parviflora* L. Pollen grains with the Renner complex *flavens* from *Oe. suaveolens* Desf. and plastids of *Oe. parviflora* L., plastome IV, are inactive (Stubbe 1960), thus preventing the transfer of the Renner complex by the pollen and causing *flavens* to behave as an egg cell complex in such a hybrid. The analysis of the latter case was made possible by the clever use of some peculiarities presented by *Oenothera*: a Renner complex *flavens* without the sporophytic lethal from *Oe. suaveolens* Desf. strain Grado, mutated white plastids of *Oe. suaveolens* Desf. with a complete disturbance of plastid pigment synthesis, transfer of plastids in the generative cell of the pollen tube, and a design for crossing experiments that gave as the result homozygotic *Oe. flavens·flavens* chimaeras with green (plastome IV) and white sectors (mutated plastome II). The flowers on green branches with *parviflora* plastids had inactive pollen and those on white branches with the *suaveolens* plastids had fertile pollen with oblong starch grains. Fertility of the egg cells was not influenced.

Another character, the sporophytically determined degeneration of ovules in the hybrid *Oe. albicans·*^{h}*hookeri* with plastome IV, also depends on the interaction between genome and plastome. Only the rate of multiplication of the plastids themselves is a genuine character of the plastome which cannot be changed by the genome (Stubbe 1960). The order is *Hookeri* I > *suaveolens* II > *Lamarckiana* III > *argillicola* V > *parviflora* IV, with the last one having the slowest multiplication rate. This order was found in all cases investigated, irrespective of the genome of the cell in which the plastids were combined (Stubbe 1960).

Similar types of pale and variegated hybrids have been observed in the subgenus *Raimannia* (Binder 1938; Haustein 1938; Sturm 1938; Schwemmle and Zintl 1939; Arnold 1967b).

4.5.2 Plastome Mutations

In *Oenothera* variegated leaves with white, yellow or light green margins or sectors or whole branches of aberrant colour are repeatedly observed by all scientists involved in experiments with this genus. When such parts do not bear flowers, then nothing can be said about the nature of the abnormality. Stubbe succeeded in raising progeny from such aberrant parts in *Oe. suaveolens* Desf. and later also in other species. This confirmed the assumption that mutations of genes located in the plastids had occurred, causing a deficiency in the synthesis of plastid pigments (summary in Stubbe 1989a). The mutations are named according to the number of the plastome in which they occurred, followed by a Greek letter, in their order of appearance, e.g. I-*alpha* (Stubbe and Herrmann 1982).

In contrast to the chlorophyll deficiency of plastids which results from incompatibility between plastome and genome and which is expressed only in certain combinations, the mutated plastids lack the ability to synthesize chlorophyll and eventually also to synthesize carotenoids, irrespective of the genome involved (Stubbe and von Wettstein 1955). All spontaneous plastome mutations are directly involved in characters expressed in the plastids, such as structure, pigments and proteins involved in photosynthesis. This type of mutation is different from those which must have occurred in the evolution of the plastomes. They give proof, however, of the character of the genes and the mutability of the genetic material located in the plastids. In this respect there is no difference to the chromosomal genes. Whether other plastome genes will mutate too is unknown. Similar mutations have been observed in the sect. *Raimannia* (Schwemmle 1940a).

A special feature of the mutability of the plastome was discovered by Epp (1973; Epp et al. 1987). In *Oe. Hookeri* Torr. & Gray strain *Johannsen* a recessive nuclear gene *pm* (plastome mutator) induces an increase in the frequency of plastome mutations. They are, as far as the morphology and submicroscopic structure are concerned, of the same types as those occurring spontaneously (Epp and Parthasarathy 1987) and exhibit the same genetic behaviour. This can be used to create plastome mutants for further analysis of the structure and function of the plastome.

Starting with the assumption that when there are genes that do mutate spontaneously it must be possible to induce mutations experimentally, several attempts were made to obtain artificial mutagenesis in the plastome. These experiments did not succeed (Arnold and Kressel 1965c; Kutzelnigg 1968; Kutzelnigg and Stubbe 1974). Considering the difficulties involved in constructing an appropriate experimental design and a concept for observation of the results when there is a multitude of mutable units in a cell, maybe inadequate planning was the main reason for the lack of success. The question of experimental mutability of the plastome is still an open one.

4.5.3 Molecular Biology of the Plastome

In contrast to the chromosomal genes, the molecular analysis of the plastome has been investigated in detail, but mainly by one group of scientists centred around Stubbe and Herrmann.

With the help of the proteins in the thylakoid system three plastome groups can be identified: I/II, IV/V and III. According to this analysis, however, the plastomes are not uniform but exhibit a certain variability. In plastome I, at least, five subtypes are found which cannot be distinguished between by genetic methods (Herrmann et al. 1980).

In *Oenothera* the circular DNA of the plastids (ptDNA) has the same basic structure as is found in other plants (summary of the first results in Herrmann et al. 1980). It is constructed of four regions, a large sequence and its inverted duplicate of about 15 MDa, separated by two single-copy regions, a large one and a small one of 57 and 14 MDa respectively (Gordon et al. 1981). In *Oenothera* this was first analyzed for plastome IV (parviflora), the oldest and most primitive one in *Oenothera* according to Stubbe (1959). The structure of the plastid DNA molecule in the other plastomes is similar to plastome IV in most characteristics (Gordon et al. 1981, 1982). Judging from the position of restriction sites, 11 size differences can be assumed between the five plastomes. These are mostly located in the larger of the single-copy regions. The ptDNA polymorphism induced by the mutator *pm* is analyzed too (Chiu et al. 1988).

When differences occur in the duplicated regions they are always the same in both copies. This suggests the existence of an editing mechanism that transposes every change in one copy to the appropriate site in the repeat (Gordon et al. 1982). When this works as a repair mechanism it may offer an explanation for some of the difficulties in experimental mutagenesis of plastid genes.

The mutants can be designated to one of six classes. They are characterized by defects in the photosystems I or II (Hallier 1966, 1968; Fork and Heber 1968; Hallier et al. 1968), in the cytochrome system, the ATP synthase, or RuBPcase, or are pleiotropic mutants. Several of the more than 40 genes, all identified as spontaneous or *pm*-induced plastome mutants, can be located on the DNA molecule of plastome I (hookeri). The loci for the ATP synthase subunit *alpha* and for pre-apocytochrome f lie close together. The nucleotide and amino acid sequences of the latter have been analyzed (Tyagi and Herrmann 1986). More genes, e.g. those for cytochrome b-559, are located in relation to the these sites (Carrillo et al. 1986). The mutant influencing cytochrome f has received special attention (Johnson and Sears 1990). Some other mutants are defective in their proteins (Schmitt 1978). The gene for the large subunit of ribulose bisphosphate carboxylase/oxygenase (rbcL) has also been mapped. Its mutant I-*sigma* differs from the normal gene by a deficiency of five nucleotides (Hildebrandt et al. 1984; Winter and Herrmann 1988).

Particular problems arose with the mutation I-*iota* which affects at least two subunits, *beta* and *eta*, in the polypeptides of the chloroplast ATP synthase. The two genes are located near one another and are cotranscribed. Following in vivo translation, the gene product can be clearly identified as different from the normal product. After in vitro translation of the RNA, however, the polypeptides of the mutated and the normal gene cannot be distinguished. A post-transcriptional event, probably a defect in a translation signal, must be involved in the expression of the mutant phenotype (Sears and Herrmann 1985).

Upto now, *Oenothera* has contributed so many interesting details to the analysis of the ptDNA in angiosperms that more and surprising facts can be expected from future research.

4.5.4 Evolution of the Plastomes

Differences between the plastomes are an important factor in the differentiation of species in the genus *Oenothera* (summary in Stubbe 1959). Such differences cannot be based on mutations of single genes. The incompatibilities must be part of a co-evolution of the genome and plastome. The common ancestor is most probably plastome IV which is compatible with genome C and more of the remaining combinations than the other plastomes. Genome differentiations could have arisen with this plastome.

Moreover, a simple consideration gives an additional indication of the most probable line of evolution. There is a strong competition between plastids of different genetic constitution in mixed cells of a hybrid (Meyer and Stubbe 1974). Eventually this extends to the complete elimination of one of the plastomes from the meristematic cells. A newly arising type of plastid, a new plastome, can only be evolutionarily successful when it is able to survive this competition. It must be stronger than the old one. That means it must have the higher multiplication rate. All other plastomes fulfil this condition when compared with plastome IV. The latter cannot therefore have evolved from any of them.

The first to arise from plastome IV were plastomes II and III which are less tolerant in the cooperation with different genomes. They could have evolved only after the genomes that were compatible with plastome IV were present. The last to appear were plastomes I and V. When a genome is compatible with different plastomes, the one found in the appropriate species always has the higher multiplication rate.

The individual steps in the co-evolution of genomes and plastomes have not yet been identified and the sequence of events is still unknown too. They cannot involve relatively simple changes of plastid function as they are observed in the spontaneous plastome mutations. They must act on another level and influence the expression of otherwise normal genes in the genome and plastome which usually work on the basis of mutual cooperation. Clarification of the nature of the disturbances found in the incompatible genome–plastome interaction will be a most fascinating task for future investigations.

4.5.5 Mitochondria

The genetic continuity of mitochondria has meanwhile been established for *Oenothera*. Inheritance is strictly maternal. Transfer through the pollen is not observed (Brennecke and Schwemmle 1984; Wasmund et al. 1987).

The gene content of the mitochondria is similar to that of other plants. The nucleotide sequence is generally the same, although there are some characteristic exceptions. A speciality of *Oenothera* is a difference in the nucleotide sequence, with the codon TGG for tryptophane found in other plants being replaced by CGG for arginine. However, a divergence is found when the nucleotide sequence is determined in cDNA from mRNA. Here the codon for tryptophane is present. The consequence is that the proteins are more similar to homologous proteins of other species. The explanation presented is that the DNA is edited in the mRNA. Some other editing is also observed (Hiesel et al. 1990). It seems to be a method for correcting changes in the genomic DNA to obtain highly conserved proteins.

In the sect. *Raimannia* some evidence has been found that in addition to the recognized plastome there could be another heritable plasmatic (extrakaryotic) factor (B. Schwemmle 1962; Schwemmle 1968a, 1970). This plasmon is now identified as the genetic material in the mitochondria (Mulcahy and Chapmann, pers. comm.).

4.5.6 Still More Extrachromosomal Factors

Plastids and mitochondria cannot be the only extrachromosomal sites of genetic material in *Oenothera*. The character *falcifolia*, crippled plants characterized by sectors with abnormal sickle-shaped leaves, is observed repeatedly in progenies of the hybrid *Oe. velans·flavens* from crosses between *Oe. Lamarckiana* and *Oe. suaveolens* (Harte 1948; Stubbe 1970). There is a basis in the genome, but the inheritance clearly does not follow Mendelian segregation. At least one, most probably even two, extrachromosomal factors must be involved. They are called determinants and are brought in by both parents. They cannot be located in the plastome because plastid mutations are inherited independently from the determinants. Transmission through the male gametes has a special importance in backcross generations (Stubbe 1970, 1989a,b; Basler 1973). This observation excludes the mitochondria as carriers.

On the other hand, the *fal* factor can be lost following several generations of backcrossing. In some cases genotypes were obtained which were genetically identical with the original parents but lacked the *fal* determinants (Stubbe 1989b). These determinants need a material basis but this cannot be located in chromosomes, plastids or mitochondria. As with all problems in biology, this is either due to black magic or chemistry. The first assumption should not be considered when the second one has not been excluded because not all possibilities have been tested. As there are no cell organelles left to carry the obscure determinants, could it be that they are transported as transposable elements, like those observed in maize and other plants? And maybe even the cooperation of two of them, one from *Oe. suaveolens* Desf. and the other from *Oe. Lamarckiana* de Vries, is necessary for the expression of the *fal* character? There is no proof for this hypothesis, but it does not contradict the observations of Stubbe. Transposable elements are still missing in the genetics of *Oenothera*. It would be strange if they did not exist. It would be nice to have them, with the obligatory *Oenothera* peculiarity. The case of *falcifolia* still leaves some problems to be solved.

4.5.7 The Model of Extrachromosomal Genetics in *Oenothera*

In *Oenothera*, in addition to the chromosomal genes, the cell organelles plastids and mitochondria contain genetic material that is transferred by the gametes to the next generation. The plastome is inherited through both sexes. The mitochondria are transmitted only through the egg cell. Most extrachromosomal genes which characterize the plastomes are active in cooperation with the chromosomal genome. Mutations of the genes located in the plastids occur, influencing the ability for synthesis of plastid components irrespective of the genome. Furthermore, two heritable factors which have not yet been localized in any known cell organelle but which are undoubtedly present and active, are transmitted in the female and male lines.

5 The Species Concept in *Oenothera* and Evolution of the Genus

5.1 Genetics and Taxonomy

The observations on the phenotype of species hybrids in the genus *Oenothera* compared with recognized species are not compatible with the classical concept of "pure species" in taxonomy. All the investigations on species hybrids confirm the conclusion that was reached by Renner (1916, 1917), namely, that most *Oenothera* species are constant hybrids. From these considerations the question arises whether the classical species concept makes sense in *Oenothera*, or whether the Renner complexes are the real taxonomic units of this genus. They can be combined in different ways to give species. When treated in the usual taxonomic manner these seem to be "good species" for taxonomists. New species can arise whenever hybridization between existing species occurs (Cleland 1944; Renner 1946).

The consequences which are derived from the genetic behaviour of species of *Oenothera* in relation to general problems of taxonomy, are discussed by Davis (1922) in a most interesting context that is still valid today. Just as in this genus, there exist in other genera a number of "impure" species which are well-defined members of the flora in a region but for different reasons do not correspond to the strict definition of a taxonomic "pure species" and botanical nomenclature. As the plants will not adjust themselves to fit the rules, it seems advisable to adjust the rules to fit the plants and use taxonomic names in the genus *Oenothera* only as a means of identification in the field without any underlying philosophy.

Shortly after his first experiments with *Oenothera*, Renner made the statement that in *Oenothera* species hybrids do not exist, only hybrids between the gametes of two species which are hybrids themselves, and the new hybrids can behave as new species. To a large extent, evolution in the genus *Oenothera* proceeds in this way.

The evolution of *Oenothera* in the New World, and especially of the subgenus *Euoenothera* in North America, has been described in detail by Cleland and his students in many publications. The main factors of evolution in this genus are reciprocal chromosome translocations, gene mutations, incompatibility systems and hybridization. On this basis Cleland could build up his excellent analysis. As this is presented by himself with many details (Cleland 1972) and is still up-to-date, in spite of publication 20 years ago, it is not the intention to rewrite it here as a short-cut version. The results are mentioned and the reader is referred to the original publication.

Recently, a new species, *Oe. wolfii*, evolved in California from *Oe. elata* subsp. *hookeri*. The Renner complexes are genetically nearly identical but have different chromosome formulas as demonstrated by the ring of 14 chromosomes in meiosis. In the evolutionary process repeated reciprocal translocations are involved as well as gene mutations changing minor characters, the build-up of a system of balanced sporophytic lethals and gametophytic lethals causing degeneration of pollen grains. This species clearly demonstrates the course of evolution in the genus (Wasmund and Stubbe 1986). More details on the evolution and population structure of the translocation heterozygote *Oe. laciniata* are present by Ellstrand and Levin (1980a,b,c, 1982).

Investigations on *Oe. grandiflora* l'Her. elucidated some of the evolutionary processes in this species. The populations consist of both homozygotic and heterozygotic strains. Some of these must have evolved recently. The Renner complexes of the heterozygotes are genetically similar but became differentiated by new reciprocal translocations of the chromosomes which, together with lethals, gave rise to new constant heterozygotes. These proved to be also interesting for investigations of the genome–plastome interactions (Steiner and Stubbe 1984, 1986; Schumacher et al. 1992).

More problems of the evolution in the genus can only be discussed when the developmental systems which are of importance have first been presented.

5.2 Some Aspects of Taxonomy of the Genus *Oenothera*

In recent years the taxonomy of the Onagraceae, especially of the genus *Oenothera* in North and South America, has been investigated by several groups of taxonomists. As these publications are mainly concerned with problems of taxonomy and plant geography of species which are, up to now, not included in research outside these fields, they are mentioned here for use by those scientists who want to look for new problems in new species.

An outline of the systematics of the genus has been presented by Raven et al. (1979). The *Oenothera* species of South America have been extensively investigated by Dietrich (1977). Keys for identification of the species found in Brasilia (Dietrich 1984) and Argentina (Dietrich 1988) are available. Species from North America and taxonomic problems related to them are discussed by Stubbe and Raven (1979a), Dietrich et al. (1985) and Dietrich and Wagner (1988). Munz (1949) described the *Oe. hookeri* group. Problems arising from investigations of some species are discussed by Dietrich and Wagner (1987a,b). The same has been done for *Oe. grandiflora* l'Her. by Steiner and Stubbe (1984). Wagner (1983, 1986) describes new taxa and species. The interesting biology of *Oe. biennis* L. as a weed in Canada is discussed by Hall et al. (1988) including an extensive list of literature on this species.

5.3 *Oenothera* in Europe

When discussing the evolution of the genus in Europe it has to be remembered that *Oenothera* is not indigenous in this part of the world. The species that arrived after 1500 A.D. came either as officially invited guests to botanical and private gardens and walked out silently into the countryside when the show was over, without asking permission, or came as unregistered passengers accompanying other seeds and plants, and sought to make a living as soon they came ashore. In any case there were small samples of a few seeds, each being a representative of a large group of species. In terms of population genetics this means an extreme reduction of genetic variability in the settlers compared with the population in the home country. At the same time, this gave the exploding populations in their new home the chance for evolution of a genetic variability that could not be identical with the *Oenothera* population in the country of their origin. The process of evolution of European *Oenothera* species needs to be investigated by somebody trained in mathematical genetics and population dynamics to develop a model for this behaviour.

The books describing the floras of European countries mention usually only a few species, *Oe. biennis* L. and *Oe. muricata*, *Oe. Lamarckiana* de Vries and sometimes *Oe. suaveolens* Desf. Even according to the elaborate *Flora of Central Europe* (Hegi 1935) only two species exist, although Renner already found this to be far from true. Now, in every region where somebody becomes interested and looks for *Oenothera* in the field, new species are discovered. Renner and his students described in a series of publications the *Oenothera* species of northern and central Germany (Hoeppener and Renner 1929; Rudloff 1929; Renner 1937a, 1938a, 1942b, 1950, 1956; Rossmann 1963b). In the course of his observations he came to the conclusion that in *Oenothera* the methods of classical taxonomy are not sufficient to describe the species. It is impossible to discriminate between species on a purely morphological basis without genetic and cytological analysis of the Renner complexes (Renner 1942b, 1943a).

On the contrary, Gates, who described *Oenothera* species in Great Britain, favoured the classical opinion that taxonomy had just the task to describe the species by morphological criteria (summary in Gates 1958). The analysis of the European *Oenothera* populations by Rostanski (1985) and collaborators depends on the principles laid down by Renner.

The genetic analysis of European species of *Oenothera* supports the assumption that in Europe only a few species are observed which can be derived directly from immigrants without further genetic changes being involved. But even the species *Oe. biennis* L., which was doubtless introduced as one of the first, cannot be identified with any of the known species from North America (Rostanski 1982), even though the *biennis* group is most abundant there (Cleland 1972). Considering morphological criteria and cytogenetic methods as well, all the investigators came to the conclusion that nearly all European species are generated by hybridization, followed by reciprocal chromosome translocations and gene mutations, including the appearance of new zygotic lethals. These are exactly the same factors as are active in the native country of the genus. Maybe the first immigrants did not survive for long because of climatic conditions, but their hybrids were better adapted to the new environment. They spread out with astonishing speed, using the banks of rivers and canals, roads, railway tracks and disturbed sites like ruined cities as their predetermined path.

The evolution of the European species is still continuing. At present, the increase in genetic variability in the genus *Oenothera* is not directed but playful (Renner 1946). In general, the differences and similarities between the newly observed species and those known from earlier observation in the region give proof of their hybrid origin and the continued occurrence of chromosome translocations and gene mutations. Some examples will be discussed.

The number of species in southwest Germany is increasing by immigration from adjacent regions and by hybridization. The Renner complex *albicans* is very stable and *flavens* is extremely variable (Kappus 1957). *Oe. chicaginensis*, observed in Freiburg i.Br., is there surely as an escapee from the experimental garden of the Botanical Institute (Kappus 1960). In a new species with a ring of 14 chromosomes, *Oe. Oehlkersii* Kappus, one Renner complex is *gaudens*, the other is similar to *flavens* but differs by some alleles and reciprocal translocations of the chromosomes (Kappus 1966). Some years later, the new species was observed in Poland (Rostanski et al. 1989). It crossed borders at a time when people could rarely do so! In *Oe. conferta* n.sp. (Renner 1956; Renner and Hirmer 1956) both Renner complexes are similar to those of *Oe. Lamarckiana* de Vries but not identical with them.

In field populations of *Oe. biennis* L. and *Oe. suaveolens* Desf. the frequency of *sulfurea* plants is approximately 0.2 to 0.4%, but this does not allow any conclusions to be drawn on the frequency of crossing-over of the allele *s* from the Renner complex *albicans* to the partner *rubens* or *flavens* as an evolutionary event because the homozygotic *s*/*s*-plants, as soon as they arise, become a part of the population and are maintained by self- and cross-pollination (Kappus 1958).

The Renner complex *velans* from *Oe. Lamarckiana* de Vries, which arose in Europe, is so similar to ^{h}*franciscana* that the hybrid *Oe. gaudens*·^{h}*franciscana* cannot be distinguished from its female parent *Oe. Lamarckiana* de Vries using morphological criteria (Davis 1915, 1926a). The same type is observed from hybridization of *Oe. biennis* L. with *Oe. franciscana* and too from *Oe. biennis* L. × *Oe. grandiflora* l'Her. (Davis 1911). This means that a phenotype which cannot be distinguished from a well-known species can be obtained by hybridization of one species as the female parent with pollen from either one of two other species!

From the analysis of further experiments it can be concluded that the Renner complex *flavens* is not identical with *acuens* from *Oe. grandiflora* l'Her. but it is very similar to it. This opens the possibility that *Oe. Lamarckiana* de Vries and *Oe. suaveolens* Desf. originated in Europe by hybridization between the early introduced *Oe. biennis* L. and the later arrivals *Oe. franciscana* and *Oe. grandiflora* (Hoeppener and Renner 1928). According to this assumption, *Oe. Lamarckiana* de Vries received from *Oe. biennis* L. the Renner complex *rubens*, which was named *gaudens* before the identity of both was recognized. For *Oe. suaveolens* Desf., however, *Oe. biennis* L. could have been the female parent, giving the Renner complex *albicans*, and *Oe. grandiflora* l'Her. the male parent, contributing *acuens* which changed to *flavens* by reciprocal chromosome translocations and some gene mutations, including the lethal that established the balanced lethal situation. However, when in addition to the genomes the differences in the plastomes are considered, it seems that the evolution was not so straightforward, but intermediates, now extinguished, could have been involved (Stubbe 1953, 1989c). Considering the differences in chromosome formulas and morphological differences between *Oe. suaveolens* Desf. from different parts of Europe, as observed by Stubbe (1953), and in the otherwise stable Renner complex *albicans* (Stubbe and Diers 1958), the latter hybridization could have occurred repeatedly. This would make the species *Oe. suaveolens* Desf. polyphyletic, thus creating a new treat for taxonomists.

Davis (1940) observed in Great Britain a new species, *Oe. cantabrigiana*. A Renner complex similar to *albicans* is active in the pollen and the egg cell complex resembles *rubens*. This was seen as proof that *Oe. biennis* L. can continue evolution. However, it was not tested (and could not be considered at the time) whether this new species is identical to or derived from a newly introduced *Oe. biennis-2* of the *beta-alpha* type described later by Cleland (1972).

In France new species, their genetics and cytology have been described by Linder (Linder 1957a,b,c; Linder et al. 1957; Linder and Jean 1969). A key for the *Oenothera* species in France is provided by Linder (1957b).

Another group, with Rostanski as the senior author, investigated locations in several parts of Europe. These publications are concerned with the taxonomy and distribution of *Oenothera* species and are presented complete with Latin diagnosis of the morphological characters of the species, the site of herbarium specimens, synonyms, ecology of the locations, keys for the species and hybrids in different regions, and extensive literature lists which provide access to publications on taxonomy and floristics. In some cases, as far as is possible, the Renner complexes and their relationships are mentioned in addition. The regions investigated are: Austria (Rostanski and Forstner 1982), Britain (Rostanski 1982) and Wales (Rostanski and Ellis

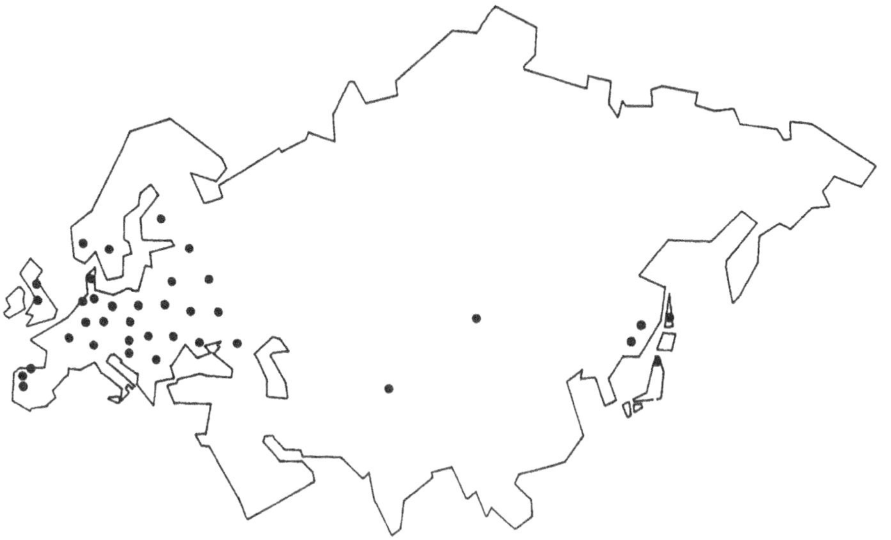

A

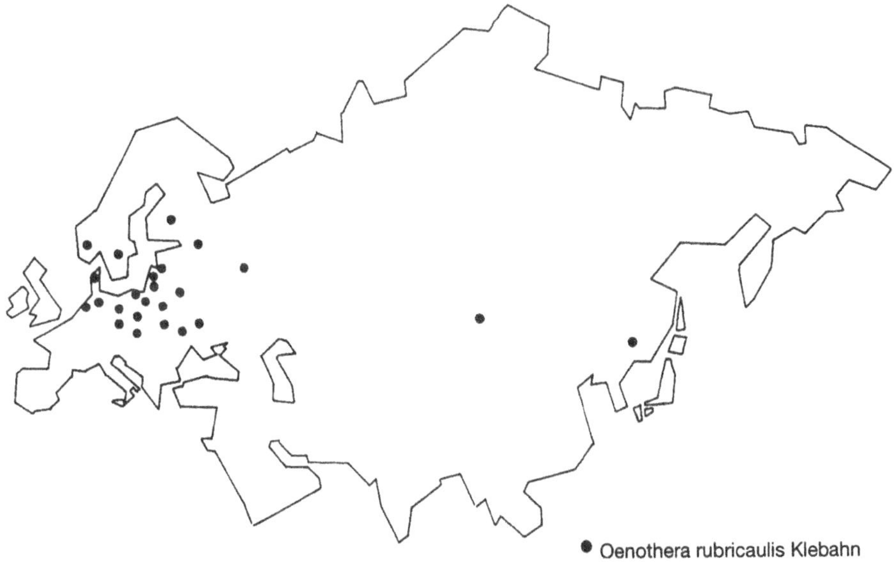

B

Fig. 16. Distribution of *Oenothera* in Europe and Eurasia (maps contributed by K. Rostanski). **A** *Oenothera biennis* L., s.str. **B** *Oenothera rubricaulis* Klebahn (syn. *Oe. muricata* L., s.str., nom.conf.). **C** *Oenothera suaveolens* Desf. (ex Pers.); *Oenothera erythrosepala* Borbas. **D** *Oenothera parviflora* L.; *Oe. subterminalis* Gates (syn. *Oe. silesiaca* Renner)

C

D

Fig. 16. *Continued*

1979), Czechoslovakia (Jehlik and Rostanski 1979, 1980), some parts of eastern Germany (Rostanski and Kloss 1965; Gutte and Rostanski 1971, 1981), Hungary (Rostanski 1966), Italy (Rostanski and Soldano 1981), Poland (Przwywara and Rostanski 1980; Rostanski and Wika 1988; Rostanski et al. 1989b; Sokol and Rostanski 1989) and especially the province Silesia (Rostanski 1965a; Rostanski and Serwatka 1968; Rostanski and Szotkowski 1973). In every region now included in the search for *Oenothera* new species and hybrids will be detected (Fig. 16).

6 Developmental Biology of *Oenothera*

6.1 General Aspects

The genetic analysis of *Oenothera* hybrids left some problems which cannot be solved by that method. These are, in fact, not problems of genetics but of development. The solution can only be found by a shift of the investigations to the appropriate methods of developmental biology.

For many aspects of development in *Oenothera* the knowledge is based on repeated and accumulated observations by the scientists working with this genus and not on experiments performed to solve problems of development. Nevertheless, there are enough facts known and, at the same time, enough problems left open for future research to make the development biology of the plants in this genus most interesting.

6.2 The Sporophyte

6.2.1 Embryogenesis

In *Oenothera* embryogenesis, both from the morphological and physiological points of view, maturation of the seeds and germination have not received much more attention from botanists than any other common plant. Normal embryogenesis was described by Renner (1914). The ovules in which no fertilized egg cell is present soon degenerate and are found in the mature fruit as "fine dust" (Weidner-Rauh 1939). Fertilization can be verified microscopically by the remnants of the pollen tube in the micropyle. The polar nucleus starts division immediately after fertilization, developing at first a single layer of endosperm cells surrounding the cavity of the embryo sac before mitosis begins in the fertilized egg cell. Later, the endosperm grows and becomes multilayered, but disintegrates before the seed maturation. However, in *Oe. lutescens* (*flavens·flavens*), which lacks the lethal allele in one Renner complex, ovules were observed with an embryo but without endosperm (Hiorth 1926).

The zygote enlarges and starts cell division to form the pro-embryo several days after fertilization. The suspensor is short; the haustorial cell at the basal end remains small. The globular and the heart-shaped stages and the initiation of the cotyledons do not show any peculiarities. The embryo enters the resting stage in the mature seed with a radicle facing towards the micropylar end of the ovule and with two cotyledons, but without differentiation of leaf primordia at the embryonal vegetation point. In the mature seed the embryo contains a large amount of fatty substances as reserve material. However, the steps in metabolism leading to this accumulation are not known in detail. Recently, the *Oenothera* oil, gamma-linolenic acid extracted from the seeds, has been used in cosmetics and is believed by some people to have certain astonishing nutritional and pharmaceutical values (Hall et al. 1988; Jean 1989).

Haploid embryos occur in some homozygotic species (Gates 1929; Gates and Goodwin 1931; Lelieveld 1932). They have never been observed in *Oe. Lamarckiana* de Vries and other species with lethal factors in the Renner complexes (Davis 1931; Renner 1941). In controlled progenies with genetic markers both androgenic (Haustein 1961) and gynogenic haploid plants (Davis and Kulkarny 1930; Stubbe 1953) were observed. This means that diploidization of the egg cell nucleus is not required to start embryogenesis. Following fertilization, either of the two gametic nuclei in the egg cell can degenerate before fusion takes place and in this case the remaining one, either the female or the male nucleus, can start embryogenesis. In experiments where a large number of haploids with genetic markers were found, both cases were observed to occur with approximately equal frequency (Harte 1973d). Because haploid embryos are rare, they have never been observed in situ and embryological observations on the endosperm in the ovules with haploid embryos are missing. In comparison with observations on other species, it seems possible that the polar nucleus is fertilized by the second male gamete provided by the pollen tube and that a normal endosperm is present.

In rare cases seeds with two embryos occur. They always develop from two embryo sacs. Their frequency is very variable in different years (Harte and Sniezko 1988).

Embryogenesis can be disturbed by recessive sporophytic lethals which, in the homozygotes, influence the order of cell divisions at different stages in the young embryo. These disturbances are followed by degeneration of the embryo. The stage in which they become apparent is different and characteristic for each lethal. In *Oe. Lamarckiana* de Vries, the two homozygotes can be distinguished by the type of disturbance of embryogenesis. Different investigators do not agree on the exact stage in which the abnormality becomes obvious, but all find two types of degenerating embryos, both of which die before the heart-shaped stage is reached (Renner 1914, 1916; Hiorth 1926). In *Oe. grandiflora* Ait., the homozygotes *Oe. truncans·truncans* die later, just after the appearance of the cotyledons. The stage that can be reached by this genotype depends, however, on the plasmone (Gerhard 1929). The homozygotes *flavens·flavens* from *Oe. suaveolens* Desf. are different from all these (Hiorth 1926). As haploids of Renner complexes containing lethal factors are never observed (Davis 1931), it can be concluded that the disturbance of embryogenesis occurs in the same way in the haploids with lethal alleles as is observed in the corresponding diploids and that it is caused by the lack of expression of the normal alleles and not by other influences, like a special activity of the mutated gene or incompatibility of genomes, between plasmone and genome or between mother plant and embryo. Attempts to save lethal embryos by in vitro culture after dissection from the ovule before or at the onset of the disturbance have not succeeded (Harte unpubl.) but should be tried again using the modern knowledge on special nutritional requirements of embryos in different stages of development.

In some species of the sect. *Raimannia* lethal factors, which cause the death of the homozygotic embryos, have been observed (Oelkrug 1934; Schwemmle and Lohmeyer 1963). In hybrids between *Oe. Hookeri* de Vries as the female partner and pollen of several species of the sect. *Raimannia*, however, all the embryos die at different stages of development, but in every case before 22 days after pollination. Here, the lethality is not initiated by a single lethal gene but is the consequence of incompatibility between the hybrid nucleus and the *hookeri* plastids in the zygote (Kistner 1955).

In *Oe. Lamarckiana* de Vries the development of adventitious embryos from cells of the nucellus has been observed by Haberlandt (1921, 1927, 1938) in ovules which were damaged by external influences, such as squeezing or wounding the ovary, or which developed from the last ovaries at the end of the flowering period. Vegetative cells near the chalaza or from the nucellar tissue adjacent to the side of the embryo sac expanded into the cavity of the degenerating embryo sac. By some cytological criteria, such as the absence of a vacuole, the presence of relatively more cytoplasm than in the neighbouring cells, or a large nucleus with a nucleolus, these vegetative cells resembled zygotes. Observation at later stages suggested that many malformations occurred, but some of these cells developed to become nucellar embryos.

It is evident that in *Oenothera* undisturbed embryogenesis requires the expression of the normal alleles of several genes at certain stages of development. In the sect. *Euoenothera* at least four different loci are involved, most probably more, each becoming active at a different stage of embryogenesis and controlling crucial steps. As yet, it is not known what these genes do exactly at the molecular level. Here a most intriguing field for research by modern methods is open (Fig. 17).

6.2.2 Development of the Seed

Fertilization initiates the development of the embryo and the endosperm. This induces the maternal tissue of the ovule to grow to the size of a normal

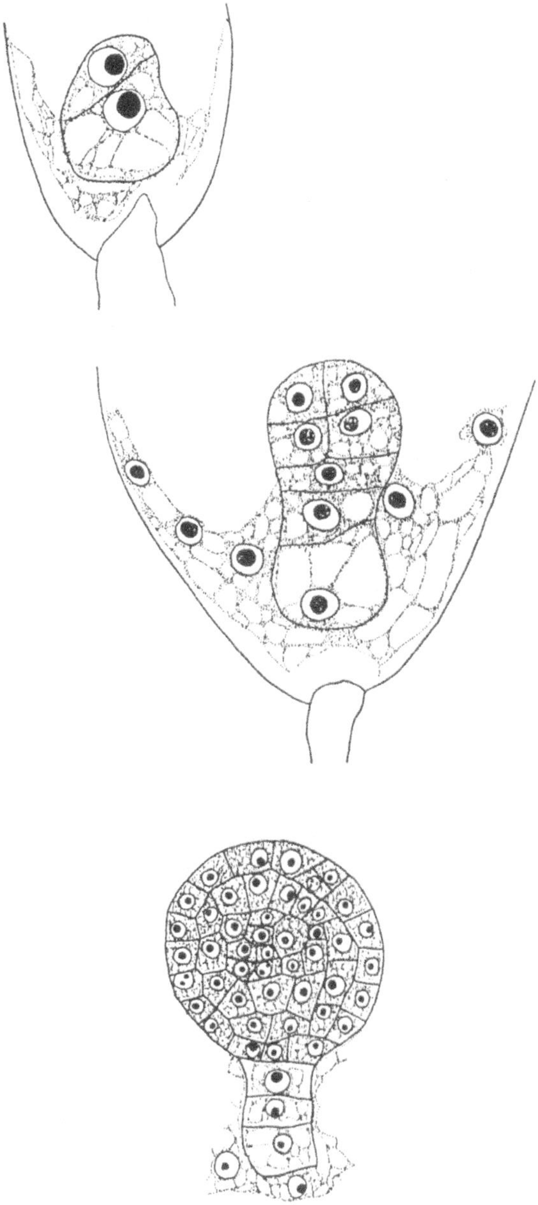

Fig. 17. Embryogenesis in *Oenothera*. Embryos from the progeny of *Oe.* (*Lamarckiana* de Vries × *Oe. biennis* L.) *fallax* × *Oe. biennis* L. The hybrid *Oe. fallax* contains the Renner complexes *velans·rubens*. Two types of embryos have been observed 9 days after pollination (Langendorf 1930). **A** Two-celled lethal embryo, supposed to be *rubens·rubens* (600×). **B** Normal embryo in globular stage with short suspensor, supposed to be *velans·rubens* (600×). **C** Vital embryo 21 days after pollination (450×)

mature seed. For some reasons, such as homozygosity for lethal alleles, the embryo can die early in development. The death of the embryo does not inhibit the growth of the seed coat if this has already started. At the end of its development there is a seed of normal size, filled with an opaque undifferentiated tissue, but without an embryo or containing only a very small undifferentiated structure that could possibly be the degenerated embryo (Renner 1916). This means that the development of the ovule after fertilization is initiated by influences from the embryo or the endosperm as an inductive process. After the start, the presence of a growing embryo is neither needed for the continuation of growth of the seed nor for determining the end of growth and differentiation of the structure of the seed coat as soon as the size of the seed corresponds to the normal mature embryo. The fact that growth of the seed coat stops at the same size in fertile and embryo-less seeds proves that the limitation of seed growth is not caused by maturation of the embryo. The size of the mature seeds depends on the species. *Oe. Hookeri* de Vries has smaller seeds than *Oe. Lamarckiana* de Vries. Investigation of reciprocal hybrids has revealed that the size of the embryo and seed mainly depends on the genotype of the mother plant and not on the genotype of the embryo (Krumbholz 1930). After induction, the development of the maternal tissue is an autonomous process both for continuing and for finishing growth. The first description of seed development was given by Taeckholm (1915).

In the exceptional case of *Oe. lutescens*, *flavens·flavens*, where, in one of the two otherwise identical Renner complexes, the lethal allele is lost, ovules with embryos but without endosperm are observed. The nucellus is not dissolved but is pushed aside by the growing embryo (Hiorth 1926). From this observation it can be concluded that the disintegration of the nucellus during seed development in *Oenothera* is neither a question of ageing of this tissue nor of starvation because the growing embryo takes the nutrients from the nucellus. Instead, it is initiated by influences from the endosperm. The growth of the embryo does not depend on substances from the disintegrating nucellus.

In some species of the sect. *Raimannia*, the degeneration of the embryos with homozygotic lethals occurs early in development, before the induction of growth of the ovule by the embryo or endosperm is completed. The fertilized ovules without embryos stop growing and can be observed in the mature fruit as "coarse powder" between the seeds with normal embryos and the "fine powder" consisting of ovules which degenerated before fertilization (Oelkrug 1934).

In summary, it can be concluded from the observations on embryogenesis and seed development that, besides the influence of genes on many processes in early development stages of the embryo, a considerable number of developmental steps in the seed are directed by induction from adjacent cells and tissues. Interactions between the embryo, endosperm and vegetative tissues of the mother plant are most important at every stage in early development.

6.2.3 Germination

After harvesting the seeds cannot germinate immediately but need some time of after-ripening. In the Northern Hemisphere, the germination of the seeds, measured as the percentage of germination and average time needed for emergence of the root from the seed coat, has an optimum with fast germination of a high percentage of the seeds in spring, from April to June, and reaches a minimum with slow germination in early winter between October and December. When stored in a dry room at moderate temperatures from 12 to 18 °C, the viability of the seeds, measured as percentage germination under optimal conditions, does not decrease during the first 6 or 7 years after harvesting. After this time the germination does not increase again after the winter minimum and the embryos seem to be dead.

The behaviour of *Oe. Berteriana* and *Oe. odorata*, sect. *Raimannia*, is different. With ageing of the seeds, the percentage of seedlings that die at or shortly after germination increases until from old seeds none survive (Schwemmle 1940b).

Germination does not take place in complete darkness. Illumination, with either natural or artificial light, even of low intensity, for at least several hours per day after soaking of the seed coat is necessary according to the general experience, obtained with species of the sect. *Euoenothera* and experiments on sect. *Raimannia* by Schwemmle (1960). Germination is best at moderate temperatures of about 20 °C (18 to 25 °C). Low temperatures of around 5 °C, or even freezing of the soaked seeds for 12 hours or longer, considerably increases the percentage of germinating seeds in a sample and reduces the average time and variability of the germination process. The result is fast germination of nearly all viable seeds of a sample treated in this way within 2 or 3 days after the root of the first seedling has perforated the seed coat. In segregating progenies of hybrids of several species of the sect. *Euoenothera*, including *Oe. Hookeri* de Vries, *Oe. suaveolens* Desf. and *Oe. strigosa* de Vries, the frequency of some genotypes was different in samples of early- and late-germinating seeds, demonstrating the influence of the genotype on the rate of germination (Harte 1949a). Similar observations have been made on species of the sect. *Raimannia* (Schwemmle 1952c).

6.2.4 Development of the Seedling

When the cotyledons become free from the seed coat, they unfold and enlarge. First, they are raised above the surface of the soil by stretching of the hypocotyl. The first leaves are initiated at the vegetation point without development of a stem. Then, the behaviour of the hypocotyl changes. The young rosette is soon brought down to the level of the substrate by shrinkage of the hypocotyl, and this is directed by the cotyledons (Renner 1948b). The leaves of the rosette are initiated at the vegetation point in a spiral.

Heterophylly is observed. In *Oe. Hookeri* and in some mutants which differ in the form and size of their leaves, the first leaves of the rosette are different from those initiated later (Baumgarten and Harte 1964, 1965a,b; Harte 1971). In most species of the *Euoenothera* number, size, form and anatomical structure of the leaves in the rosette are modified by the photoperiodic conditions (Renner 1948b).

In *Oe. Hookeri* de Vries two mutants were observed with a disturbance in the IAA metabolism of the seedling. The vegetation point between the cotyledons develops first as a callus upon which leaf primordia and vegetation points of shoots later appear. From these a quite normal-looking plant can grow, in spite of some abnormalities of the leaves. By treatment of normal seedlings with high concentrations of IAA, phenocopies with similar disturbances in the development of the young vegetation point can be obtained (Baumgarten and Harte 1965b).

Early in development the raphide cells start to differentiate in still meristematic tissues (Kowalewicz 1956b; Schötz et al. 1970). They are already enlarging while neighbouring cells are still dividing, and can reach a final length of 1 mm. In homozygotic recessives of the gene *kr/kr* (*Knirps*), segregating from hybrids with the Renner complex *stringens* (Oehlkers 1922; Harte 1942), an exceptionally large number of raphide cells is found in all tissues (Oehlkers pers. comm.), confirmed by observations of Harte (unpubl.) suggesting a metabolic disturbance correlated with this genotype and expressed from the first stages of growth of the seedling onwards.

6.2.5 Development of Shoot and Leaves

Some genes are known to influence growth. The height of the shoot is limited by the genes *nanella* from *Oe. Lamarckiana* de Vries (Davis 1926b) and *dwarf* from *Oe. franciscana sulfurea* (Emerson 1929a,b, 1931a; Davis and Kulkarny 1930; Davis 1937); and also by *Zwerg* from *flavens* and *Knirps* from *stringens*, both found in the progeny of the hybrid *Oe. suaveolens* Desf. × *strigosa* de Vries (Oehlkers 1922; Cleland and Oehlkers 1930; Harte 1942, 1948). Dominant alleles influencing the geotropic reactions of the growing shoot, causing plagiotropy of the top or curvature of the stem, are known from several segregating hybrid progenies with the Renner complexes *curvans* of *Oe. muricata* (gene *Cu*) and *flectens* of *Oe. atrovirens* (gene *Fl*) (Renner 1925, 1933a, 1942a). Several mutants with a reduced shoot length and one with a changed geotropic behaviour, causing a plagiotropic growth of the stem, are observed in progenies of regenerated plants from tissue cultures of a stock of *Oe. Hookeri* de Vries with small *sulfurea* flowers (Harte unpubl.).

The mutant *pitys* of *Oe. biennis* L. (Renner 1959a) and a phenotypically similar mutant of *Oe. Hookeri* de Vries (Harte 1973b) develop disturbances during the growth of the inflorescence, in which parts with normal internodes, bracts and flowers are interrupted by regions with extremely short internodes and bract-like leaves without flower buds. In both cases, fluctuations in the metabolism or activity of growth hormones, either auxin or gibberellins, can be suspected. The same holds for another mutant, *Gi-1*, with malformations of the shoot (branching of the shoot by partition of the vegetation point, resembling fasciations) (Harte 1973b).

In the rosette the final size of subsequent leaves increases and the form, measured as the length/width relation, changes gradually to a steady state in which the size of the new leaves at the end of their growth does not increase any further (Baumgarten and Harte 1964, 1965a,b; Harte 1971). This stage will last during the course of initiation of several leaves. Then, the final size of the newly formed leaves decreases until the growth of the shoot starts. From an analogy to the behaviour of other plants (*Antirrhinum*, Harte 1983) the decrease in final leaf length can be interpreted as the first sign of induction of the flowering process. During the development of the shoot, the final size of the leaves decreases, and the form, now described by the length/width relation and the relative position of the broadest part of the leaf, changes continuously in a transition zone to the bracts.

In the publications by Renner (e.g. 1925, 1933a, 1941, 1942a) on segregating progenies, many genes are mentioned which influence size and form of the leaves. The great variability of these characters in species and in segregating progenies (photographs in Hoeppener and Renner 1929) suggests the activity of a polygenic system with quantitative influences on leaf growth. Similar results have been obtained by Harte (1948).

In *Oe. Lamarckiana* de Vries two genes are known to cause abnormalities of the leaves. These are *acutifolia* (Brittingham 1931a) and the mutant *funifolia* in which the margins of the leaves roll downwards (Shull 1921). The phenotype is nearly identical to the mutant *formosa* from *Oe. pratincola*. For none of these mutants has there been an investigation of their developmental features compared with those of the normal leaves.

Several mutants with small and narrow leaves or with very long and extremely narrow leaves are observed in mutation experiments with *Oe. Hookeri* de Vries (Harte 1972c, 1973a) and also originating from tissue cultures (Harte unpubl.). The investigations on leaf development in these mutants gave some insight into the mode of gene action influencing morphogenesis. The leaves are initiated at the apical vegetation point in the subprotoderm. At the sides of the primordium a marginal meristem is formed, from which the lamina of the leaf develops. A predetermined sequence of the orientation of cell divisions in this meristem cannot be observed. The new cell walls are arranged in a way that gives undisturbed development of a flat structure. The final number of cell layers in the lamina is not determined in the marginal meristem but by the orientation of the cell walls after mitosis at a short distance from the margin (Baumgarten and Harte 1964).

The veins in the leaves are initiated in the laminar meristem near the margin. The width of the leaves depends on the number of cell divisions between the veins and is influenced by several genes. The mutant *bullata* (Shull 1928a, 1930) has very crooked leaves, caused by a disturbance of the growth of the mesophyll in relation to the veins. In mutants of *Oe. Hookeri* de Vries, compared to the normal type, an influence on the growth of the mesophyll between the veins, without a change in the basic pattern and number of veins in the lamina, is observed. In the mutant *b* (broad leaves) an increase in cell number between the veins is found, in contrast to the mutant *Sp* (Spitzblatt) which is characterized by a decrease. This indicates an influence of these genes on cell division in special regions of the meristematic leaf after initiation of the veins (Baumgarten and Harte 1964). By investigation of morphogenesis (Baumgarten and Harte 1965b) and repeated measurements of the leaves during the growth period, quantitative data were made available to analyze the growth curves for length and width (Baumgarten and Harte 1965a) and to determine the parameters of the developmental function which describe the relative growth (Harte 1971). Quantitative differences in developmental processes, as a basis for the differences in form and size of the leaves in the mutants, were revealed. This series of investigations demonstrated that the manifold differences between the mutants are secondary effects. The primary action of the genes involved in the study is to influence growth by cell division in the primordium and leaf meristems, in combination with differences in the growth rate and duration of growth.

The *falcifolia* type, characterized by asymmetrical leaves and general growth disturbances and which has been investigated by Stubbe (1953, 1970, 1989a,b; Basler 1973), requires special attention. The observations on development and anatomy of the sickle-shaped leaves stress the importance of cooperation between epidermis and mesophyll for normal growth of the leaves.

A great number of genes are involved in the morphogenesis of the leaves, changing size, form, structure and growth relations in the lamina. Nothing is yet known about the sequence of their expression in leaf development and about the physiological processes involved in leaf morphogenesis in *Oenothera*.

For several mutants it has been demonstrated that a disturbance of the auxin content is correlated to the mutant phenotype. The first of these is the mutant *helix* from *Oe. biennis* L., named according to the form of the leaves on the stem which are rolled into a spiral. The leaves contain considerably more auxin than the normal leaves. Similar disturbances in leaf development can be found after treatment of the seedlings with 2,4-D (Renner 1953; Chrometzka 1955). In *Oe. Hookeri* de Vries two mutants, *Sm-4* and *Sm-7*, both characterized by very long leaves with a narrow, interrupted lamina, were observed. The vegetation point of the seedling becomes a callus-like structure upon which buds and shoots later regenerate. Phenocopies can be obtained by treatment of normal seedlings with auxins (Baumgarten and Harte 1965b). These and some other mutants of *Oenothera* could be interesting objects for investigations of the genetics of growth hormones.

6.2.6 The Flowering Process

The physiology of flowering has not yet been properly investigated in *Oenothera*. After germination, all *Oenothera* plants grown until now in experimental gardens, even the annuals, develop a rosette of leaves before shoot elongation starts. The biennials normally germinate in their natural habitats in late spring and survive the winter as a rosette of leaves, but under experimental conditions most of them can be grown as annuals. In all cases the elongation of the shoot is correlated with the induction of the flowering process. The species from Europe and North America are mostly long-day plants which start flowering when the rosette of a certain size is influenced by long days (Renner 1948b). The biennials germinate in late spring under long-day conditions and form a rosette with a considerable number of leaves, but the inflorescence develops only in the following year. The photoperiod cannot be the only factor for induction of flowering. It can be suspected that the *Oenothera* species are either short-day–long-day plants (according to the definition by Salisbury 1963) or are temperature-dependent plants. Some biennials need low temperatures after the rosette has reached a certain size, but it is evident that the temperature requirements are very different in species from different geographical regions. Many biennials can be grown as annuals, germinating and flowering in the same year, when the seeds are sown between January and March and the plantlets are brought out of the greenhouse in time for them to be influenced by at least some cool nights. Other species, e.g. *Oe. strigosa* de Vries, will form a shoot and start flowering only when sown early and given cold temperatures of near 0 °C in the rosette stage. The species from the south of North America and Central America are day-neutral or short-day plants.

The initiation of new flower buds and growth of the very small ones stops when the first seed capsules at the base of the inflorescence reach maturity. This stop in the growth of the apical meristem can be prevented by removal of the fertilized ovaries, so that no mature or nearly mature fruits are present on the plant. After maturation of the seeds from the summer flowers, growth of the shoot will continue by activation of the apical vegetation point and initiation of new flower buds and, in some species, a second flowering period may occur in the autumn but under the climatic conditions of middle Europe the seeds from these flowers will not ripen before winter. When the initiation of new flower buds has ceased and the experimenter needs fresh flowers, it is not necessary to wait for the spontaneous activation of the vegetation point. Renewed growth can be enforced within a few days at any time by removing all mature and a sufficient number of still-growing fruits from the main stem (Harte unpubl.). These observations give proof of a feedback system between the apical meristem of the main shoot (inflorescence) and the fruits. It is not clear whether this is based on the transport of photosynthetic products, needed for the development of both the embryos and seeds and also in the apical meristem for initiation and growth of flower buds, or whether the transport of hormones which could be provided by or just used by the embryos are involved. In any case, something indispensable to growth is missing at the top when in lower regions of the inflorescence growing fruits in the last stages before maturity are present.

In *Oe. Hookeri* de Vries there exists an early-flowering monogenic mutant (Harte 1973a). The seedlings of the homozygous recessive mutant do not develop a leaf rosette but form a shoot with flower buds immediately after germination and initiation of a few leaves. These plants differ physiologically from the normal ones by having a high content of gibberellins immediately after germination. In the seedlings and rosettes of *Oe. Hookeri* de Vries, which are homozygous for the normal allele at this locus, gibberellins cannot be found and the same amount of gibberellins as in the early-flowering mutant is only observed when the shoot starts elongation (Roscher 1973, unpubl.).

In the sect. *Raimannia*, most species develop a rosette stage. Only in *Oe. odorata* does the elongation of the shoot start directly after germination. *Oe. Berteriana* is a long-day plant which needs a photoperiod of at least 14 h light per day for induction of flowering. In contrast to this behaviour, *Oe. odorata* is a facultative long-day plant which can start flowering in a photoperiod of 6 h daylight, although flowering is promoted by longer photoperiods (B. Schwemmle 1962).

6.2.7 Development of the Flower

6.2.7.1 General Aspects

At the vegetation point in the axil of a bract that will develop into a flower, two sepals are initiated at the sides perpendicular to the axis shoot–bract. Then the other two sepals follow. The position of the petals is alternating with the sepals, and then two alternating circles of four anthers each, the first opposite the sepals, are initiated. The primordia of the carpels are seen opposite the inner circle of the stamens. The ovary is inferior. Its development needs special consideration. In the young flower bud the sepals are growing fastest. They become connected at the margins by outgrowths of the epidermis. The petals soon stop growing and remain very small until meiosis in the microspore mother cells in the anthers is completed. The same holds for the ovary.

In all species, but more often on plants homozygous for *s/s* (*sulfurea* flower colour), some flower buds, especially in the beginning of the flowering period, will grow only upto a size when meiosis in the microspore mother cells in the anthers should occur, then they die and fall off. Cytological studies indicate that in the anthers of these buds the development of the PMCs is disturbed at a very early stage and is followed by degeneration of the contents of the anthers and death of the whole bud. The process is different from the disturbances which occur in cases of male sterility in that it affects only single buds on an otherwise fertile inflorescence. As the event is singular and cannot be predicted, only accidental observations are available from material collected for cytological purposes. Therefore, nothing can be said about the sequence of events leading to the loss of the flower bud. The first step either could be an early developmental mistake in the flower, causing a disturbance in the development of the anthers as the first observable fact in the process of death of the whole organ, or the young archespore may be essential for development of the flower at this stage and after its destruction, for unknown reasons, the useless organ is discarded (Harte unpubl.).

6.2.7.2 Sepals and Petals

Sepals and petals are those organs of the flower whose origin as transformed leaves is easily recognized in their development. The most prominent part of the sepals is the unifacial tip, developing from the uppermost part of the primordium. The growth in width of the sepals depends on a group of initial cell rows in the subepidermis of the margin (Kowalewicz 1956a).

The primordia of the petals appear at the vegetation point of the flower in the first stages like the sepals, but their development is different from the beginning. The unifacial tip is missing or, in some species, is reduced to a tiny triangular tip. The marginal meristem consists of only one row of subepidermal cells and in the last stages of meristematic growth of the petal a marginal region of two layers of epidermal cells without mesophyll develops. The size of the mature petals corresponds exactly to the size of the bud and is determined by several genes and inhibitors which were first described by Renner (1925), who found the dominant inhibitor of large flowers, *Co/co*, and repeatedly confirmed by others. The gene *mp* (missing petals), described by Cleland (1970, 1972), inhibits the formation of petals at one or more of the sites in the second circle of the leaf primordia in the flower, while the remaining ones appear in their normal place. The gene acts by inhibiting the formation of a primordium at the site of a petal, without disturbing the pattern for localization of the remaining primordia either in this circle or of the stamens. The expression of the gene is highly dependent on conditions outside the gene locus. Early in the flowering period nearly all the petals can be missing, while towards the end of flowering many petals, and even complete flowers, may be observed on the same inflorescences. It is an open question whether this reaction depends on ageing of the plants or on the photoperiod.

The form of the petals depends on three genes: *cruciata* (*cr*), petals reduced by different degrees with uneven margins; *apetala* with extremely narrow, greenish petals; and *sepaloid* (*sep*). The development of the petals and the differences between normal and *cruciata* petals were investigated by Kowalewicz (1956a), the sepaloid petals were investigated by Rossmann (1963a,b). The normal sepals consist of more cell layers than the petals. Both organs differ in the form of the cells in their epidermis and mesophyll. The sepals have hairs and stomata at the outer surface, anthocyanin in some regions of the epidermis and chloroplasts in the mesophyll. The petals have neither hairs nor stomata, no anthocyanin, and yellow chromoplasts which develop from chloroplasts present in the young petals.

The petals of the *cruciata* type, investigated in *Oe. atrovirens*, *Oe. biennis* L. *cruciata* and several other combinations of Renner complexes, are sepaloid with the extreme form resembling sepals of only a few millimetres wide. The less extreme types have many morphological and histological characteristics of sepals, especially hair cells, stomata and chloroplasts. In some cases petals of intermediate size consist of yellow areas with a petaloid structure, combined with other regions of clearly sepaloid characters with chloroplasts which have not been transformed into chromoplasts. A correlation between the size and structure of the petals is observed in so far as the widening of the petals depends on an increase in the area of petaloid structure. In the nearly normal broad petals of some *cr*-plants the sepaloid areas can only be recognized microscopically, but they are always present.

From the combination of all the observations on the development of the petals under the influence of *mp* and *cr*, Cleland (1970) concluded that the vegetation point of the flower undergoes changes during flower development, its state determining the differentiation of the primordia. In *cr*-homozygotes the shift from the state determined to differentiate sepals, to the state "petals" is delayed, while the pattern of the primordia for flower organs develops at the normal speed. As a consequence, for the primordia of the second circle, which normally develop as petals, differentiation starts during the stage "sepal formation" or the transition to the state "petal formation". They develop the morphological and histological structures of sepaloid petals according to this delay. In the mutant *sep* apparently the delay for start of the stage "petal differentiation" is so long, or this stage is missing altogether, that the leaf primordia at the sites of the petals are always differentiated as sepaloid structures.

As soon as microspore tetrads are found in the anthers, the petals start to enlarge slowly. Their growth is only very fast in the last stage before anthesis.

When the flower opens in the late afternoon, the sepals separate at the margins. First, two openings are seen on opposite sides at the thickest part of the bud, and this is followed by separation of the sepals on the two remaining sides. Two or all four sepals can remain connected in the uppermost region just below the tips and the whole structure is pushed aside by the evolving petals (Sigmond 1930a,b). On the basis of superficial observations on the process of opening of the flowers and without controlling the development in earlier stages, it was formerly believed that *Oenothera* had only two sepals (Davy de Virville and Obaton 1923). A mutant of *Oe. biennis* L. with very small petals that cannot push aside the separating sepals completely, with the consequence that the sepals remain connected above the tip of the stigmas, is described as "*hemikleistogama*" (Stomps 1957; Renner 1958a). A similar phenotype is expressed in the mutant *apetala* and in some cases in *Oe. atrovirens* and the mutant *cruciata*. In all cases it is not an autonomous character but is correlated to the size of the extremely small or narrow petals which can expand in the space between the partially separated sepals without pushing them aside. In the extreme cases of sepaloid petals of the *cruciata* and *apetala* type the movements of the transformed petals and their orientation in the open flower resemble the behaviour observed in the sepals (Kowalewicz 1956a).

The analysis of the movement of the sepals made clear that the opening of the flower is not only enforced by the expanding and unfolding petals but also that the sepals perform active movements. Sepaloid petals of the extreme *cruciata* type resemble sepals in their behaviour during anthesis and in their orientation in the open flower. The flowers open in the late afternoon. Dehiscence of the anthers occurs during the day on which the flower opens, the time of day depending mainly on the temperature. On warm sunny days the process can start as early as 8 a.m. but on cool days it can be delayed until shortly before anthesis. Experiments by Roze (1895, 1899) demonstrated that the main factors in initiation of flower opening are not the absolute values of temperature and humidity but the decreasing temperature and increasing humidity in the evening and the change in turgescence correlated to these factors. Light has also an influence. Most important is the last light period prior to opening of the flowers. Complete darkness for 24 h inhibits the process, but when the last change from darkness to light occurs according to the normal schedule of photoperiod, then a shortening of the following light period and complete darkness at the time of anthesis cannot inhibit dehiscense of the anthers and opening of the flowers at the normal time. In *Oe. Berteriana* similar observations were made, stressing the importance of the photoperiod during development of the flower buds for the timing of anthesis (Arnold 1959).

6.2.7.3 Hypanthium

The hypanthium appears rather late in the development of the flower when the growth of the petals and sepals is enhanced after completion of meiosis in the anthers (Weinland 1941). It starts as an elongation of axial structures, raising a ring with the insertions of the sepals, petals and stamens on the top of a hollow tube above the level of the ovary. In this development there is some similarity with the development of a ring fasciation, a well-known malformation of the vegetative apex in many plants. Just as in these cases, the ring raises the leaves. The vegetation point remains in the middle at the bottom of the beaker formed by the elevation of the ring. In the flowers of *Oenothera* at this spot the primordia of the carpels are initiated before the development of the hypanthium starts. I have not found any suggestions compatible with this idea in the morphological literature, but the similarity is so striking that the interpretation of the hypanthium as a ring fasciation within the flower should be considered. The hypanthium grows slowly until a few days before anthesis. Then the cells start enlarging rapidly and growth is finished at the time of anthesis.

Experiments with excision of different parts of the flower, either the whole bud above the hypanthium or only the sepals, petals or anthers, at different stages during development of the flower bud and with eventual subsequent supplementation with auxin demonstrate that the enlargment of cells in the hypanthium depends on auxins provided by the young anthers (Weinland 1941). When the hypanthium has reached 1 cm in length, it has received enough auxin to continue growth to its final size. Excision of the anthers after this stage does not inhibit elongation. The style lengthens at the same time, its growth being correlated to the hypanthium. The source of auxin is also found in the anthers. It passes by diffusion from the walls of the anthers into the stigmas, which are enclosed by the anthers at this stage. Auxin transport in the hypanthium, filaments and style is strictly basipetal (Weinland 1941). From observations by Kumar and Hecht (1965) on *Oe. organensis* Munz. after emasculation, it was concluded that an influence of the anthers on the growth of the style and hypanthium could not be seen, but the experiments were performed shortly before anthesis on rather large flower buds in which the transport of auxin is finished, and the cells of the hypanthium elongate according to the hormones received beforehand.

6.2.7.4 Stamens and Anthers

6.2.7.4.1 Normal Development

The development of the anthers was first described by Beer (1906). More details about the histological differentiation were added by Oehlkers (1927) and Harte and Bissinger (1952). The initiation of the primordia at the vegetation point is like that of any other leaf in the flower region. The first stages from primordium to the differentiation of the archespore do not differ from those found in other objects. Soon the special differentiation as an anther can be recognized by the development of four lobular evaginations on the sides of the staminal primordium.

In a recent investigation the development of the anthers has been described as a sequence of 12 stages characterized by the development of the pollen (Noher de Halac et al. 1990). The development of the tapetum and the adjacent outer layers of the anther wall is correlated with the events in the archespore. In the meristematic anther the differentiation of the tissues starts with the archespore. It can be recognized histologically as a mass of non-vacuolate cells in the centre of the lobes. Next, the prospective microspore mother cells (pollen mother cells, PMCs) can be distinguished by shape and volume from the tapetum which consists of isodiametric uninucleate cells without vacuoles. Between the tapetum and epidermis three or four layers of meristematic cells are seen.

Parallel to the development of the PMCs the tapetal cells enlarge and become binucleate, with abundant RER (rough endoplasmic reticulum) and ribosomes, by the end of meiosis in the PMCs (Noher de Halac 1985b). The tapetal cells continue to enlarge and appear swollen. This stage is always seen only in a part of the pollen sac. The cells must pass through it very quickly before they loosen their adhesion, separate from one another and partially move into the fluid of the pollen sac (Harte and Bissinger 1952).

As soon as young microspores are present in the pollen sac, vacuoles appear in the tapetal cells and lipids accumulate in their plastids. The lipids are secreted into the lumen of the pollen sac before degeneration of the tapetal cells occurs. They disintegrate during pollen development and only some lipid drops remain. By anthesis the tapetum and surrounding cell layers have disappeared. The wall of the mature microsporangium consists of the epidermis and the subepidermis with the usual spiral thickenings of the cell walls.

The disintegration of the contents of the pollen sac, normally affecting only the tapetum and parenchymatous cells, is initiated by enzymes. In *Oenothera* these have not been investigated, but the process is so similar to that found in other species that it can be assumed to occur in the same way, sparing only the cells of the subepidermis and the young pollen grains (Noher de Halac et al. 1990).

6.2.7.4.2 Male Sterility

In *Oenothera* two genes, *fr* and *ster*, are known to cause degeneration of the contents of the pollen sac in the recessive homozygotes *fr/fr* and *ster/ster*. This occurs after the tetrad stage at the end of meiosis and leads to male sterility (Oehlkers 1921, 1923; Harte 1942, 1948). The development of the male-sterile anthers was investigated by Oehlkers (1927) for *fr* and the male-sterile *Oe. Lamarckiana* de Vries mut. *lata*, by Harte and Bissinger (1952) for *fr* and *ster*, and by Noher de Halac (1985a; Noher de Halac et al. 1990) for *ster*. The first differences to the normal development are seen in the tapetum. In the mutant *sterilis* dark-staining inclusions appear as vesicles with electron-dense deposits (Noher de Halac 1985b). It can be assumed that these are accretions of defective sporopollenin precursors synthesized in the plastids and accumulated because of a delay in secretion. There are differences between the course of degeneration in the two mutants (Harte and Bissinger 1952), but before anthesis the microspores have been completely dissolved and only a dark-staining unstructured mass remains. In both cases malfunction of the tapetum in the synthesis and secretion of sporopollenin precursors is the first step and finally leads to degeneration of the microspores (Noher de Halac et al. 1990, 1992).

In the male-sterile anthers of the two mutants the genetically determined disturbance in the development of the microspores is initiated sporophytically in the homozygotes by the malfunction of the tapetum without active participation of the genes in the haploid nuclei of the microspores.

6.2.7.5 Ovary and Ovules

The development of the flower is influenced by several genes, some of which were observed first in *Oe. Lamarkiana* de Vries. Morphological analysis of the flowers of the mutants reveals that in all cases the ovary is involved. So the development of this part of the flower must first be described. It has been investigated several times with considerable differences in the results found by the authors. Analysis of the papers, however, reveals that the conclusions depend completely on the regions of the ovary and the developmental stages included in the observations. Of historical interest is the first investigation of the development of the flower and ovary in *Oe. suaveolens* Desf. by Duchartre (1842), and the description by Weiße (1899). The most complete descriptions, avoiding the errors of earlier investigations by starting at the vegetation point of the flower and considering the whole ovary, are provided by Kaienburg (1950) and Kowalewicz (1956a).

The development of the inferior, syncarpic ovary starts with four separate carpel primordia in opposition to the inner circle of stamens. The next step is a deepening of the top of the axis between the primordia. Four partitions, which will later develop central placentas, appear in the beaker in connection with the margins of the carpels. By growth of the axis, the wall of the beaker increases in height, leaving four deep cavities in which placentas are formed in a central position. This structure becomes the lower part of the ovary, while the carpels and partitions seen first form the upper part of the ovary with the style and four stigmatic lobes. The histological differentiation and structure of the mature tissues of the ovary and placentas in both parts of the ovary cannot be distinguished in spite of their different origins. The morphological interpretation is that the ovary of *Oenothera* is an integrated structure, formed by the axis and carpels. This is in agreement with the position of vascular bundles in the wall of the ovary. The other floral structures begin at the upper margin of the hollow hypanthium and do not contribute to the wall of the ovary or to the hypanthium (Kowalewicz

1956a). Considering the distribution of vascular bundles in the wall of the hypanthium, this supports the interpretation of the hypanthium as an axial structure homologous to a ring fasciation.

The ovary contains from several hundred to a thousand ovules. These are anatropous with two integuments. In the crassinucellate nucellus usually one megaspore mother cell (MMC) is located near the chalaza and can already be recognized before the integuments develop. The MMC is surrounded by cells which can be designated as the tapetum. In rare cases the archespore consists of two MMCs. This has been observed quite frequently in the mutant *brevistylis* (Sniezko pers. comm.). The micropyle is enclosed by both integuments (Noher de Halac and Harte 1975). At the chalaza, the hypostase appears at the tetrad stage and persists until the completion of embryo sac development (Johansen 1928). From the extensive investigations by Schwemmle and his students on species of the sect. *Raimannia* it is known that many of the ovules degenerate before fertilization for unknown reasons. The same later happens to those ovules which are not fertilized. The degeneration of all cells in a megaspore tetrad, leading to degeneration of the whole ovule, has also been observed (Sniezko and Harte 1985). In the ovaries of the mutant *brevistylis* all the ovules usually degenerate (Fig. 18).

In the ovule of *Oe. biennis* L. the nucellar cells differentiate in distinct patterns at the chalaza, in the parietal layers around the embryo sac and the regular nucellus cells (Noher de Halac 1980c). EM observations revealed important differences in the fine structure of these cells during development of the MMC in the ovule from a meiocyte to the gametophyte. Lysosomes give the earliest evidence of differentiation. Tannins are present in the hypostase and integuments. The presence of several types of vacuoles, Golgi apparatus and other submicroscopic characteristics can be taken as indicators of secretory activity of the chalazal cells. The nucellus cells accumulate starch before degenerating. The degradation of the reserve material to be used in embryogenesis depends on amylases provided by other cells.

In the epidermis of the placenta near the ovules groups of cells are located which, according to their anatomical structure as seen in cross sections of the ovaries, could have a secretory function. In mature ovaries of flowers at anthesis they seem to be secreting some substance which could act in guiding the pollen tubes to the ovules.

In a mutant of *Oe. missouriensis* (sect. *Raimannia*) the anthers are transformed into carpels. The ovary seems to be normal but is sterile (Linder 1953). If the interpretation given by Cleland for the case of the *cruciata* petals is applied in analogy here, than it can be assumed that in the vegetation point of the flower the state "differentiate stamens" is extremely short or has been completely deleted. The consequence is that the state "differentiate carpels" will already have been reached when the pattern for initiation of primordia prescribes the formation of primordia at the site of the stamens, which will accordingly differentiate as carpels.

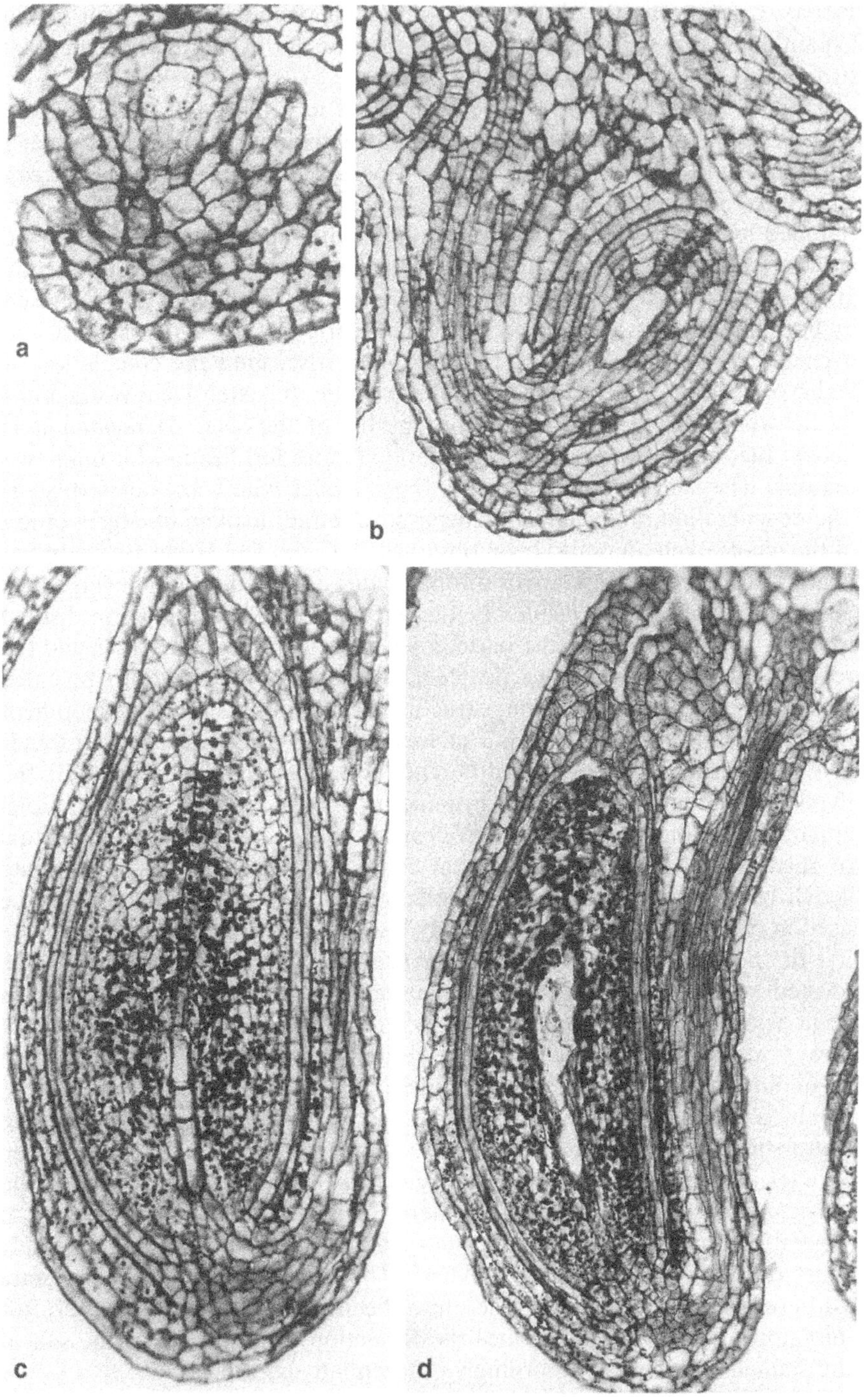

6.2.7.6 More Mutants

The mutation *pollicata* in *Oe. Lamarckiana* changes the tissue connecting the ovary and hypanthium in the upper part of the ovary. In the mutant, the inhibition of growth of this part, which is characteristic of normal development, is missing. A solid structure appears, resembling the hypanthium only in the distribution of vascular bundles and the presence of a few longitudinal air-filled spaces. The size of the hypanthium is reduced by the length of this structure. The short style lacks supporting tissues and has anatomical abnormalities at the base of the stigmas. However, pollen tubes can pass from the style to the ovary through the middle of the abnormal part (Shull 1934, 1937), indicating the presence of a functional transmitting tissue. In the mutation *supplena* (*su*) the style is shortened and additional petals develop, making a "double" flower (Shull 1928c). In the description no information is given whether the petals are additional structures or are transformed stamens, as is usually the case in double flowers of other species. In the latter case the interpretation of gene action would be a prolongation of the state "differentiate petals" combined with a deletion of the state "differentiate stamens" in the vegetation point of the flower.

In *Oe. missouriensis* another hereditary anomaly of the flowers was detected and was found to influence a basic developmental step. In plants homozygotic for the recessive allele of this gene a secondary flower develops in the ovary as a flower bud enclosed by the placenta and causing complete female sterility (Linder 1953). In other plants cases are known in which not all the cells of the vegetation point are used in the formation of the flower and the remaining ones can, under certain conditions, repeat the development of flower organs. Considering the special events in the normal development of the ovary, it seems a plausible interpretation to suppose that some cells at the base of the beaker, which is formed after initiation of the carpel primordia, do not differentiate in the normal manner but remain meristematic for a short time, then remember their determination as flower initials and start the whole sequence again. The process would be terminated by the stop of all developmental processes at anthesis of the first flower.

The mutant *brevistylis* (*br*), which made *Oenothera* famous (de Vries 1990a,b; Davis 1918, 1921; Shull 1928c; Catcheside 1954) has a short style. The stigmas are, for the greater part, enclosed in the short hypanthium.

Fig. 18. Development of the ovule (microphotographs contributed by R. Sniezko). **a** Very young ovule, integuments developing from the base of the nucellus; megaspore mother cell already visible. **b** Crassinucleate ovule with meiocyte in the chalazal part, two integuments still not completely covering the nucellus; the outer integument consists of three cell layers, the inner integument with two cell layers. **c** Mature ovule, integuments with micropyle, megaspore tetrad at the chalazal end of the nucellus, hypostase with dark-stained cells at the chalaza. **d** Mature ovule with young uninucleate embryo sac growing towards the micropyle

Malformations of the style include its dissection in several separated stripes of stylar tissue and differentiation of a placenta with additional ovules in the style and on its separated parts (Renner et al. 1952; Sniezko and Harte 1993). Other disturbances in the development of the ovary and ovules lead to nearly complete female sterility. Furthermore, the tissue separating the hypanthium and ovary after pollination is poorly developed, and in most flowers is missing altogether. In plants with normal flowers the hypanthium is discarded along with the corolla within 1 or 2 days after fertilization and unfertilized ovaries die within a few days and fall off, but in *brevistylis* the dried flowers remain on the seedless ovaries with degenerating ovules for several weeks, often until the end of the flowering season. When, however, *br* is transferred to the Renner complex ^{h}*blandina*, the *brevistylis* character, including the pleiotropic influences on leaves, buds and growth of the shoot, is expressed in the homozygotes *br/br*, but the stigmas develop nearly normally and the ovaries are fertile with a reduced number of seeds (Catcheside 1954). To a certain extent the same is also observed by transfer of *br* to ^{h}*hookeri*. In contrast, in the combination of Renner complexes ^{h}*hookeri·velans* most plants are female sterile with great variability of the malformations of the style. Only exceptional plants have some ovaries with a few seeds. Considering the pleiotropic effects of *brevistylis* the first activity of the gene seems to be an influence on the growth of the precursor tip of the leaf primordia which is followed in the carpels, where the tips of the primordia become the stigmas, by secondary reactions leading to the observed malformations. A comparison of the phenotypes of the mutants *brevistylis* and *pollicata* demonstrates that the expression of the normal alleles at these loci must be necessary to control the development of a tiny structure in the flower, namely the few cell layers which close the ovule and separate it from the primordia of the other parts of the flower. The allele $+^{br}$ gives the signal for the start to form this tissue and $+^{pol}$ is, after only a few cell divisions, responsible for the command to stop it and differentiate on top of it the style and hypanthium (Sniezko and Harte 1993). The influence of *br* on flower development can be modified by the genetic background of other genes. An influence of the plasmone or plastidome can be excluded by the behaviour of *br* in *Oe.* ^{h}*hookeri·velans*.

The combination of *su* (*supplena*) with *br* in a double recessive *su br/su br* results in an interaction of both genes, producing a completely new phenotype. The length of the style is further reduced, as can be expected from the combination of the two genes, but in addition to this, the whole flower is involved. The carpels are lacking and the ovary and hypanthium are replaced by a sterile, solid structure bearing on top a repetition, by up to fourfold, of the sequence calyx–corolla–anthers (Shull 1928c). A description of the development in earlier stages is missing. In analogy to the cases previously discussed, the interpretation could be that because the activity of two normal alleles, necessary to push forward the development at certain stages, is lacking, the changes in the state of the vegetation point are arrested completely before the state "differentiate carpels" is reached. But the initiation of primordia continues and the signal to stop the differentiation of the carpels is missing, with the result that there is a continued repetition of the floral structures normally initiated before the blockade. The elongation of the axis is independent of all other processes and proceeds as usual, but it becomes meaningless as all the other steps connected with it, like formation of the ovary and initiation of the hypanthium, are missing.

As far as can be judged from the morphological descriptions of the phenotypes, most of the genes influencing flower structures act mainly in one of two ways: either by causing an early stop of development or by lacking the ability to stop certain development processes, thereby leading to repetition of structures. In the case of "carpels instead of anthers" there is a short-cut in development. The primordia are initiated at the vegetation point in the right order, but in differentiation the step "anther" is left out and the instruction "make carpels" has already been given when the primordia of the anthers are just formed, irrespective of the subsequent initiation of carpel primordia in their proper place. Combined with the observations on sepaloid structures this means that the sequence of initiation of flower organs at set places on the vegetation point is independent of the following differentiation of these primordia. Two sequences of developmental events, initiation of primordia for flower organs at given places and differentiation of these primordia to form the structures of the flower, in a given order from sepals to carpels, are independent of one another but act in perfect coordination in normal development. The mutants demonstrate that this coordination is genetically organized and can be disturbed by mutations of genes which are responsible for starting the next step.

Because the investigators of these genes stopped their research at a time when gene banks did not exist, the stocks have been lost and the most interesting contribution they could have made to investigations on gene action in flower development, and generally to the coordination of developmental steps, is missing. Only *brevistylis* and *cruciata* remain.

6.2.7.7 Stigma

In *Oenothea* the stigma is four-lobed, with arms up to 1 cm long and 1–1.5 mm in diameter in *Oe. organensis* Munz. and somewhat smaller in most other species. In the middle of each lobe there is a vascular strand. Transmitting tissue is found on the adaxial side. The receptive surface covers the stigma lobes. In the style, the vascular strands remain separated and surround the core of transmitting tissue as a ring of four bundles. The elongated cells in this part of the style separate at maturity by secreted material of pectic polysacharides (Heslop-Harrison 1990). The stigmas are of the wet type. On a young style the secretory papillate cells form a continuous layer on the receptive surface that is later disrupted. They initially bear a very thin cuticle and a thin layer of protein through which lipids are copiously secreted. Later, aqueous globules containing polysaccharides are rapidly inserted into the lipid layer. On the mature stigma the surface is coated with a foamy emulsion in which esterase activity can be detected. The changes in the structural and chemical composition of the layer covering the stigmas indicate a succession of phases in the glandular activity correlated to a sequence of histological events in the epidermal cells (Heslop-Harrison and Heslop-Harrison 1985). They are most probably based on a sequence of gene expression.

6.2.7.8 The End of Flower Development

The flowers are open for one night. In Europe pollination is mainly performed during daylight hours in the late afternoon or early morning by bees and bumble-bees as pollinators. In natural populations in California pollination is carried out preferably during the night by hawkmoths (Gregory 1963). Following pollination, wilting starts ca. 15 h later, when the pollen tubes have passed through the style. The flower parts above the ovary have dried out nearly 20 h after opening of the flower. Only flowers which are open in wet and cool weather and have not been pollinated remain fresh for another night. Following successful pollination, a separating tissue develops at the base of the hypanthium. The preformed cells, which can be distinguished from the other hypanthium cells of the mature flower only by slight differences, enlarge and the middle lamellae in the walls disintegrate, starting in the outer cell layers of the hypanthium (Sigmond 1930a,b). A similar tissue separates the style from the ovary at the same time. The fertilized ovary starts growing immediately and 2 days later the increase in size can already be some millimetres, irrespective of the number of fertilized ovules present (Sniezko and Harte unpubl.).

Flowers which are emasculated a few hours before dehiscence of the anthers and are then isolated using paper bags to prevent pollination by insects, can survive the expected wilting time by several days. From this observation it can be concluded that wilting of the flower is not a spontaneous process of ageing but is induced by fertilization of the egg cells.

6.2.8 Still More Genes

Two genes are known to change the flower colour: *sulfurea* with pale yellow flowers and *vetaurea* with old-gold flowers (Naples yellow according to the colour tables of Ridgway). The pigments are carotenoids and their relations to plastid development will be discussed together with plastid pigments (Sect. 6.3) (Plate 1).

The vegetative parts contain anthocyanin in the epidermis. The basic gene for formation of anthocyanin is *P* (*punctata*). The recessive allele $+^P$

inhibits production of anthocyanin in all parts of the plant and was discovered in hybrid progenies of *Oe. suaveolens* Desf. (Oehlkers 1924). The general impression gained in the experimental garden is that homozygotes without anthocyanin are weaker than plants with anthocyanin, although exact data on growth and survival rates are not available. Synthesis of the red pigment in the epidermis of the green flower parts starts in each flower bud just a few days before anthesis. This indicates a very exact timing of the gene expression, responsible for the synthesis and pattern formation of anthocyanin, in relation to other developmental processes in the flower. This problem has as yet not been investigated.

6.2.9 Summary: Development of the Sporophyte

In summary, the statement can be made that in the development of the sporophyte in *Oenothera* there exists a fascinating combination of gene actions on primary developmental processes and induction of new steps in development by interactions between different parts of the plant. In these interactions growth hormones, and maybe other substances too, must play an important role, However, the research in this field has until now been more accidental rather than planned on a great scale, and has depended mainly on the special interest of some scientists and on observations made while looking at other problems. For some questions, like the development of special structures of the flower, even the knowledge of basic facts and events at the level of anatomy and histology is missing. Quantitative research is very rare, but will become necessary when a basis for a theoretical approach to models of development has to be found. These gaps must be filled before the role of gene expression on the molecular level can be investigated. Overall the state of the problems is of such a kind that the methods of classical experiments on development of plants are necessary in some fields to provide basic information for planning future experiments. It cannot be expected to gain a revolutionary new insight into general problems from these first steps, but a follow-up of such basic research by an approach with the most modern methods of molecular and developmental biology can bring the solution for many interesting problems which are presented by *Oenothera* but are not solved.

6.3 Plastomes and Development

The plastomes of *Oenothera* were discovered by their influence on plastid pigmentation, leading to pale, often lethal seedlings. The disturbance of chlorophyll synthesis in certain genome–plastome combination is interpreted as an incompatibility between the genetic information of chromosomes and plastids. The special suitability of the *Oenothera* system for investigations on genome–plastome interaction has been presented by Stubbe (1989c), who discussed the different aspects of the problem and constructed a scheme describing the incompatibility reactions between the plastomes and genomes based on pigment synthesis in the plastids.

Disturbances of plastid development and pigment synthesis can, however, have very different reasons. Incompatibility of the genome and plastome is only one of them. Mutations of chromosomal or plastomal genes involved in plastid development and pigment synthesis are the others. All are of importance to an investigation of the genetic influences on plastid development and also of the influence of the plastome on other developmental processes.

6.3.1 Plastids in Gametophytes, Embryos and Meristems

In *Oenothera* plastids are inherited by the egg cell (maternal plastids). However, the generative cells of the pollen tube contain proplastids (paternal plastids) which are included in the zygote, as was observed by Renner. The number is nearly constant (Stubbe 1963a). Their fate is decided during embryogenesis. The decision whether paternal plastids will be included in the embryo depends on their position in the cytoplasm of the zygote and on the reproduction rate of the proplastids. The position of the paternal proplastids in the zygote depends on the place where the sperm cell entered the egg cell after passing through one of the synergidae. This factor has not yet been properly investigated in *Oenothera*.

Basic information on the propagation of proplastids in the apical meristem of the shoot and in different tissues of young leaves in the hybrid *Oe. albivelutina* is presented by Schröder (1962). Considerable differences are observed in the rate of propagation of the proplastids, depending mainly on the plastome but also influenced by the genome present in the nucleus (Renner 1924; Schötz 1954, 1958a, 1968, 1974, 1975). The quantitative relation of the different types of plastids in the mixed cells and the importance of this factor to the problem of plastid competition is discussed by Schötz and Heiser (1969). However, the importance of a delay in the start of propagation of the paternal proplastids in the zygote, which also depends on the plastome, must be considered too (Chiu et al. 1988). On the other hand, the egg cell plastids can be eliminated from the embryo in extreme cases, leaving only paternal plastids, when the maternal plastome belongs to a type considerably weaker than the paternal one. A ranking of plastomes, according to their ability to be transferred by the pollen and propagated in the embryo, has been analyzed by Stubbe (1960). The highest rate of multiplication is found in plastome I (*hookeri*). In decreasing order follow plastome II (*suaveolens*), III (*lamarckiana*), V (*argillicola*) and IV (*parviflora*). In combination with the genome of *Oe. Hookeri* strain *Johannsen* the order of transmission of the plastomes is I > III > II > IV. The special problems which arise from competition between different plastids in a cell, based on differences of multiplication rate, have been investigated by Schötz (1958a, 1968, 1971, 1975).

Differences in the frequency of two plastid types in the proembryo could also occur by differential destruction of proplastids in the zygote. However, submicroscopical investigations on zygote development do not give any indication of organelle deterioration at this stage, so this hypothesis can be safely discarded (Meyer and Stubbe 1974).

When both types of plastids are included in the top cell of the proembryo, and when they differ in their reaction to the genome of the hybrid, they can be distinguished in the "mixed" cells of the embryo after germination as soon as chlorophyll synthesis starts (Hoeppener and Renner 1928). When only the paternal plastids are able to accumulate plastid pigments, the two types of pale maternal and green paternal plastids will appear in some of the seedlings in addition to the pure white or yellow plants. The mixed cells, in which the different types of plastids can be recognized, give proof of the genetic independence of the plastomes.

In some cases a segregation of the plastid types during mitosis is observed in the embryo or later in the meristems. It depends on the distribution of the plastids in the cell when the new cell wall is formed after mitosis (Krumbholz 1925). As expected, mixed cells have been observed, containing different plastids side-by-side but, in addition, chimaeric plants arise with cells that contain only one of the two plastid types. They are either sectorial chimaeras with variegated leaves or whole sectors of different colour, or periclinal chimaeras with the epidermis, containing one type of plastid, covering a mesophyll that contains a different type. In the progeny of such a cross even completely green plants with only the paternal plastome are observed in addition to the expected white and variegated seedlings.

In special cases, in which the mother plant had mixed cells with two different mutated plastids of the *suaveolens* type and the father contributed a third type, plants with three-colour variegation were obtained. Stubbe (1957, 1958) succeeded in analyzing such plants from the cross *Oe. suaveolens* Desf. (two different types of mutated plastids) × *Oe. Lamarckiana* de Vries. The mutated plastids from *Oe. suaveolens* Desf. can be distinguished: one type becomes pale yellow-green and the other cream-coloured. In the hybrid *Oe. albicans·velans*, *albivelutina*, normally the *lamarckiana* plastids are pale. Mixed cells as well as sectors with only two or one of the three plastid types are observed. In mixed cells and when a cell with *lamarckiana* plastids lies in the mesophyll in close contact above or below, but not side-by-side to, a cell with mutated *suaveolens* plastids the *lamarckiana* plastids become green (Plate 2).

This observation proves that the disturbance of pigment synthesis in the cases of genome–plastome incompatibility cannot be a direct interaction of genomic and plastomic gene products but must rely on complex metabolic processes in the cells. Some of the intermediate metabolites can be synthesized with the help of the mutant plastids but cannot be used by these plastids themselves. They are present in the cell and available to other plastids for use in their development. They can even be transported in the mesophyll through the cell walls, but only in certain directions, perpendicular to the leaf surface. These substances can be recognized by plastids of the *lamarckiana* type, which are deficient in the ability to synthesize them under the influence of the hybrid genome, but can use them in chlorophyll synthesis when they are provided by external sources (Kandeler and Schötz 1956; Stubbe 1958, 1963a).

In some hybrids with extreme disharmonious genome–plastome combinations, such as appear in some crosses between species of the sect. *Euoenothera* and species of the sect. *Raimannia*, the embryos with maternal plastids will die, leaving empty seeds. When, however, paternal plastids are included in the top region of the proembryo and enough mixed cells are present in the early stages, nearly normal embryogenesis is possible. White–green chimaeras or even wholly green plants can occur. Later segregation of the two types of plastids in meristematic cells can lead to pure white sectors on such plants, which can develop normally because they receive photosynthetic products from the green regions. Irregular leaves of the type *irregularis* will develop in periclinal chimaeras when mesophyll cells with green paternal plastids are covered by a disturbed epidermis with maternal

---

Plate 2a–e. Disturbance of chlorophyll synthesis following changes in fine structure of plastids by mutation of a nuclear gene, by a plastome mutation, by genome-plastome incompatibility and by interaction between plastomes. Contributed by W. Stubbe. **a** Example for disturbance of chlorophyll synthesis by a nuclear gene. *Oenothera suaveolens* mut "Weißherz". The recessive allele from the Renner complex *albicans* is transferred by crossing over to the Renner complex *flavens*, so it can be realized as homozygote. **b** Genomes AB with plastome I. *Oenothera albicans* Grado·*flavens* Friedrichshagen with *Bauri*-plastome, showing this special type of chlorophyll degradation caused by genom-plastome incompatibility. **c** Genome AA with plastomes IV/II beta (normal plastids type IV and mutation beta of plastom II with genetic disturbance of chlorophyll synthesis). Three leaves of a plant. From left to right: left half of the leaf white over green, right half green over white, periclinal and sectorial chimaeric structure in one leaf! – in the upper half of the leaf the structure is reziprocal to the first leaf, in the lower part on the right side a completely white sector is present, – left half of this leaf is pure white, the right half is a periclinal chimaera green over white. **d** Genome AA with Plastome IV/II beta. *Oenothera rigens*·^{h}*hookeri* with green *oakesiana* (= *parviflora*) plastids and mutated white *biennis*-(II)-plastids. This picture demonstrates the different types of periclinal and sectorial chimaeras in plants with two types of plastids. **e** Example of a three-colour chimaera, leaves of *Oenothera albicans*·*velans*. The plant contains plastids of type III from *Oe. Lamarckiana* in combination with mutated plastids of the plastome type II which remain white. The plastids of the *Lamarckiana*-(III)-type are pale *virescens* in combination with this genom, but become normal green when they are in cells which lie directly under or above cells with the mutated *suaveolens*-(II)-type plastids

a
b
c
d
e

plastids (Stubbe 1963a). Examples of pale and variegated plants are presented in plates 2 and 3.

6.3.2 The Pale Plants

Plants with disturbed pigment synthesis can have any colour from pure white through yellow and yellowish green to greyish green. The striking influence of paleness on development is the decrease in growth rate, often followed by death of the plantlet. It cannot be decided whether this is a primary effect of disturbances in gene expression or rather a secondary effect of starvation following the lack of photosynthesis in the plastids. From the observation that pale or white sectors can survive, and even produce progeny, on plants with sufficient green areas, the second hypothesis seems to have the greater plausibility. This is further supported by experiments with white seedlings which can survive for some time in sterile culture on a nutrient with a source of carbohydrate (Chiu et al. 1988).

As can be seen from the scheme of Stubbe (1989c), the phenotype of the pale plants can be either constant or variable during the life of the plant. In this respect considerable differences which depend on the genome–plastome combination are observed. In some types the seedlings are white or yellow from germination onwards, indicating a complete disturbance of plastid development. These seedlings are always lethal. The viability of hybrids which contain only the incompatible plastome is reduced (Schötz and Reiche 1957).

In other cases the phenotype changes during development of the plant. A description of the many different types observed is given by Schötz (1958b). Those types which start with green cotyledons and then undergo a change of colour are particularly interesting. The colourless proplastids of the embryo can develop a normal green colour at germination. This stage can last either for only a very short time or be maintained during the initiation of a number of leaves. Then the state of the cells in the meristems changes and in the newly developing tissues only pale plastids are found. This second stage, following the first "green" stage, can start a few days after germination, with the result that pale tissue is present at the base of the cotyledons, or appears after the development of several green, photosynthetically active leaves. The plastids in the new tissues are disturbed and the lack of chlorophyll causes a deficiency in photosynthesis. This stage can be maintained during the formation of several leaves. The plants rely wholly on the photosynthesis in the green parts formed in the first stage. In some cases these are only the tips of the cotyledons. A considerable lethality is observed during this stage. Later, when several leaves have developed during the "pale" stage, in the surviving plants again a change can occur and a second green stage follows. This stage can last until the end of the life of the plant, eventually interrupted by more pale stages, but it is not stable and is maintained in most cases only in low light intensity, like that present in a shaded greenhouse. Most plants will die in strong light in the field (Schötz 1958b).

6.3.3 The Pale Plastids

Of special interest is the behaviour of the pale plastids themselves. The first question is, how the changes in the phenotype of the plants arise. Does the change from green to pale occur by the formation of new tissues from the meristem, in a process which will cause these cells to exhibit differences in the development of the plastids, or can differentiated plastids in mature cells change their state from "green" to "pale" and vice versa? Both cases are observed. The character "pigment deficiency" of a plastid is not constant.

In some cases of incompatibility, the development of the proplastids in differentiating cells depends on the developmental stage of the plant. This means that the cooperation between the plastome and genome can change. These changes are not distributed at random throughout the life span of the plant but follow a given sequence, characteristic for each plastome–genome combination (Schötz 1958b, 1959)

Pigment synthesis in disturbed plastids was investigated by Schötz and his collaborators. When the lack of plastid pigments depends on a mutation of plastomal genes, then the normal development of the plastids is irreversibly disturbed in all tissues of the plant and only pale plastids, varying in colour from white or yellowish white to light green, are observed. These mutants can only be maintained as chimaeras. However, when the disturbance depends on the mutation of a nuclear gene, degeneration and regeneration of the plastids is observed. The pale plastids of incompatible combinations can react in both ways.

In the case of paleness caused by mutation of a nuclear gene, e.g. *Weißherz* (white heart) from *Oe. suaveolens* Desf., reversible degeneration of green plastids occurs (Schötz and Senser 1961). During the first green period of the rosette the plastids have the normal submicroscopic structure and pigment content. At the start of the pale phase, degeneration of the structure occurs, the grana deteriorate. Several weeks later the structure of the chloroplasts is reconstructed and pigment content becomes normal again During the pale phase, first carotenoids are synthesized, then some grana appear. The development of these plastids is similar to that of the chromoplasts in the petals. Not only is the pigment content reduced but the quantitative relation of the pigments is changed. In the petals, and also to a small extent in the sepals, a new pigment, carotenoid-N, appears. It is composed from at least two xanthophyll esters, one of which is derived from violaxanthin. In the petals of the mutant the overall decrease in pigment content during growth occurs as in the leaves. Only the amount of carotenoid-N continues to increase during the last day before anthesis. It seems that the

Plate 3a–d. Disturbance of chlorophyll synthesis as consequence of changes in fine structure of plastids on the basis of genom-plastome incompatibility, demonstrated on chimaeras with two different types of plastids. Contributed by W. Stubbe. **a** Genomes AC with plastids type IV/I. *Oenothera albicans* Grado·*percurvans* with green plastids type IV from *Oe. ammophila* and plastids of the type I from *Oe. Bauri*, which became pale, type *diversivirescens*, by incompatibility with the combination of genomes in this hybrid. **b** Genomes BB with plastids type II/I. *Oenothera gaudens·flavens* with green plastids of the type II from *Oenothera suaveolens* and plastids of the type I from *Oe. Hookeri* which are white because of the incompatibility with genome B. **c** Genomes CC with plastids type II/V. *Oenothera argillicola* with green *argillicola*-plastids of type V and type II from *Oenothera suaveolens*, which stay white by the incompatibility reaction, demonstrating the inhibition of growth in the white sector of the rosette. **d** Genomes BC with plastids type II/V. *Oe. gaudens·*^{h}*argillicola* with light green *argillicola*-plastids type V and white plastids type II from *Oe. suaveolens*, demonstrating the differences in disturbance of pigment synthesis in both types by the incompatibility reaction. Comparison of **a** and **b** demonstrates the different reactions of plastom I in combination with different genomes. Comparison of **b** with **c** and **d** shows examples of reaction of plastome II to different genom combinations. Finally comparison of **c** and **d** gives an example for the complicated reactions that are possible when the same combination of plastomes is under the influence of different genom combinations

new compound is protected against the deterioration of the plastid pigments caused by the mutation (Schötz 1926b).

The structure and pigment content of the pale plastids differs from that of the normal green ones in several points. The pale hybrids have the normal content of carotenoids in the petals. In contrast, the mutants *sulfurea* and *vetaurea* are characterized by the presence of different carotenoids in the petals, but the chloroplast pigments are not involved. The disturbance in expression of these genes, caused by the mutation, is restricted to the petals.

In hybrids with an incompatible combination the phenotype of the plants, first green and later becoming pale, is similar to *Weißherz* but the development of the plastids is different. An example is *Oe. albivelutina* with *lamarckiana* plastids (Schötz and Senser 1961; Schötz 1962b; Schötz and Batheld 1964). In all cases of pale hybrids the colour of the plastids is correlated to disturbances of plastid structure. It depends only on disharmony between the genome and plastome and not on deficiencies in the genetic structure of one of the partners (Schötz 1964, 1965; Schötz and Diers 1968; Senser and Schötz 1964).

The combination of genome AA with plastome III also leads to deterioration of plastidal pigments but the course of events is again different and, moreover, slightly variable between the five types belonging to this group, depending on the Renner complexes involved. Together, they can be described as the *albivelutina* type of pale hybrids. With low light intensity the plants can survive for some time. Under conditions of higher illumination the plants lose up to 90% of their pigments while, at the same time, the relation of chlorophyll a:b and of carotene:xanthophyll changes (Schötz and Batheld 1964, 1965a). In the white parts of the leaves the fine structure of the plastids is changed. In the space between the inner and outer membranes of the plastids, thylakoids and sometimes even grana are present. Their origin is seen in evagination of the inner plastid membrane; this is in contrast to the normal structure where only invagination of this membrane occurs (Schötz and Diers 1966).

In the hybrid *Oe. velans·^{h}hookeri* (genome AA with *lamarckiana* plastids type III) the differentiation of the thylakoid system is disturbed. The membranes are formed and grana appear with approximately a 10-fold increase in the diameter compared to normal grana. At the same time the distance between the membranes decreases to the point where membranes stick together. The pale plastids are still active in photosynthesis. Starch grains can be formed, even when the membrane system is drastically disturbed (Schötz and Diers 1968; Schötz et al. 1968; Schötz 1970).

The course of plastid development and degeneration is also different in another type, *pictirubata*, *Oe. pingens·rubens* from the cross *Oe. atrovirens* × *Lamarckiana* de Vries with paternal plastids (Schötz and Senser 1964, 1965; Schötz and Batheld 1965a,b). When the same genome is derived from the cross *Oe. atrovirens* × *biennis* L., the degeneration of plastids is characterized by a three-phase development of the plants with instability of the second green phase. At low light intensity the pigment content can be even higher than in normal green plastids and in the field it is still 60% of the normal level, but during ageing of the leaves the pigment content decreases, just as in the leaves of the pale period of the seedling. The relation of chlorophyll a:b decreases, the relation of xanthophyll:carotene increases, and the quotient a+b:x+c decreases to such an extent that eventually the carotenoids contribute more to the pigmentation than the chlorophylls (Schötz and Batheld 1965b). The microscopic and submicroscropic structure of the degenerating plastids differ from that of the other types of pale plastids (Schötz and Senser 1964).

In *Oe. rubens·flectens*, from the cross *Oe. biennis* × *atrovirens*, or in the same combination of Renner complexes obtained as a rare exception from a *flectens* egg cell as a twin of *Oe. pingens·rubens*, the *atrovirens* plastids develop a normal structure and green colour, while the *biennis* plastids remain yellowish white (Schötz and Senser 1965). A direct comparison of this type with the normal development is possible in one cell, when in the hybrid mixed cells containing plastids from both parents are present. The development of the pale plastids up to the formation of the central body is normal. This holds also for the beginning of grana formation. Then, the development of the grana structure is disturbed. At the end, the disturbed plastids disintegrate completely (Schötz et al. 1965).

In summary: it can be stated that the paleness of the plants with incompatible genome–plastome combinations depends on the degeneration of plastids after a normal start. In most cases the disturbances are not absolute but variable during development according to a characteristic time pattern. Considering the plants as well as the individual plastids, the course of events in the hybrids is different according to the genome and plastome involved. It cannot be a general disturbance but there must be specific reactions in every case which need further elucidation (Fig. 19).

6.3.4 Plastid Development in Plastome Mutants

Information of a quite different kind is achieved from observations on the development of plastome mutants. The first stages of development of disturbed *biennis* plastids are similar to those in disharmonious hybrids. In sectorial chimaeras of *Oe. rubens·^{h}hookeri* with mutated and normal *biennis* plastids both types can be compared in the same plant and, eventually, even in sectors of the same leaf. The normal central body develops primary grana. Chlorophyll synthesis starts when the diameter of the grana is approximately 3 µm. In the degenerating plastids lipid droplets appear initially at the margin, the plastids flatten and the development of the grana is interrupted. The central body degenerates, but the plastids themselves remain visible in the cytoplasm (Schötz et al. 1965) (Fig. 20).

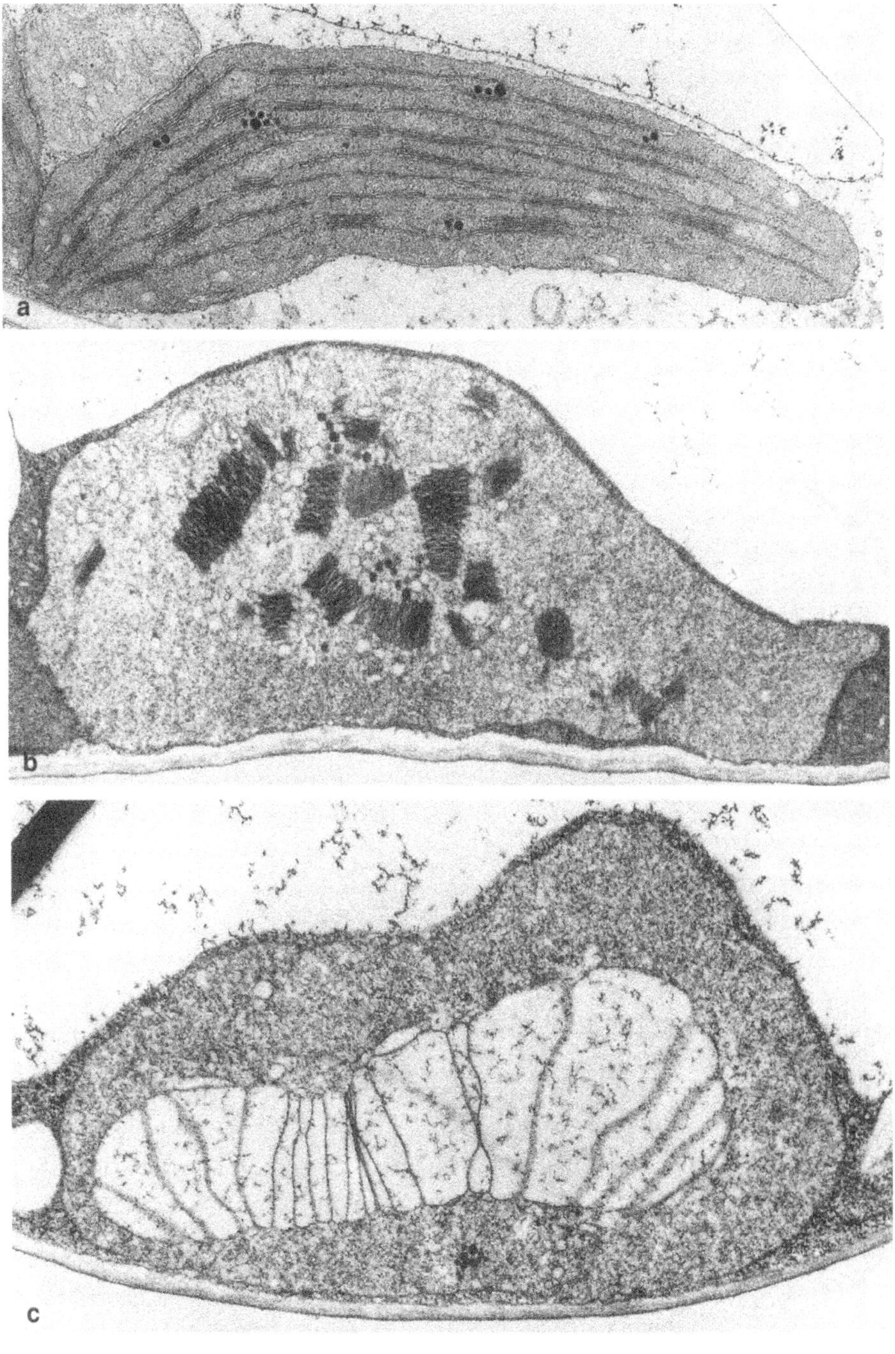

Fig. 19. *Continued*

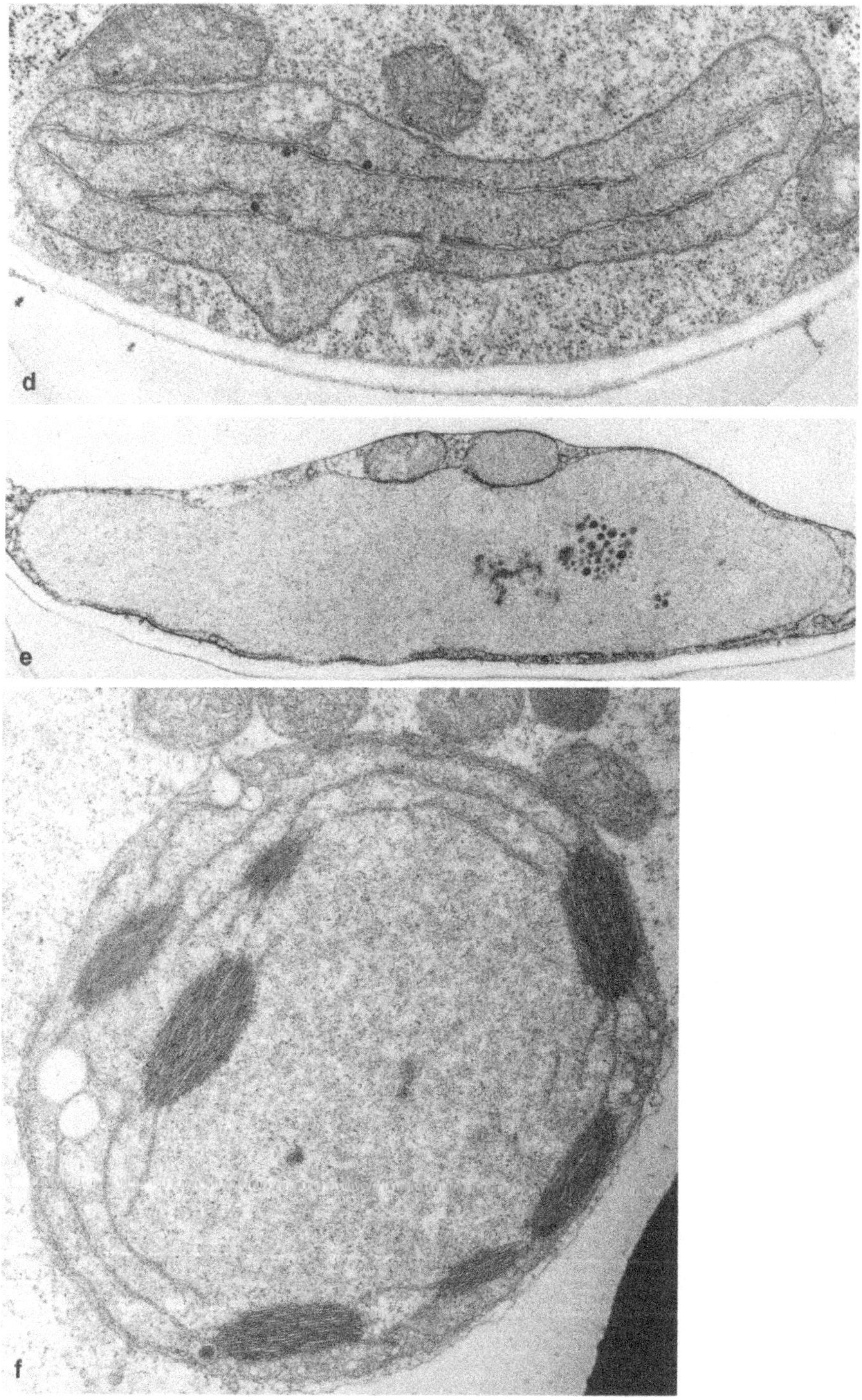

Fig. 19. *Continued*

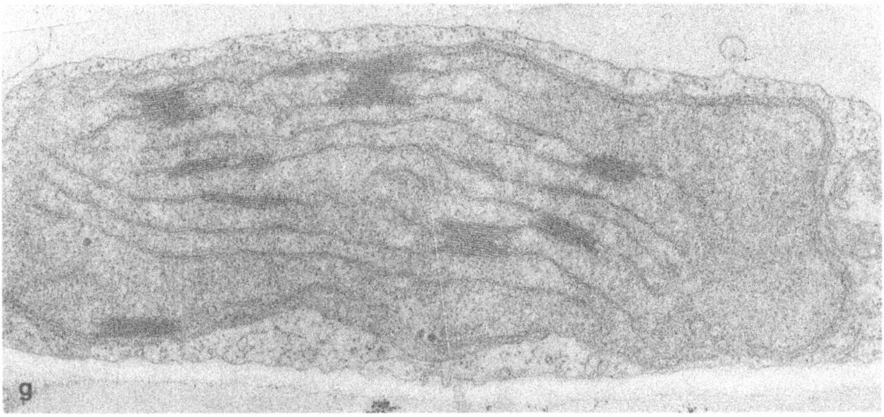

Fig. 19. Ultrastructure of plastids in *Oenothera* (contributed by F. Schötz). Fixation glutaraldehyde, followed by osmium tetraoxide contrast by uranyl acetate and lead citrate, ×25000. **a** Hybrid *Oenothera atrovirens* × *biennis pingens·rubens*, normal chloroplast from a green leaf with plastids of the female parent *Oe. atrovirens* which are fully functional. This type represents the compatible combination of genome BB with plastome II. **b** Same hybrid as **a**, degenerated plastid from a pale leaf area with plastids from the male parent *Oe. biennis* L. The leaf areas with this plastid type lose chlorophyll and become pale depending on age and light conditions. The structural defects connected to paleness are very variable and exhibit nearly all types which have been observed in other disharmonic genome–plastome combinations and as a result of plastid mutations. This type represents the incompatible combination of genome BB with plastome II. **c** Same hybrid as **b**, other type of degenerated plastid in pale leaf area; grana completely missing, some abnormal thylakoid structures present. **d** *Oenothera suaveolens* Desf. mut. "Weißherz", degenerated plastid from a pale leaf area. The mutation, which causes a chlorophyll degeneration in certain developmental stages in the homozygotes, is located in the genome. **e** Degenerated plastid without any membranous structures. **f** Degenerated, inflated plastid, reduced number of grana, thylakoids lie parallel to the outer membrane and determine the position of the grana in a circle. **g** Plastid with slightly abnormal structure from a pale, light-green leaf area in a chimaeric plant; reduced number of thylakoids and grana, small primary vesicles

Some of the spontaneous plastome mutants from plastomes I (*hookeri*) and II (*suaveolens-biennis*) could be maintained by passage through tissue culture from leaves with the mutated plastids (Stubbe and Herrmann 1982). Others could be preserved as chimaeras with the mutated plastome in the subepidermis from which the sporogenic tissues are derived (Kutzelnigg and Stubbe 1974). Embryos from the egg cells of flowers appearing in the axils of bracts with white margins have the mutated plastids. These have been investigated to determine the type of disturbance which is characteristic for the mutant.

Investigations on the development of the fine structure of the plastids in 26 plastome mutants gave a fascinating picture of the different phenotypes. The mutated plastids contain ptDNA and also plastidal RNA. The mor-

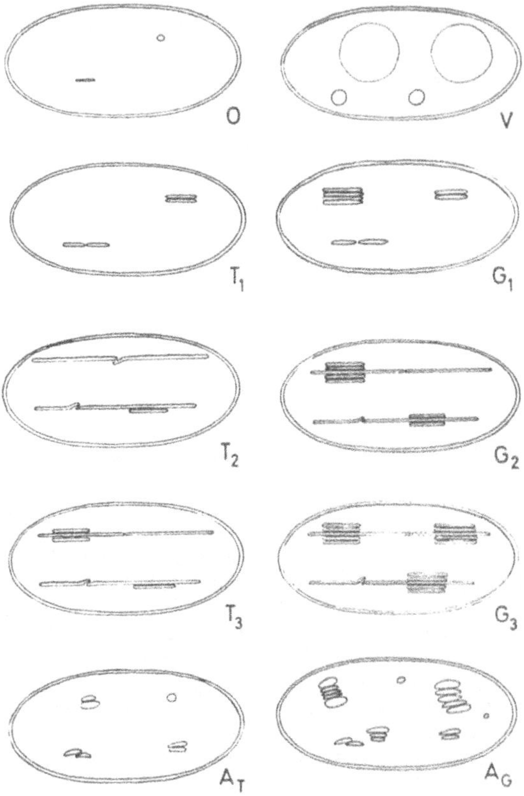

Fig. 20. Schematic presentation of the structure of the thylakoid system in normal and mutant plastids of *Oenothera* (Kutzelnigg et al. 1975a); compared to the green plastids, mainly the internal structure of the grana involved in the changes which are observed in the mutant plastids. The degeneration pattern of type *T* is characterized by a considerable reduction in the number and size of the grana. *O* Thylakoid system reduced to a few lamellar or vesicular structures; T_1 reduced internal structure of the plastids, few isolated short thylakoids; T_2 membranes and few thylakoids, rarely in small packages; T_3 in several places packages of two to four thylakoids present, but no normal grana; A_T early degenerating, vesicle-like thylakoids; *V* no thylakoids, only large and small vesicles present; G_1 isolated thylakoids, rarely in small packages; G_2 reduced thylakoids; G_3 normal structure of green plastids with many large grana; A_G degenerating G-type, grana enlarged to form compounds of vesicles

phology is mostly the same as the lens-shaped wild-type plastids, but the outline can be round or irregular. Sometimes, size and shape are determined by vacuoles and inclusions. The thylakoid system is reduced or missing. The variability of different types can be large, even within a cell. During the lifetime of a leaf the structure of the green plastids usually remains the same. In contrast, in most of the mutants considerable differences are observed between young and old leaves.

According to the developmental changes the mutants can be allocated to three main groups. In group 1 the development, as observed in leaves of ca. 1 cm length or more, starts with a reduced thylakoid and grana system and ends with vacuolation and complete destruction of the lamellar system. In the young leaves of group 2 the early development must be the same, as far as can be concluded from the structure of the plastids in these young leaves. This stage is maintained until the end of leaf growth. Then, degeneration starts and proceeds in essentially the same way as observed in much younger leaves of group 1. In group 3, however, the development of the plastids in growing leaves is similar to group 2, but then the build-up of the thylakoids continues. In the old leaves the fine structure is similar to the wild type, although fewer lamellar structures are found. In fact, the main difference from the normal type is a slowing-down of plastid development (Kutzelnigg et al. 1975a,b). Until now the different types of development of the mutated plastids cannot be directly correlated to the physiological defects caused by the mutated plastidal genes.

6.4 Developmental Biology of the Haploid Generation

6.4.1 Why Are the Haploids So Important?

One of the results of the genetic analysis of *Oenothera* is that the maintenance of constant heterozygosity, characteristic for many species in the genus, is not only a question of genetics but rather of the developmental processes involved in sexual propagation. The developmental steps leading to the formation of the gametes and fertilization are therefore of special interest. Higher plants are characterized by a generation cycle with alternating diploid and haploid generations, the sporophyte and the gametophytes, respectively. The diplont is an asexual sporophyte. On specialized leaves which, in spite of the asexual nature of the plant, are sexually differentiated as the carpels and stamens, sporangia with an archespore are formed, namely, the megasporangia as ovules and the microsporangia as pollen sacs in the anthers. The meiocytes of the archespore are differentiated as the megaspore mother cells (MMC) or as the pollen mother cells (PMC). In a correct terminology, the latter should be called microspore mother cells, the microspores really being the pollen mother cells sensu strictu, as these develop into the pollen grains. However, the abbreviation "MMC" could lead to confusion as it would not be clear which sex was being discussed. Furthermore, the term PMC is so widespread in the cytological literature that a change would not be an improvement.

The process of meiosis does not lead to the formation of gametes. Instead, tetrads of asexual but nevertheless sexually differentiated spores (megaspores and microspores) appear as the result of meiosis. From this stage onwards we have to look for developmental processes which lead eventually to the elimination of certain genotypes. The spores develop into the female and male gametophytes, the embryo sacs and pollen grains, respectively, which produce the gametes.

The gametophytes exist in a very reduced form of a few cells. The megaspore is never released by the mother plant and the female gametophyte, the embryo sac, is wholly dependent on the sporophyte and can until now not be isolated without lethal damage. The microspores are set free, but nevertheless the male gametophyte depends for its development on two sporophytes: one from which it originated and the other on which it will germinate after a short resting stage and finish its life as a pollen tube. The only function of the gametophytes is to produce the gametes and bring them together. In the process of fertilization, the gametes of both sexes are completely passive until the last moment of cell fusion. When the result of fertilization is not according to the expectation on the basis of random meeting of the gametes, the cause has to be sought and found in the behaviour of the gametophytes. These are the reasons why, in *Oenothera*, the development of the haploid generation must be considered in order to understand the genetic results of sexual propagation. General aspects of gametophyte ontogeny based on *Oenothera* have been presented by Noher de Halac and Cismondi (1992), a summary of investigations on *Oenothera* is given by Harte (1984).

6.4.2 Development of Megaspores and the Female Gametophyte

6.4.2.1 Ovules and Embryo Sacs

In the nucellus usually one megaspore mother cell (MMC) is located near the micropyle. In rare cases, two MMCs are seen side-by-side (mentioned by Haberlandt 1927; Gates 1928; Rudloff and Schmidt 1932) and more observations have been made by Noher de Halac and Harte and also by Sniezko and Harte in the course of investigations on embryo sac development. Two MMCs frequently occur in the ovules of the mutant *brevistylis*. The MMC elongates in the direction of the longitudinal axis of the ovule and becomes the meiocyte while the number of cell layers in the nucellus increases. By the orientation of the meiocyte in the ovule, a monoaxial organization with a micropylar pole and a chalazal pole can be recognized (Noher de Halac and Harte 1975). The development of the megaspore tetrad was first described by Geerts (1909).

Meiosis and megasporogenesis are not synchronous in the ovules of an ovary. The most mature stages are to be found in the middle part of the ovary. At the onset of meiosis, the MMC is surrounded by a layer of callose, which is thicker at the end towards the chalaza in homozygotic and heterozygotic isogametic species. The distribution and ultrastructure of the callose deposits in *Oe. biennis* L. and *Oe. Lamarckiana* de Vries have been investigated by Jalouzot (1971) and Jalouzot and Buvat (1970). Special aspects of the ultrastructure of the nucleus during meiotic prophase in the MMC of *Oe. Lamarckiana* de Vries (Jalouzot 1973) and in the embryo sac (Jalouzot and Buvat 1975) have been discussed.

After the first meiotic division, a dyad is formed. The haploid cells are orientated according to the long axis of the meiocyte and separated by a thick callose wall. The second meiotic division can be simultaneous in both dyad cells. In *Oe. Hookeri* de Vries and in some other cases the second meiotic division starts first in the micropylar dyad cell and is delayed in the sister cell. In some cases the preferential position of the micropylar end of the dyad even leads to a suppression of the second meiotic division and degeneration of the chalazal part of the triad before the tetrad stage is reached. In *Oe. biennis* L. and *Oe. suaveolens* Desf. synchrony of the second meiotic division is found in all tetrads. In *Oe. Lamarckiana* de Vries and the hybrid *Oe. ^{h}hookeri·flavens* ovules with synchronous and asynchronous tetrad development can be observed side-by-side in the same ovary (Noher de Halac and Harte 1975; Sniezko and Harte 1984a).

After meiosis in the MMC, the four tetrad cells, the megaspores, are found in a linear arrangement parallel to the long axis of the ovule, the top cell lying near the micropyle. To facilitate the description, the tetrad cells are numbered, starting with number 1 at the micropylar end of the row. Numbers 1 and 2 are sister cells from the second meiotic division, as are also numbers 3 and 4. Immediately after each meiotic division, first the dyad cells and then the tetrad cells, the haploid megaspores, are isolated by callose in the cross walls (Noher de Halac and Harte 1975, 1977; Noher de Halac 1980b,c; Sniezko and Harte 1984a,b, 1985). The pattern of callose deposits in *Oenothera*, in comparison with that in other plant genera, has been investigated by Rodkiewics et al. (1976).

Callose is not the only substance to point to polarization of the meiocyte and tetrad. Plastids with starch grains are present in the meiocyte at the poles and later in the two megaspores at the ends of the tetrad or even only in the one in the preferential position (Rodkiewics et al. 1976; Noher de Halac and Harte 1977, 1985; Sniezko and Harte 1985).

In *Oe. Hookeri* de Vries, the micropylar megaspore enlarges, while the three other megaspores, or megaspore number 2 and the remaining dyad cell, degenerate. Degeneration can start at any of the positions 2 to 4, but preferentially begins in the chalazal cell at position 4. In the homozygotic species of the type *Oe. Hookeri* de Vries, the mode of development of the meiocyte and megaspores does not have genetic consequences. This situation is different in constant hybrids. The two Renner complexes of the constant heterozygotes are separated in the first meiotic division. The two dyad cells are genetically different. In the linear tetrad the two micropylar cells at positions 1 and 2 are genetically similar sister cells, and the same holds true for the two other megaspores at positions 3 and 4. If orientation of the

Renner complexes during metaphase I is at random, then the two Renner complexes are expected to be represented in the micropylar position at equal frequency, each in 50% of the tetrads and of the embryo sacs developing from the micropylar megaspore at position 1.

Following growth of the micropylar megaspore in the direction of the axis of the former meiocyte, two haploid mitoses occur. The megaspore develops into a monosporic, four-celled embryo sac, the female gametophyte. This has the same polarity as the tetrad, with the preferential developmental position towards the micropyle. In addition to the egg cell and two synergidae at the micropylar end, the embryo sac contains one haploid polar nucleus, situated nearly in the middle of the embryo sac. There are no antipodal cells. In the terminology of polarity, this four-celled embryo sac can be described as initiated by a continuation of the polarity in the ovule, causing a very strong suppression at the chalazal end of the structure which has already led to the degeneration of the chalazal megaspores and now prevents any development at this site, even the entrance of a nucleus into this region, which could give rise to an antipodal cell group (Noher de Halac and Harte 1975). When the archespore consists of more then one meiocyte, then they will all undergo meiosis and from one megaspore in each tetrad an embryo sac can develop.

After fertilization by the two generative cells of the pollen tube, both the zygote from the egg cell and the endosperm mother cell from the polar nucleus are diploid and have the same genetic constitution.

The embryo sacs are mature at the time of anthesis. They are susceptible to fertilization only during a relatively short period. Exact data are not available, but observations on emasculated flowers with too early or delayed pollination and with double pollinations give estimates of 2 to 3 days during which the egg cells can be fertilized. From the same experiments it can be concluded that maturation and ageing of the embryo sacs start in the basal part of the ovary (Schwemmle 1957a,b, 1958), proceeding upwards with a difference of less than 1 day between the first and the last embryo sac. The loss of competence for attracting a pollen tube during the process of ageing can be derived from the seed set that occurs after very late pollination only in the upper part of the ovary. In the sect. *Raimannia* it seems that the genetic constitution has some influence on the ageing process, because following late pollination certain genotypes are found less frequently than after early pollination (Schwemmle 1957a,b, 1958).

The plastomes have an influence on the development of the ovules, as is demonstrated by the hybrid *Oe. biennis* L. × *Hookeri* de Vries, *albihookeri*. With the plastomes I or II from the parents, the hybrid is fully fertile, but when the same genome is combined with plastome IV, many ovules degenerate during development of the MMC, megaspore tetrad and embryo sac. The degeneration starts in the cytoplasm with changes in the structure of the membranes of organelles. The plastids are destroyed. Only a few embryo sacs reach maturity. Later, degeneration of the embryos, especially

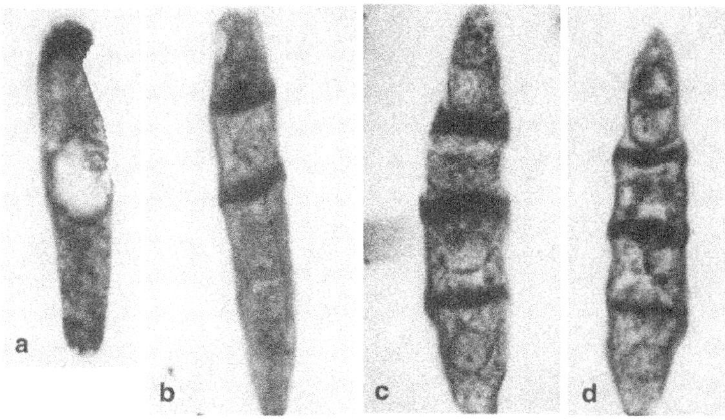

Fig. 21. Development from meiocyte to tetrad (microphotographs contributed by R. Sniezko). **a** Isolated meiocyte, large nucleus with nucleolus shortly before onset of meiosis. **b** Isolated triad from an extremely heteropolar meiocyte with thick cross wall resulting from the first meiotic division and similar cross wall separating the two cells from the second meiotic division in the micropylar dyad cell. In the chalazal dyad cell the second meiotic division is delayed and will probably not occur. **c** Isolated heteropolar tetrad with thick cross wall between the dyad cells at the end of telophase I and a similarly thick cross wall which separates the two micropylar tetrad cells. The second meiotic division is delayed in the chalazal dyad cell as is indicated by the still very thin cross wall between the two chalazal tetrad cells. **d** Isolated homopolar tetrad; all three cross walls have nearly the same thickness, regardless of whether they arise after the first meiotic division in the meiocyte or after the second meiotic division in both dyad cells (micropylar end on top)

those of the type ^{h}*hookeri*·^{h}*hookeri*, can occur in the same way. Some embryos of this type survive when they have enough paternal plastids. This behaviour is initiated by the incompatibility between genome and plastome which is expressed in the hybrid only during development of the ovules (Stubbe et al. 1978).

6.4.2.2 Competition Between Megaspores and Embryo Sacs (or Catch As Catch Can)

In the heterogametic species nearly all egg cells contain the same Renner complex, in contrast to the Mendelian expectation. Based on the genetic experiments, this phenomenon was formally described as due to competition in the female line which would eliminate one genotype from the population of active egg cells. But, just as in all other cases of unusual genetic results in *Oenothera*, the development basis of the observations has to be elucidated.

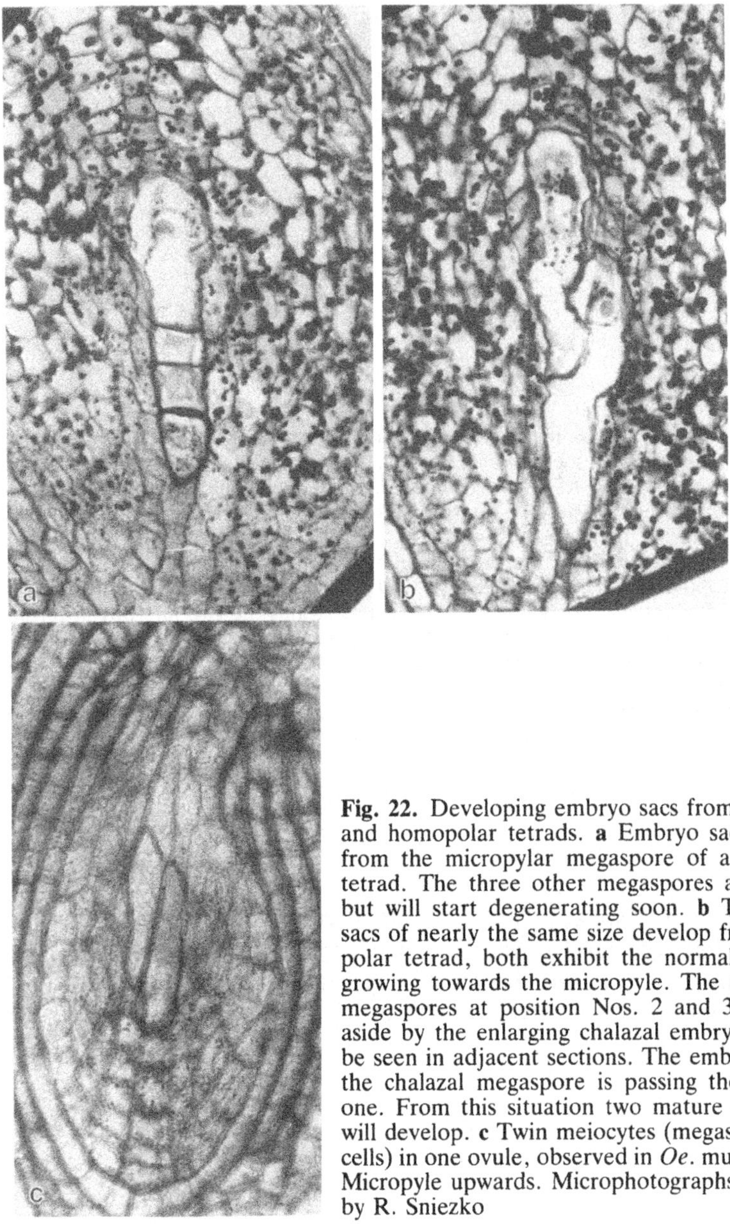

Fig. 22. Developing embryo sacs from heteropolar and homopolar tetrads. **a** Embryo sac developing from the micropylar megaspore of a heteropolar tetrad. The three other megaspores are still alive but will start degenerating soon. **b** Twin embryo sacs of nearly the same size develop from a homopolar tetrad, both exhibit the normal polarity by growing towards the micropyle. The degenerating megaspores at position Nos. 2 and 3 are pushed aside by the enlarging chalazal embryo sac as can be seen in adjacent sections. The embryo sac from the chalazal megaspore is passing the micropylar one. From this situation two mature embryo sacs will develop. **c** Twin meiocytes (megaspore mother cells) in one ovule, observed in *Oe*. mut. *brevistylis*, Micropyle upwards. Microphotographs contributed by R. Sniezko

The question is: At what stage between the differentiation of the meiocyte and fertilization of the egg cell does this elimination take place and what is the developmental mechanism by which the goal of genetically uniform egg cells will be reached?

To find the solution, the development of the MMC and its descendants, the megaspores, in the cases of genetic female competition must be compared with the situation in the homozygotic and isogametic species. The first investigations were carried out by Renner (1917, 1921). Several differences can be observed. The common developmental step is meiosis in the MMC. But already before meiosis, the first difference in the polarity of the meiocyte can be detected by the distribution of callose in its wall (Noher de Halac and Harte 1975, 1977; Sniezko and Harte 1984a,b, 1985, 1986). As examples *Oe. biennis* L., some of its hybrids and *Oe. muricata* will be presented. Although the meiocyte is elongated in the same direction as in *Oe. Hookeri* de Vries, callose deposits in the outer wall of the meiocytes are not different at the poles, but very thin and evenly distributed or callose is missing altogether. The second meiotic division is simultaneous in both dyad cells, demonstrating their developmental equality. The two cross walls between the megaspores after the second meiotic division are of nearly the same thickness. The four megaspores do not exhibit morphological differences until degeneration starts in one of the cells at position 2 or 3, while the two remaining megaspores at the end of the tetrad enlarge (Figs. 21, 22).

All these observations indicate an important difference in the development of polarity in the ovules. A strong polarity is expressed in *Oe. Hookeri* de Vries and related species by the development of callose deposits and by the rate of development of both the meiocyte and young megaspores. This polarity is changed in the other cases. The clear preference for further development at the micropylar end of the tetrad and early degeneration at the chalazal pole is called heteropolarity of the tetrad. The other case, when two megaspores at opposite ends of the tetrad start development, is called homopolarity (Noher de Halac and Harte 1975; Sniezko and Harte 1986). In the homopolar tetrads, degeneration of the megaspores does start in the middle positions, leaving the two megaspores, number 1 at the micropylar end and number 4 at the chalazal end of the tetrad, for further development.

The chalazal megaspore can, with some delay, still degenerate, but, in the same way, degeneration can proceed to the micropylar end of the tetrad, leaving the chalazal megaspore alive to form the embryo sac. Even both megaspores at the polar positions in the tetrad can start development by enlarging and each eventually ends up as a complete embryo sac. The polarity of the ovule in the first stages of its development is the same in all cases, as can be judged by its morphology and the position of the meiocyte. Degeneration of all megaspores in a tetrad is observed with a rather high frequency of up to 25% of the ovules. It does not exhibit any relation to callose pattern or to other morphological characters of the ovule. Most probably, this type of degeneration is caused by irregularities during meiosis or disturbances of unknown origin (Sniezko and Harte 1986).

When degeneration starts in the megaspore at the chalazal position and the micropylar megaspore develops directly without delay into the embryo sac, the mode of development is called primary heteropolarity. However, when first a homopolar tetrad is formed and with some delay, caused by the simultanous start of growth in the two megaspores in the polar positions and later degeneration of one of them, only one embryo sac develops, the phenomenon is called secondary heteropolarity (Sniezko and Harte 1986). A chalazal megaspore developing into an embryo sac, irrespective of whether it does so alone or as a twin simultaneously with the megaspore in the micropylar position, always immediately regains the original polarity of the ovule and enlarges in the direction of the micropyle, eventually passing alongside the degenerating megaspores, as has been described by Renner (1917), Haberlandt (1927), Gates (1982) and Rudloff (1929). At the end of a true homopolar development two four-celled functional embryo sacs exist, the twins lying side-by-side in the nucellus. The polar orientation in the embryo sacs, irrespective of their origin from megaspores of different positions in the tetrad, is always the same: the egg apparatus with the egg cell and synergidae lies at the micropylar end.

Observations on the differences between the homogametic and heterogametic species in the development of the megaspore tetrad can be interpreted as a disturbance of polarity in the ovule in the heterogametic case. This does not involve meiosis and the linear arrangement of the megaspores in the tetrad but prevents an immediate difference in the development of the cells in the polar positions. The first sign of this disturbance is seen as a difference in callose deposition in the wall of the meiocyte, and it is maintained during meiosis and early stages of megaspore development until the start of development of the embryo sac. At this stage, the original polarity is restored and the growing embryo sacs have the same orientation as in *Oe. Hookeri* de Vries with the site of preferential development at the micropylar end and complete suppression at the chalazal pole. A schematic presentation of the different pathways from megaspore tetrad to embryo sac is given by Noher de Halac and Harte (1975, 1977) (Figs. 23, 24).

6.4.2.3 Genetics of the Development of Polarity

In considering polarity and the pattern of callose deposits in the ovule, everything looked good as long as only species and F_1-hybrids were investigated. All the plants of a population behaved in the same way with respect to the general type and even to the variability of the phenomena between the ovules in a plant. The unsolved problem is the relation between callose deposits and polarity of the tetrad. A correlation between the polarity of callose deposits and degeneration pattern of the megaspores has been found

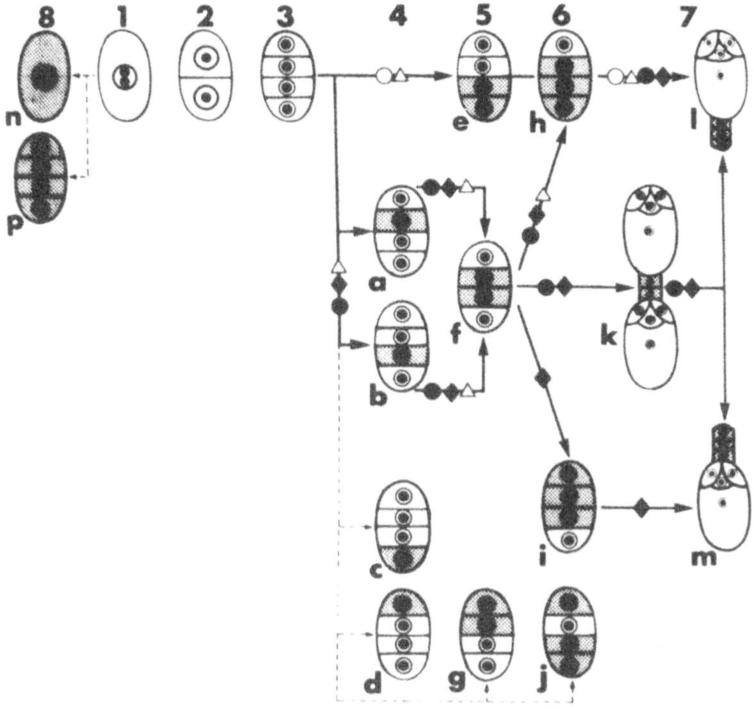

Fig. 23. Megasporogenesis and development of the embryo sac in *Oenothera*, homopolarity and heteropolarity in the ovule. The diagram illustrates the different ways of development of polarity in the ovule (Noher de Halac and Harte 1974). In the ovule a strong polarity is usually expressed with preference to the micropylar end, leading to polar development of the meiocyte, the megaspore tetrad and the embryo sac. The micropylar megaspore can develop directly to the embryo sac, combined with early degeneration of the three other megaspores (primary heteropolarity). The other modus of development of the megaspore tetrad starts with a temporary breakdown of polarity in the ovule and as a consequence the development of the two megaspores in the polar position to form twin embryo sacs and degeneration of only the two megaspores in the middle position (homopolarity). Later, one of the two young embryo sacs can still degenerate, leaving only one functional embryo sac (secondary heteropolarity). As soon as the development of the embryo sacs starts, the polarity in the ovule is restored. The embryo sacs are always oriented in the long axis of the ovule with the egg apparatus towards the micropyle

by several investigators for many plant species. The pole of the meiocyte, which is covered by a thick callose cap, indicates the site where degeneration of the megaspores starts. It is interpreted as an isolating factor related to degeneration phenomena. In contrast to most other plants, in *Oe. Hookeri* de Vries and also in *Antirrhinum* the callose cap and the start of degeneration are on the chalazal side and the micropylar megaspore develops into the embryo sac (Jalouzot and Buvat 1970; Jalouzot 1971;

Rodkiewicz 1970; Rodkiewicz and Kuran 1971; Kuran 1972; Noher de Halac and Harte 1975, 1977; Sniezko 1976). There is certainly a relationship between callose and degeneration. Callose deposits are present in the walls of the degenerating megaspores in *Oe. Hookeri* de Vries and also in the degenerating embryo sacs of the homopolar *Oe. biennis* L. The ultrastructure of the callose is similar in both cases (Noher de Halac 1980b). Investigations of Jalouzot (1971) on *Oe. biennis* L. and the homopolar *Oe. Lamarckiana* de Vries are interpreted in the same way (Jalouzot and Buvat 1970).

These observations are taken as further proof for a relationship between isolating callose and degeneration of meiospores. Furthermore, in those species of *Oenothera* which have homopolar megaspore tetrads, the polarity of callose deposits is missing. Taken together, these observations were interpreted as a causal relationship: the isolation by the thick callose deposits is the reason why the cells will degenerate. *Oenothera* provides additional confirmation, because the familiar degeneration pattern is disturbed when the callose pattern is changed. That's clear, isn't it? It is not. But it is a very good example of a logical fault, not uncommon in biology, "post hoc, ergo propter hoc" (it occurs after it, therefore because of it!), as demonstrated here by *Oenothera*.

→

Fig. 24. Callose pattern during tetrad development in the ovule. Microphotographs of ovules with meiocytes, dyads and tetrads, fluorescence staining of callose which demonstrates the differences in callose patterns in homopolar and heteropolar tetrads (contributed by R. Sniezko). **a** and **b** Meiocytes with different callose deposits. **c** Dyad with thin callose layer in the outer wall and thick callose deposits in the cross wall. Typical tetrads of different plants in a segregating F_2-progeny of *Oenothera* ^{h}*hookeri·flavens* demonstrate the independence of callose deposits in the outer wall of the original meiocyte and the polarity of the tetrad as seen in the delay of the second meiotic division and the differences in thickness of the cross walls in the micropylar and chalazal part of the tetrad. In **d–g** tetrads with callose deposits in the outer cell wall and chalazal callose plug are shown. **d** Heteropolar tetrad with callose in the outer wall and thick callose plug at the chalazal end; slight delay of the second meiotic division in the chalazal dyad cell as demonstrated by the thin cross wall in this part of the tetrad. **e** Heteropolar tetrad with chalazal callose plug and callose deposits, especially in the outer wall of the three tetrad cells which will later degenerate, exhibiting extreme differences in thickness of the callose walls. **f** Heteropolar tetrad, callose completely missing in the outer wall of the micropylar megaspore, chalazal cross wall sligtly thinner than the cross wall in the micropylar part of the tetrad. **g** Tetrad with extremely thick callose deposits in the outer wall of the chalazal part and thin callose at the micropylar end, in combination with nearly similar thickness of the cross walls separating the megaspores which indicates homopolarity of the tetrad. Tetrads without callose deposits in the outer wall derived from the meiocyte are shown in **h–j**. **h**, **j** Homopolar tetrads shortly after the second meiotic division which occurred simultaneously in both dyad cells. **j** Homopolar tetrad with cross walls of nearly the same thickness. **k** Heteropolar tetrad with delay of the second meiotic division in the chalazal dyad cell, as demonstrated by the very thin cross wall compared to the thick one which separates the two micropylar megaspores. **l** Extremely heteropolar triad, the second meiotic division omitted in the chalazal dyad cell (micropylar end to the left)

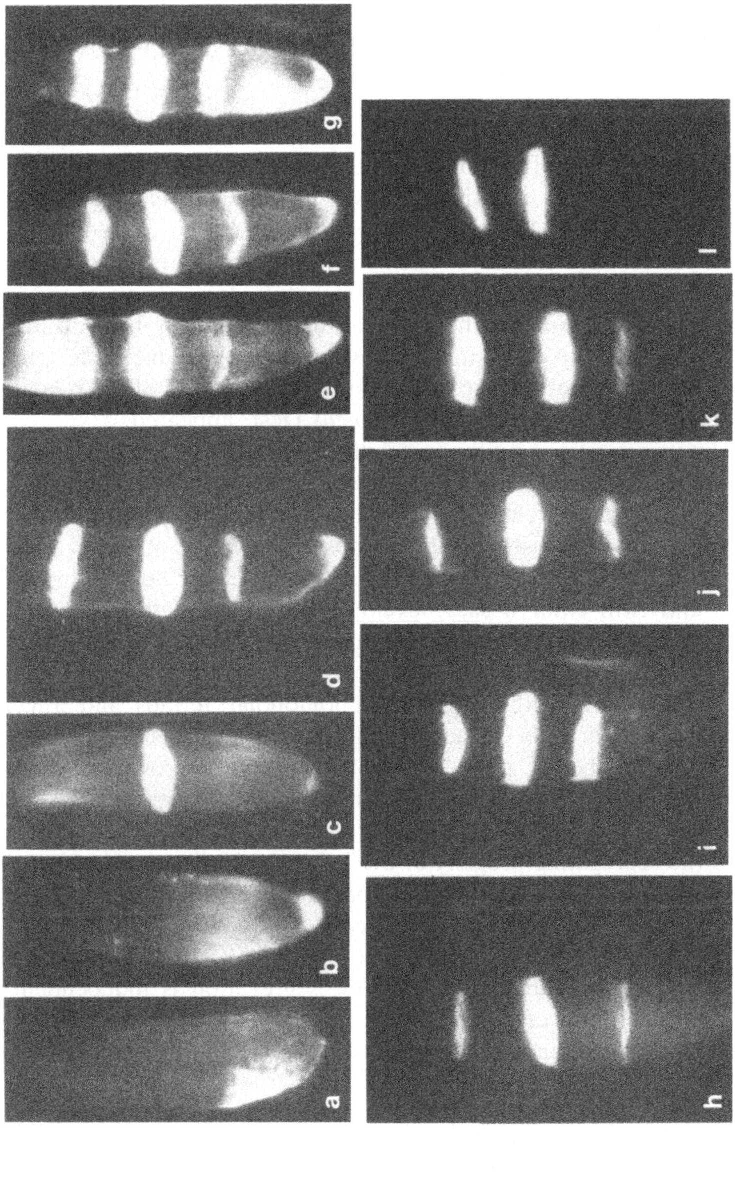

Fig. 24.

The first indication that the relations were maybe not so simple was found by Rudloff (1929). *Oe. purpurata* has many homopolar tetrads and competition occurs between megaspores and young embryo sacs in spite of its homozygosity. Renner (1940a) also reported the observation of homopolar megaspore tetrads in homozygotes. Callose was not investigated at the time.

When plants of segregating F_2-progenies, obtained by self-pollination of the hybrids *Oe. flavens*·^{h}*hookeri* and *Oe. albicans*·^{h}*hookeri*, were considered in an investigation of the development of tetrads and embryo sacs in individual plants on a quantitative basis, the picture changed (Sniezko and Harte 1984b, 1985, 1986). The correlation between development of the callose pattern and polarity, which was earlier interpreted as a causal relationship, is in fact a perfect and meaningful coordination of independent processes. This coordination can break down following hybridization. New combinations of the characters "callose pattern of the meiocyte/tetrad" and "degeneration pattern of the megaspores in the tetrad" of the parent species *Oe. Hookeri* de Vries and *Oe. suaveolens* Desf. appeared.

The investigations on the variability of polarity phenomena in the ovule in segregating F_2-progenies of the *Oenothera* hybrids *Oe.* ^{h}*hookeri*·*flavens* revealed that the development of polarity in the ovule is genetically determined. Because at least four phenotypes of plants are observed which differ in the frequency of types of megaspore tetrads, either homopolar or heteropolar, and twin embryo sacs developing from one tetrad, the number of genes involved must be greater than one. As, at the same time, the two extremes, namely plants with only heteropolar tetrads like *Oe. Hookeri* de Vries, and others with nearly only homopolar tetrads and a very high frequency of twin embryo sacs, were observed in a small F_2-progeny of the hybrid, the number of genes acting on polarity in the ovule cannot be very large, probably two would be enough to explain the observed variability of phenotypes (Sniezko and Harte 1984b) (Fig. 25).

In the segregating progeny of the hybrid *Oe. albicans*·^{h}*hookeri*, consisting of only two types similar to the parent *Oe. Hookei* de Vries (^{h}*hookeri*·^{h}*hookeri*) and the F_1-hybrid *Oe. (albicans*·^{h}*hookeri)* with recombination of a few characters, plants with only heteropolar tetrads and others with 80% homopolar tetrads were oberved. This segregation is independent of the genetic constitution of the plants as determined by the Renner complexes and other segregating characters. It can be explained by the recombination of genes influencing the development of polarity in the ovules. The extreme form with nearly only homopolar tetrads, observed in the progeny of *Oe. flavens*·^{h}*hookeri*, is missing, indicating that *flavens* differs from ^{h}*hookeri* in probably one more gene locus than *albicans* (Sniezko and Harte 1985).

In this progeny some plants were observed which had nearly no callose in the wall of the meiocyte, like the grandparent *Oe. suaveolens* Desf., but which developed extreme heteropolar tetrads, like the other grandparent *Oe. Hooker* de Vries. In the ovules of another plant, also without callose around the meiocyte, virtually only homopolar tetrads with twin embryo sacs were observed (Sniezko and Harte 1984b). This character is more extreme than that found in *Oe. suaveolens* Desf. and probably corresponds to homozygosity for alleles from the Renner complex *flavens* which are not present in *albicans*. A segregation and recombination for the characters "callose pattern in the development of meiocyte/tetrad" and for "polarity of the degeneration pattern in the megaspore tetrad" is observed, in which both are independent of one another and different combinations of the

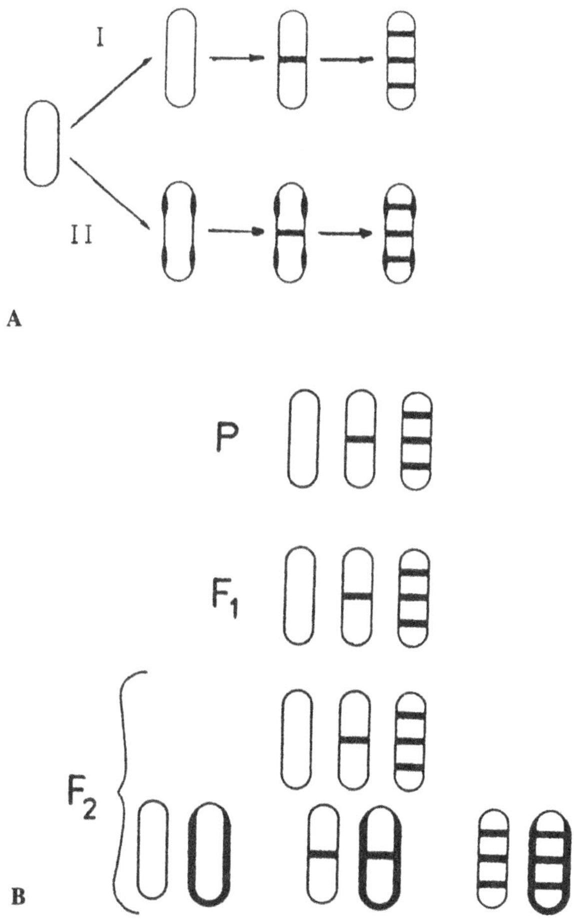

Fig. 25. Distribution of callose during megasporogenesis. **A** Schematic representation of the two types of callose distribution with and without callose deposits in the outer cell wall separating the meiocyte from the nucellar tissue, from the young meiocyte to the homopolar megaspore tetrad. **B** Examples of the variability of callose distribution in the outer cell wall of homopolar meiocytes and tetrads in a progeny segregating for genes influencing the callose pattern (contributed by R. Sniesko)

parental characters are possible. This points towards a genetic basis with several genes directing the developmental processes involved. Considering the number of phenotypes observed, the number of genes must be greater than one for each of the two characters but cannot be very large because both parental types were found in a population of limited size. Probably a difference between the parents in the alleles of two genes for each character will be sufficient to explain the number of types and the frequencies with which they were observed.

The observations on developmental processes in *Oenothera*, leading from the meiocyte to the megaspore tetrad and embryo sac, reveal another case, similar to that found in flower development, where a sequence of events which look as though they belong together and are interpreted as based on causal relationships, in fact point to coordination in the timing of two independent series of developmental steps, each governed by its own genetic influences. When, in such a situation, investigations are started with the aim of clarifying the supposed causal relations, failure is included in the programme. Instead, the goal must be to break down the correlations and to elucidate the genetic mechanisms which form the basis of the perfect coordination in the course of independent developmental events that looks so natural and logical to us at first sight.

6.4.2.4 Polarity in the Megaspore Tetrad and Genetic Competition

Now the genetic basis of development in the megaspore tetrad is known, the connection between the polarity phenomena in the ovule and the genetic observations on competition in the female line of gamete formation can be discussed. The development of embryo sacs from megaspores at the chalazal position in the heterogametic species *Oe. biennis* L. and *Oe. muricata* had already been observed by Renner (1917) and interpreted as being connected to the genetic phenomenon of competition in the female haploid phase. However, this needs more considerations which could not be seen by Renner at the time. Later, Renner, (1940a) observed homopolarity also in homozygotes and concluded that this must not always be connected to genetic competition.

When the distribution of the Renner complexes to the two poles of the meiocyte in meiosis occurs at random, then both Renner complexes should appear in the egg cell of embryo sac which develops from the micropylar megaspore with a frequency that deviates only at random from 50%. The same proportion of the two genotypes in the embryo sacs will result when the decision for development of the embryo sac from the micropylar or chalazal megaspore in homopolar tetrads occurs at random without considering the genotype of the competing megaspores.

When, however, the genetic constitution of a megaspore influences this decision by giving it a different chance for developing an embryo sac, depending on the genes of the Renner complex in its nucleus, then a random distribution of the two Renner complexes in the developing megaspores, and consequently in the egg cells, can no longer be expected. In primary heteropolar tetrads, only the position of the megaspore decides its fate, either forming an embryo sac, if situated at the micropylar pole, or resulting in death if it is in one of the other positions. The genetic constitution has no chance to influence the development. If, on the other hand, in a homopolar tetrad the micropylar preference is abolished and two megaspores start further development, then the megaspore with the "stronger" genotype can win. The spore with the other Renner complex will have no chance, irrespective of its position.

The consequence of this is that the heteropolar tetrads will give rise to embryo sacs which contain the two Renner complexes of the species with equal frequency, but in embryo sacs from the homopolar tetrads only one of the two Renner complexes will be present (Noher de Halac and Harte 1975, 1977). The twin embryo sacs, developing from two megaspores of one tetrad, will always contain different Renner complexes. After fertilization, they will give rise to non-identical twins (Harte and Sniezko 1988).

In the species and F_1-hybrids investigated upto now, all the ovules can be heteropolar, or a mixture of ovules with heteropolar and homopolar tetrads will occur, but in no case is complete homopolarity in all ovules of an ovary observed (Sniezko and Harte 1984a). Now it becomes clear that in species and constant hybrids with strong, undisturbed polarity in the ovule during its development, both Renner complexes will be represented in the egg cells at random. When the species is a complex–heterozygote, the egg cells will be ditypic. However, in heterozygotic species with the described disturbance of polarity in the development of the tetrads, either both Renner complexes can be present at random or one Renner complex will be represented in the egg cells in most cases, while the other will appear only with a frequency equal to one-half of the frequency of heteropolar tetrads in the ovules. The result will be nearly monotypic female gametophytes and transmission of the second Renner complex by the egg cells with only very low frequency. The decision between these two possibilities is made by the sporophyte and depends on the genotypes of the two Renner complexes combined in the species or hybrid. It is regulated by the alleles of the two sets of genes influencing the development of polarity in the ovules.

This expectation corresponds exactly to the observations on tetrad development in the ovules and to the origin of the embryo sacs from megaspores in different positions in the tetrad in species with monotypic and ditypic egg cells. In the species and hybrids in which most ovules contain homopolar tetrads of megaspores, those in the two polar positions enlarge to develop an embryo sac in approximately one-half of the ovules each. Some ovules, varying between plants from a few to nearly all, later contain two functional embryo sacs, one from each of the two megaspores in the polar positions.

When the two Renner complexes differ quantitatively in their ability to develop an embryo sac from a megaspore, then in those ovules where two megaspores start enlarging, further development will be determined by an interaction of preference of position and preference of genotype. As the preferential position is weakened by a disturbance of polarity in the decisive stage, the megaspore with the appropriate genotype will have a better chance of winning the competition. This means that in the potentially homopolar tetrads the decision will be in favour of that megaspore which contains the stronger Renner complex, and this will be located at the two poles of the tetrad with equal probability. From this it follows that development of the megaspores in the two positions is expected to occur with equal frequency.

The logical form of the conclusion, however, must be kept in mind. When the genetic model of competition between the megaspores or the female gametophytes is accepted on genetic evidence from breeding experiments, then this provides a valid interpretation of the histological and developmental facts observed in embryo sac development from homopolar tetrads of the appropriate species, but the conclusion cannot be reversed. The conclusion from the occurrence of homopolar tetrads with equal frequency of further development for the two polar megaspores to the expectation of genetic competition phenomena in the female line is not possible. The equal chance of development of megaspores in two different positions can just as well be observed when the decision is at random and independent of the genetic constitution of the competing megaspores.

In investigations on diplarrhenic hybrids, containing two pollen complexes, Renner (1940a) found that the ovules were mostly sterile. Megaspores with a pollen complex seem to be deficient to a great extent not only in their strength to compete with others in the megaspore tetrad but basically in the ability to develop an embryo sac at all. They are not completely unable to do so, but their lack of success in the competition has its cause at least partly in their own weakness and not so much in the activity of their partner. The result is that in many ovules of the diplarrhenic hybrids no megaspore in the tetrad is able to develop. It would be very interesting to look at the developmental processes of the meiocyte, meiosis, megaspores, embryo sacs and polarity in the ovules of such hybrid genotypes with modern methods of histological investigation.

The competition between Renner complexes in the female haploid line, as shown in the genetic experiments, is based on genetically determined influences of two different processes: first, the sporophytic component of polarity in the ovule, influencing the megaspore tetrad, and, second, in the case of homopolarity the genetic constitution of the megaspores, determining their competitive ability when they get the chance to compete in a homopolar tetrad.

Homopolarity of the megaspore tetrad is very unusual in angiosperms. In homozygotic species it would be a useless complication of a developmental process, the selection of a megaspore from a tetrad for embryo sac formation, and would introduce a delay by a stochastic process where an early determination is the shortest route with the smallest risk of failure and is, therefore, the best solution. But in hybrid species it is, in combination with differences of the genetically determined competitive ability of the megaspores, an efficient method for avoiding the production of unused embryo sacs and lethal embryos. It is a perfect cooperation between two generations. The genetic competition between haploid genotypes is made possible by the genes for polarity in the ovule, active in the sporophytic tissue, and by the genes active in the haploid megaspores and young embryo sacs, determining their competitive ability, which makes the final decision.

We can summarize the conclusions: when the decision which megaspore in the tetrad will develop into the embryo sac is made by position of the megaspore only, in a constant hybrid both Renner complexes have the same probability of being in the right position and should be observed in the egg cells with equal frequency. If, on the other hand, the decision about the development of the micropylar or chalazal megaspore in a homopolar tetrad is made at random, then the same expectation results. Only in the case where homopolarity of the megaspore tetrads is found in the majority of ovules and the decision about the future fate of a megaspore depends on its genotype, will the result be a population of monotypic egg cells on a heterozygotic parent.

6.4.3 Development of the Pollen Grain and the Male Gametophyte

6.4.3.1 From Meiocyte to Pollen Grain

Meiosis in the pollen mother cells (PMC) and the final structure of the pollen wall have attracted the interest of investigators for a long time, but apart from these two points the development has found only little interest, mainly in relation to genetically determined male sterility. The genetic studies were started by Renner (1919b). A description of the development of pollen grains and the male gametophyte, including the sterile pollen grains, is given by Jean (1974). The most recent paper with a description and presentation of a "normal table" of anther development makes a clear delimitation of 12 developmental stages possible (Noher de Halac et al. 1990).

The archespore is differentiated from the other tissues in the anthers at a very early stage in flower development, when the lobular evaginations are formed at the sides of the stamen primordia. Soon, the innermost cells, the archespore proper, can be distinguished as non-vacuolate enlarging cells from the surrounding cell layer which becomes the tapetum. While cell division in the somatic cells of the anther goes on for several cycles, mitosis stops in the prospective PMCs. The correct name would be "microspore mother cells", the microspores being the real pollen mother cells, but introducing this nomenclature will lead to confusion with the abbreviation MMC used for the megaspore mother cell. The old name will be maintained for practical reasons.

The meiocytes are closely connected with one another during the premeiotic phase and up to mid-pachytene. Then, the cell connections break down, the cells separate and become globular. A callose wall is formed which increases in thickness until the end of meiosis. This is best seen in squash preparations of anthers. At the end of meiosis, the tetrad cells immediately separate and develop their own cell wall without contact between the cells. The free space inside the old wall of the meiocyte and the tetrad cells is filled with callose, isolating the young microspores completely.

It seems that the disintegration of the old callose walls starts immediately afterwards. The dissolution of the callose proceeds very rapidly. This can be concluded from the fact that in squash preparations of anthers the tetrad cells slip out of the callose cover, lying near the characteristic structure with the four compartments and disrupted thin outer walls. This stage is sometimes seen only in a part of an anther which contains complete callose-covered tetrads at one end and free microspores without remnants of the callose wall of the tetrad at the other end.

The young microspores have no vacuoles and a moderate number of round starch grains near the central nucleus. After dissolution of the callose the three apertural chambers start their development, regularly spaced along the equator of the microspore. At this stage the sporoderm consists only of the endexine. As the next step, the vacuole becomes visible. The nucleus is lying at the side, surrounded by starch grains. The apertural chambers are fully developed and the microspores stop growing. The ektexine develops with a paracrystalline beaded structure and the viscin threads, later connecting the mature pollen grains, are formed outside the sporoderm and connect the microspore with the tapetum. This stage marks the end of the microspore (Figs. 26, 27).

The process of pattern formation involved in the development of the apertural chambers on the pollen grains is disturbed in the diploid pollen of tetraploid plants and in pollen grains with an irregular number of chromosomes on triploids. Instead of three, there are four or even five apertural chambers. They are located as usual at equal distances along the equator of the young microspore. This phenomenon was already detected before the connection between the form of the pollen grains and the chromosome number was known (Boedijn 1925b) and interpreted at the time as part of the phenotype of certain mutants.

The male gametophyte starts its development inside the sporoderm with the first pollen mitosis of the haploid nucleus when the anthers are ca. 6–7 mm long (Jean 1974). From this time onwards the young pollen grain is no longer a microspore. The generative cell contains proplastids (Diers 1963a), is separated by a thin cell wall and lies free in the surrounding cytoplasm of the vegetative cell, which accumulates lens-shaped starch grains, (Diers 1963b). The pollen grain is a two-celled gametophyte, enclosed by a sporoderm of special structure, that is formed on the basis of the microspore cell wall. Its origin is partly from the microspore itself and partly contributed by the sporophyte. The ultrastructure of the mature sporoderm, consisting of intine, endexine and ektexine, has been described by Lepouse and Romain (1967), Jean (1974) and Noher de Halac et al. (1990).

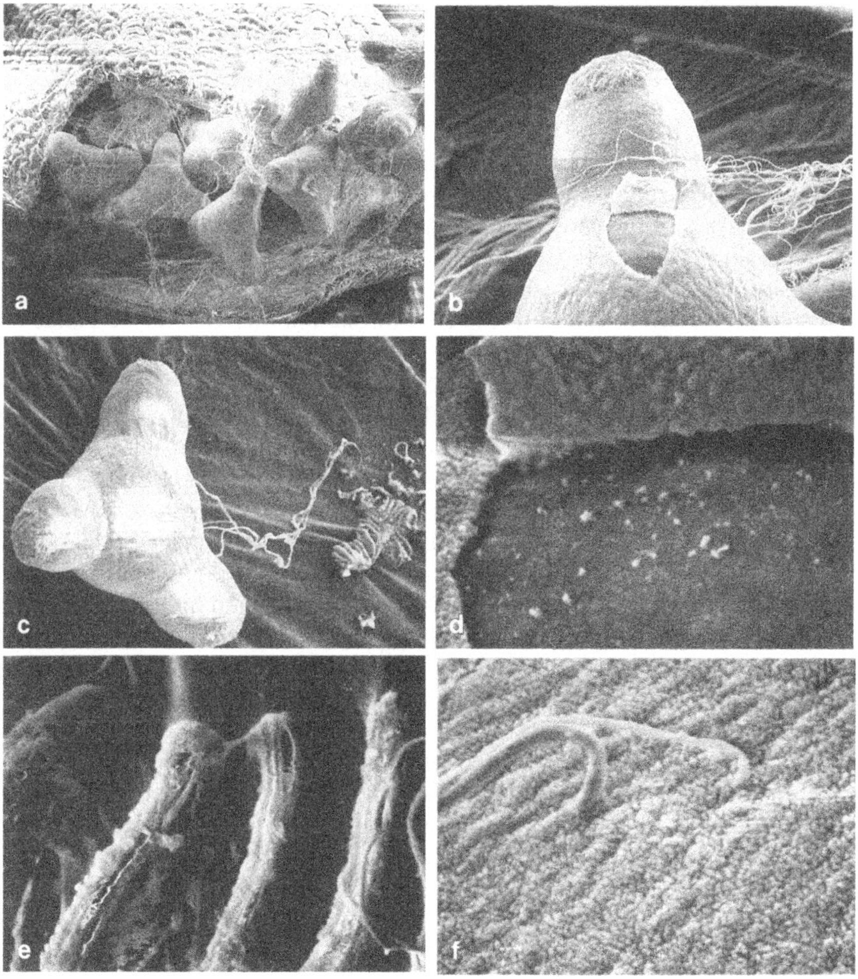

Fig. 26. Scanning electron micrographs of pollen and anthers of *Oenothera recurva* Dietrich from Catamarca/Argentina (contributed by I. Noher de Halac). **a** Anther with pollen and viscin threads; ×320. **b** Ektexine exposing endexine, viscin threads; ×1500. **c** Pollen with viscin threads; ×750. **d** Detail from **b**, exposed endexine; ×10000. **e** Viscin threads; ×10000. **f** Ektexine with attached viscin threads; ×10000

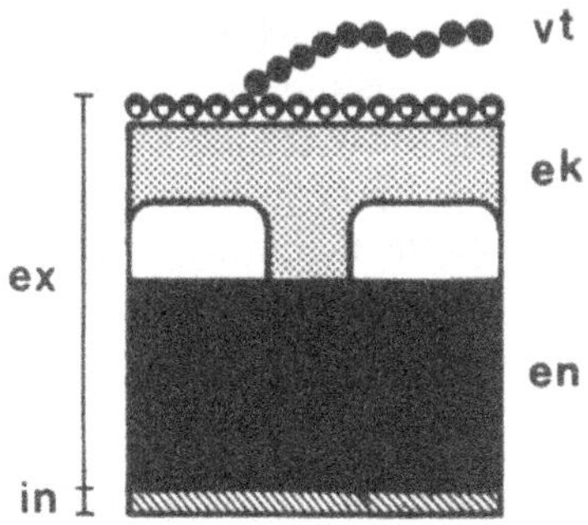

Fig. 27. Structure of the pollen wall, mature sporoderm, in *Oenothera. in* intine; *ex* exine; *ek* ektexine; *en* endexine; *vt* viscin threads (contributed by I. Noher de Halac)

The development of the microspores is correlated to developmental changes in the structure and function of the tapetal cells. The precursors of the outer layers of the sporoderm are produced not by the microspore or the pollen grain but by the tapetum and are therefore of sporophytic origin. The synthesis and secretion of substances in the tapetum and the formation of the sporoderm layers can be connected to one another (Noher de Halac et al. 1990).

In the young anther the tapetal cells can be recognized by their size and staining behaviour on histological slides and by their submicroscopic structure (Noher de Halac 1985b). Up to the end of meiosis in the PMCs the tapetal cells enlarge, contain one nucleus, but have neither vacuoles nor lipids. When meiosis is completed in the PMCs and the tetrads have been formed, the tapetum has become binucleate. Submicroscopically, laberinth is observed during the late tetrad stage but disappears afterwards. When vacuolating microspores are present, vacuoles appear and lipids are accumulated in the tapetum. While the pollen grains are growing, the lipids disappear from the tapetal cells, which soon degenerate, and are found in the locular fluid. There the amount decreases while the sporoderm layers are formed.

That the parallel activity of the tapetum and the formation of the pollen coat is not an otherwise meaningless time correlation or coordination of independent processes, as is found in other cases, but is in fact a causal relationship, is demonstrated by the developmental processes in male-sterile anthers (Oehlkers 1927; Harte and Bissinger 1952; Noher de Halac et al. 1990).

6.4.3.2 Development of Pollen in Male-Sterile Anthers

In *Oenothera* several types of genetically determined male sterility are described, but in most cases only the effect is stated without considering the development of sterility. The causes are very different. Degeneration of the contents of the anthers can be determined sporophytically. The first case found was *Oe. Lamarckiana* de Vries mut. *lata* (Pohl 1895), which later turned out to be a trisomic. In progenies of hybrids with the Renner complex *stringens* from *Oe. strigosa* de Vries the gene *fr* was found (Oehlkers 1922) and later the gene *sterilis* was detected in the Renner complex *velans* from *Oe. Lamarckiana* de Vries *brevistylis* (Harte 1948). In these cases, after a normal meiosis and microspore formation, the contents of the anthers degenerate completely before pollen mitosis. At the time of anthesis the anthers remain closed and seem to be empty. By microscopic investigation, the remnants of the pollen grains can be found as a mass of dark lamellar material (Oehlkers 1927; Harte and Bissinger 1952; Noher de Halac 1985b; Noher de Halac et al. 1990) (Table 1; Fig. 28).

The degeneration process is somewhat different in the three cases, demonstrating that the two known genes influence different steps in the synthesis and secretion of the ektexine sporopollenin precursors in the tapetum. The malfunction causes an aggregation of the defective precursors in the tapetal cells and the absence of the ektexine from the microspores. On the now unprotected microspores the endexine is dissolved, probably by an enzyme, and the naked protoplasts are attacked and digested by the enzymes present in the locular fluid, otherwise responsible for the dissolution of the tapetal cells and other cell remnants in the thecae.

Apart from these cases little is known about male sterility in *Oenothera* (Kaul 1988). In several other cases genetically determined male sterility is observed, but only the effect is stated, and observations of details on the development of pollen and anthers or the process of degeneration are missing. In *Oe. fruticosa* a case of monogenic inherited male sterility has been mentioned by Linder (1954, 1961). At anthesis the anthers are empty, as observed for *fr* and *ster*. Schwemmle (1929) found a segregation of male-sterile plants in an F_2-progeny of the hybrid *Oe. Berteriana* × *odorata*. The development is normal up to pollen mitosis, then the cell contents degenerate and at anthesis the anthers contain only empty and shrunken pollen grains. The tapetum seems normal, as far as can be judged from the histological methods used at the time, and the cause for the degeneration is unknown. It must be a sporophytically determined influence from the mother plant because it affects all the pollen grains and, in contrast, the pollen grains of the heterozygotes are all fertile, irrespective of their genetic constitution.

Generally, the trisomic types of *Oenothera* are partically pollen sterile. Pollen grains with supernumeray chromosomes degenerate or are inactive. In trisomics of *Oe. Berteriana* the microspores or pollen grains with the extra chromosome degenerate between meiosis and anthesis, in most cases before pollen mitosis (Arnold 1955). In *Oe. odorata* the trisomic type *tenella* is pollen sterile. At anthesis all pollen grains are inactive, irrespective of their chromosome number (Arnold and Bina 1957).

Table 1. Development of normal and two types male sterile anthers, homozggotic for the alleles *fr* or *ster*

	Normal	Alleles fr	Alleles ster
1. Sporophytic diploid pollen mother cells	Callose special wall. "Pattern" genes transcribed. Products transferred to cytoplasm and plasma membrane		
2. Meiosis	Organelles undergo cycle of de-and re-differentiation ER membranes enclosed in DMI's	no callose special wall	
3. Tetrad of microspores	DMI's release contents. Membranes participate in wall morphogenesis through initiation of primexine	no primexine formation	
	Synthesis and deposition of sporopollenin-like polymer in primexine. Endexine formation	no endexine formation	
4. Young spore period	Tapetal cells synthesize sporopollenin, and transfer it to surface of microspores to form the paracrystalline structure of the ektexine	no further accretion of exine polymer. No paracrystalline structure of ektexine	
5. Late spore period	Intine formation	no intine formation	dissolution of endexine
6. Young pollen period	First mitosis	no first mitosis	
7. Late pollen period	Starch deposition	no starch deposition, death	
8. Mature pollen period	Anther opens	remains of tapetum and spores stick together	

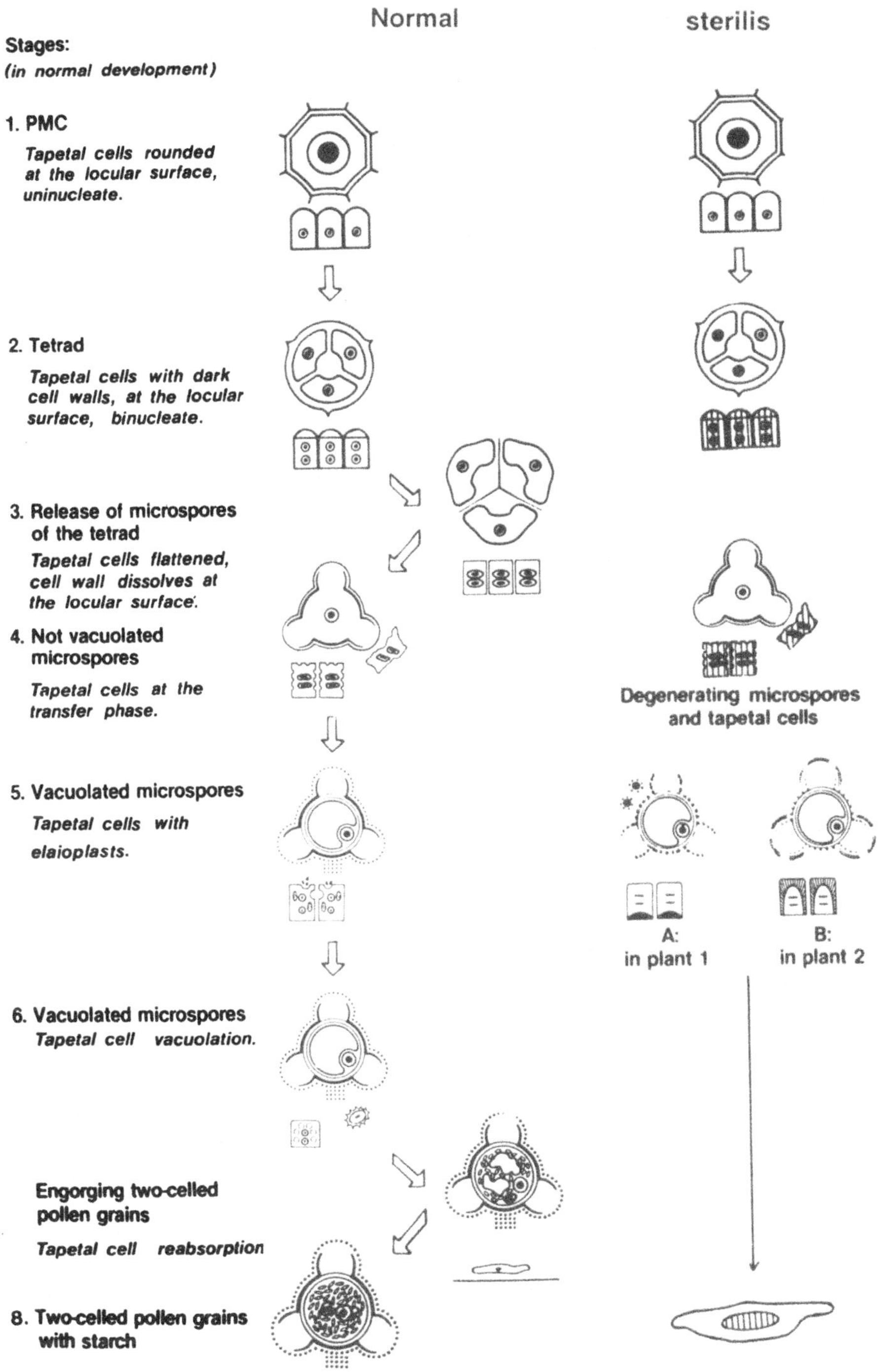

Fig. 28. Schematic presentation of pollen development in fertile pollen and in male-sterile anthers of the genotype *sterilis* (*ster/ster*) (contributed by I. Noher de Halac)

In only a limited number of cases is the degeneration or inactivation of pollen grains known to occur under the influence of the genes in the microspore itself, leading to only 50% vital pollen grains. In *Oe. fruticosa* such cases of 50% pollen sterility are observed (Linder 1954). The shrunken, empty pollen grains have been described (Linder 1967; Linder and Jean 1968). They usually do not have an intine, indicating early degeneration of the contents before pollen mitosis as observed by Jean (1974). In some species of the North American *Oenothera biennis*-1 species pollen lethals are observed, contributing to the elimination of one Renner complex from the male gametes (Steiner 1956). In trisomics generally the pollen grains with the supernumerary chromosome will die.

Quite another phenomenon is observed in *Oe. biennis* L. and its hybrids containing the Renner complex *albicans*. The pollen grains with this genotype are alive, as demonstrated by their physiological behaviour, but cannot germinate. They are inactive, causing partial male sterility that becomes effective in this species and its hybrids only for pollen grains of this genotype. At the same time they can be distinguished by the globular form of the starch grains, in contrast to the lens-shaped form in the active pollen grains (Renner 1919b). Similar differences were observed in *Oe. muricata* (Renner 1919c). The trimorphism of pollen grains in *Oe. biennis* has still to be explained.

6.4.3.3 Metabolism in Developing Pollen Grains

For a modern interpretation of the observations, the physiological development of the meiocytes to pollen grains has to be considered. From the end of the meiotic prophase to the tetrad stage starch accumulates in the meiocytes. The microspores contain round starch grains which lie near the nucleus. As soon as the pollen grains become binucleate, starch accumulates in the vegetative cell until this is filled nearly completely, but now the grains are lens-shaped, indicating a different structure. Small lipid droplets can be seen in the meiocytes from late meiotic prophase onwards, but by the tetrad stage they have disappeared. Only shortly before anthesis do larger lipid drops appear in the pollen grains (Noher de Halac et al. 1990, 1992).

A detailed description of the metabolic changes during pollen development is presented by Noher de Halac et al. (1992). The observed changes in lipid and starch contents are indicators of metabolic changes during meiosis and pollen development. The inactive pollen grains have round starch grains while the mature active pollen grains have spindle-shaped starch grains. The interpretation now given to the differences in the form the starch is that the inactive pollen, with round starch grains, stopped development at a juvenile stage, before the metabolism changed to form the new starch grains found in mature pollen. Round starch grains are also found in sporophytic tissues and in immature microspores. This type is replaced by a special gametophytic type of starch during the last stages of maturation in pollen grain development (Noher de Halac et al. 1990) (Table 2; Fig. 29).

In *Oe. organensis* Munz. a gene family of about six to eight members has been detected that is expressed at high levels only in the pollen. The protein products of these genes are at least 38 kDa and present in mature pollen and in growing pollen tubes in the style and in vitro as well. Nucleotide and amino acid sequences are similar to the enzyme polygalacturonase, suggesting a function in depolymerizing pectins during the development of pollen and pollen tubes (Brown and Crouch 1990). This is the first case investigated to such a depth in *Oenothera*. The authors stress that many genes are known to be active in developing pollen and that a considerable number are only active here and not in sporophytic tissues.

At anthesis, the pollen grains lose water. The nearly globular vegetative cell becomes flattened and the pollen enters a resting stage with reduced metabolism, comparable to the state of mature seeds. For further development induction by the stigma is necessary. The pollen grains do not have pollen kitt but are connected to viscin threads which facilitate the transport of a large amount of pollen by pollinating insects. The viscin threads have been investigated by Skvarla et al. (1978) and Hesse (1983). Recently Takahashi and Skvarla (1990) stated that the fibres connecting the exine with the tapetal cells are seen immediately after dissolution of the callose walls of the microspores, but the origin of the viscin threads is still unknown.

6.4.3.4 Pollen Grains and Embryos, Comparable Developmental Stages?

It seems that development has something in common in embryos and young pollen grains. The transformation of a zygote into an embryo and of a microspore into a pollen grain can both be seen as embryonic processes, starting from a single cell that is considered to be undifferentiated but is really highly specialized for its task. Both cells are meristematic as they start further development with cell division, and in both cases the first part of the development leads to a resting stage, characterized by the accumulation of metabolites as reserve material, loss of water and increased resistance against otherwise unfavourable environmental influences. To continue development, both embryos and pollen grains require an induction from external sources. The first step is water uptake, and finishing the resting stage is irreversible. It would not be very surprising to find that the metabolic processes during this development have many steps in common too. It is not known whether the lipid reserves in seeds and pollen grains are chemically the same, but the sequence of events, of first producing starch, then forming another type of starch and finally accumulating lipids, is the same.

In this respect, the situation in *Oenothera* is not exceptional. As far as is known to this author, a comparison of the metabolism and reserve materials

Table 2. Normal ontogenesis of pollen in *Oenothera*. Data based on material collected both during the day and at night. Duration of development: one to several weeks, depending mainly on the temperature (Noher de Halac et al. 1990)

Stage Ontogenesis	Characteristics[a]										
	Reproductive cells						Tapetum				
				Lipids	Starch	Vacuoles	No. of nuclei	Cell wall	Laberinth	Lipids	Vacuoles
Microsporogenesis	Callose										
1 Archesporium	–			–	–	–	1	+	–	–	–
2 Early pre-meiotic PMC	–			–	–	–	1	+	–	–	–
3 Late pre-meiotic PMC	+			(+)	–	(+)	1	+	–	–	–
4 Meiotic PMC	+			(+)	+	(+)	1	+	–	–	–
5 Early tetrad, cytokinesis	+			(+)	+	(+)	2	+	–	–	–
6 Middle tetrad stage	+			(+)	+	–	2	+	–	–	–
	en	ek	in								
7 Late tetrad stage	+	–	–	(+)	+	–	2	–	+	–	–
8 Non-vacuolate microspores	+	–	–	–	+	+	2	–	+	–	–
Pollen development											
9 Vacuolating microspores	+	+	–	–	+	+	2	–	(+)	++	+
10 Vacuolated pollen	+	+	+	–	–	+	2	–	(+)	++	(+)
11 Engorging pollen	+	+	+	–	(+)	(+)	Degenerate	–	–	–	–
12a Bicellular pollen with starch	+	+	+	–	++	–	Degenerate	–	–	–	–
12b Bicellular pollen with starch and lipids	+	+	+	+	++	–	0	–	–	–	–

[a] en, Endexine; ek, ektexine; in, intine. ++, abundant; +, present; (+), traces; –, absent

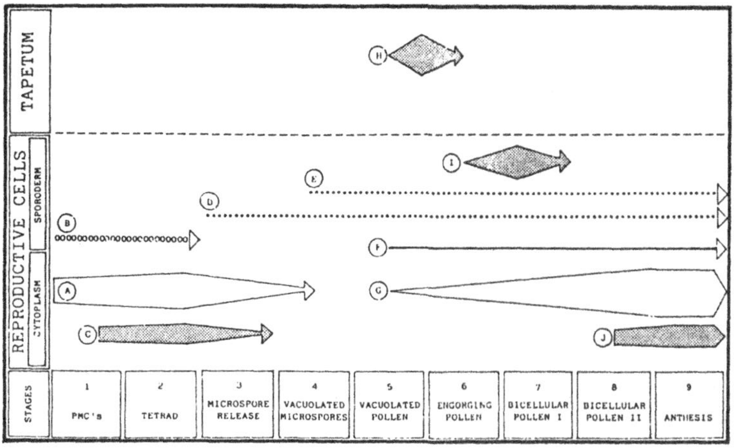

Fig. 29. Changes in the metabolic products during development of the anther. *A* round starch grains in PMCs; *B* callose; *C* lipid bodies in PMCs and tetrads; *D* sporopollenin in the endexine; *E* sporopollenin in the ektexine and viscin threads; *F* callose in the intine; *G* oval starch grains in the vegetative cell of the pollen grain; *H* lipid bodies in the plastids of the tapetal cells; *I* lipids in pollen kitt; *J* lipid bodies in the vegetative cell (I. Noher de Halac et al. 1992)

found in pollen grains and embryos does not exist for other species either. Nevertheless, it would be interesting to look at it.

Starting at the tetrad stage at the end of meiosis, the microspores and young male gametophytes represent a new generation which differs from the sporophyte not only by the haploid state of its nuclei but also by its metabolism and morphogenetic development. These processes are independent of the sporophyte and are directed by genes which are, at least in some cases, active only during pollen development. On the other hand, the development of the pollen grain depends on the sporophyte not only for nutrition by the thecal fluid during the first stages of microspore growth but also for the formation of the sporoderm that protects the developing gametophyte in the anther and on its way from the anther to the stigma.

Here, the cooperation between the cell wall of the microspore and the precursors of sporopollenin, delivered by the tapetum and used to build up the outer sporoderm layers, is evident but needs further investigation. The male-sterile mutants of *Oenothera* which make defective molecules that cannot form a complete ektexine will be of great help in elucidating the synthesis and structure of sporopollenin.

6.4.3.5 Germination of Pollen Grains and Growth of Pollen Tubes in the Style and Ovary of Oenothera

To make fertilization possible, the male gametophytes have to meet the female gametophytes which are hidden in the ovules in the ovary. The pollen grains, which are young male gametophytes in a resting stage, contain a vegetative nucleus in the cytoplasm and the generative cell. They are transferred by pollinators onto the stigma. There they germinate in the secretion layer of the stigma. The role of the secretion in germination of the pollen grains has not yet been investigated. However, on young, dry stigmas, germination is not possible. It cannot be just the fluid medium, as water or a mineral nutrient will cause the pollen grains to burst. For in vitro germination a very high sugar content in a solid agar or gelatine medium is needed. In the germination process, exine and intine are destroyed, probably by an enzyme provided by the pollen grain itself after induction by the secretion layer of the stigma, or by the nutrient in vitro. At one of the three preformed lobes a pollen tube appears which enters the style between the papillary cells of the stigma, intruding into the transmitting tissue and growing from there into the central part of the style (Dickinson and Lawson 1975; Heslop-Harrison 1990).

Here, the pollen tubes grow between the cellulose walls of the cells (Renner and Preuss-Herzog 1943; Kaienburg 1950). They prefer long walls and as cells of the transmitting tissue in the centre of the style are elongated parallel to the axis of the style, the pollen tubes, following the path of least resistance, will arrive at the base of the style. There they do not enter the cavity of the ovary but grow in the central placenta, following the transmitting tissue, until they reach the base. Here the orientation of the cells is different, and with the same strategy of preferring the long cell walls, the pollen tubes change their direction of growth and turn upwards. During the first part of their development the pollen tubes only elongate but branching of the pollen tubes occurs often in the last part of their growth, as though they were searching for the right way. Recent observations have shown that this can also happen in the style. Then the prescribed pathway will guide them to the cavity of the ovary. As this is crowded with ovules, the pollen tubes will enter always near the micropyle of an ovule.

In the first investigations of this behaviour it was concluded that the growing pollen tubes dissolve the middle lamellae between the cells. More recent investigations made clear that in mature styles the middle lamellae between the long walls of the cells in the transmitting tissue are already dissolved before the pollen tubes arrive. The intercellular cavities are filled with a fluid, containing pectic carbohydrates, that is easily lost in the course of preparing the style for microscopical investigation (Heslop-Harrison 1990). The pollen tubes do not actively choose their way but take the path prepared for them in the transmitting tissue. This path is first in the intercellular fluid of the alveolar cell layers in the stigma and then in the moisture covering the cell walls of the intercellular spaces in the transmitting tissue of stigmas and style (Dickinson and Lawson 1975).

In species of the sect. *Euoenothera* the first pollen tubes enter the ovary at the base. They will enter the micropyle of one of the first ovules which they pass, and this will be located naturally in the lower part of the ovary.

Those pollen tubes which arrive later find the cell walls occupied by other pollen tubes and will turn towards the cavity higher up in the placenta, again entering one of the first suitable ovules encountered. When pollination was scarce, the seeds are found only in the basal part of the fruit (Kaienburg 1950). This has been confirmed by observations made during many years of experimental crossings and in certation experiments (Harte 1969c and unpubl. observ.). In experiments with repeated pollination by small batches of pollen of different genotypes, it can be expected that in the progeny those genotypes which correspond to the first pollination are found in the lower part of the ovary. The pollen tubes which arrive later will find unfertilized ovules only in the upper part.

Glenk (1964) could not confirm this observation but stated that fertilization starts in the middle or even upper part of the ovary. The differences can be attributed to the fact that his experiments are performed with the sect. *Raimannia*. There are differences in the histology and development of the ovaries between the two subgenera, especially in the arrangement of the solid central part of the ovary and the conducting tissue, which can be the cause of different behaviour of the pollen tubes. The abnormal behaviour of pollen tubes of *Oe. Hookeri* de Vries in styles of species of the sect. *Raimannia* mentioned by Glenk (1964) as proof of this interpretation is not convincing because the infertility of crosses between subgenera does not allow conclusions to be made on the reactions of pollen tubes in the stylar environment of their own species.

The reserve of starch in the pollen is considered to be insufficient for growth through the relatively long styles. In young styles starch is accumulated in the cells. Following anthesis, this starch is hydrolyzed before the pollen tubes arrive and even without pollination. The oligosaccharides can be used by the pollen tubes. Perhaps free boric acid is of some importance in this process. It is accumulated in the style and in vitro experiments show better growth of the pollen tubes when boric acid is present in the nutrient (Glenk 1964; Kwack 1965).

For the last part of their route in the cavity of the ovary, the behaviour of the pollen tubes demonstrates that a mechanism of recognition must exist. This can only be a chemotropic attraction of the pollen tubes by the ovules, most probably by the micropyle. In the ovary gland cells are located on the placenta in the epidermis near the micropyle. Excretion by these cells will be the same throughout the whole ovary, so it can only act as a guide to come near to a micropyle but not a mean of discriminating between mature and immature or degenerating ovules or between ovules with embryo sacs of different genotypes. However, it is also known that in the micropyle of mature ovules a small drop of fluid can be seen, which could play a role in attracting pollen tubes. If this is the case, then it should contain a pheromone, which acts as a messenger to announce the presence of a fertilizable egg cell and is responsible for attracting the pollen tube. For this function it is unimportant whether it is produced by maternal tissue or by the embryo sac, either the egg cell or one of the synergidae. Only when the messenger is required to give information about the genotype of the mating partner must the secretion be produced by the gametophyte.

The attraction of pollen tubes by mature ovules has been demonstrated by Schneider (1956). Pollen grains of different species of the sect. *Raimannia* were put in a line opposite mature ovules of the same or related species at a distance of ca. 2 mm on a slide coated with sugar-gelatine, also containing aneurin and boric acid. The growth of the pollen tubes towards the ovules occurred with significantly greater frequency than could be expected from random orientation. In similar experiments by Schwemmle and Koepchen (1953) the attraction of pollen tubes by the ovules of different genotypes reflected the results of the crossing experiments. In the cross of *Oe. campylocalyx* × *Hookeri* de Vries no seeds were obtained. Pollen of *Oe. Hookeri* de Vries germinated normally on the stigma of *Oe. campylocalyx* and could grow through the style uninhibited but fertilization did not occur. The in vitro experiment demonstrated that ovules of *Oe. campylocalyx* (sect. *Renneria*) attract pollen tubes of the same species while pollen of *Oe. Hookeri* de Vries (sect. *Euoenothera*) will germinate under the experimental conditions, but the direction of growth does not give any indication of attraction by the ovules of *Oe. campylocalyx*.

When a chemotropic attraction guides the pollen tube towards the egg cell, then in any case the pollen tube must have receptors for the active agent. This means that there must be two genes or gene systems involved in the last steps of the development of the gametophytes, one for the production of the pheromone by a fertilizable embryo sac and one for the receptors in the wall at the tip of the pollen tube. Both are members of the sets of genes responsible for the development of the gametophytes and are active in one sex each, either in the mature embryo sac or in the tip of the pollen tube growing in the ovary. Each is active only during a short phase immediately before the last step in the fertilization process occurs.

In summary: the pollen tubes are growing on their whole route from the stigma through the transmitting tissue in the style and central placenta, in the ovary and into the micropyle in fluid media provided by the sporophyte and by the gametophyte. It requires a whole set of pheromones and of receptors adjusted to them to follow this path, but only the facts are known, not the molecules which perform this task.

For germination and growth of pollen tubes in vitro the secretion of the stigma is not sufficient (Renner 1919c). The nutrients used by different investigators for this purpose are slightly different. All have as their basis an agar or gelatine coating on slides and a rather high level of sucrose, sometimes supplemented with boric acid or other substances which are supposed to improve the germination rate or growth of the pollen tubes (Schwemmle and Koepchen 1953; Schneider 1956; Kwack 1965).

6.4.3.6 Elimination of Male Gametophytes

It is evident that in many species of *Oenothera* the efficient method of maintaining heterozygosity is to reduce the loss of zygotes, preventing the formation of homozygotes of one genotype by elimination of the microspores or of the resting stage of male gametophytes of a certain genotype before the pollen tubes can develop. Here two different mechanisms are found: (i) death of microspores or pollen grains of the unwanted genotype and (ii) inactivation of living pollen grains before they can germinate.

The first case, selective and genetically controlled death of microspores and young pollen grains as the result of gametic lethals, is very rare and found only in some American species as mentioned by Cleland (1972) and in a few European species (Linder 1961). Empty pollen grains are found with variable frequency in all species. Probably they have defective sets of chromosomes, caused by occasional meiotic irregularities, and have no importance for general considerations. An extensive discussion of many aspects of pollen lethality in *Oenothera* is presented by Jean (1974) with details on the morphology and submicroscopic structure of the pollen coat in viable and empty pollen grains (Fig. 30).

The second case is realized in the semi-heterogametic *Oe. biennis* L. As has already been demonstrated, the egg cells are ditypic with the Renner complexes *albicans* and *rubens*. The problem in this species is the monotypic pollen which inherits only the Renner complex *rubens* in controlled self-pollinations and in species crosses as well. Why is the Renner complex *albicans* missing in progenies from crosses with *Oe. biennis* L. as the male parent? Why are there only two types of fertilized ovules after self-pollination, those with apparently healthy embryos, which all prove to be of the *biennis* type at germination, and those which are much less frequent with degenerating embryos of the *rubens·rubens* type? The investigation of the ovary does not reveal fertilized but degenerating ovules. In no stage of embryogenesis or seed development can any types be identified which could correspond to *albicans·albicans* homozygotes. In the progeny of species crosses with pollen of *Oe. biennis* L. hybrids of the *albata* type are never found.

When the mother plant is heterozygotic, the two Renner complexes should be separated in meiosis and two types of pollen should be present. When only one type can be found in the progeny from pollen and when, furthermore, by observations the loss of embryos of the expected genotype has been excluded, then the only conclusion remaining is that somewhere on their way between meiosis and the formation of the male gametes the microspores with the *albicans* complex, or their descendants, are eliminated and prevented from reaching fertilization.

In the development of the anthers not more degenerating microspores or pollen grains are found than in homozygotic species. In mature anthers,

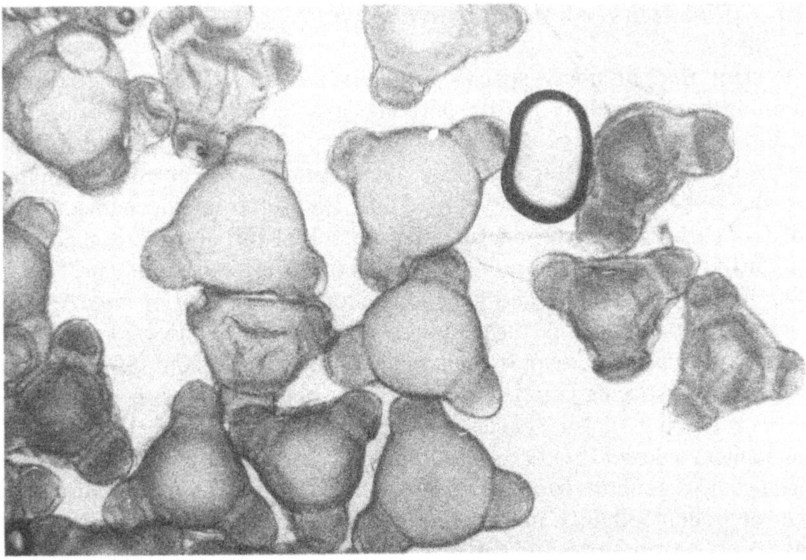

Fig. 30. Mature pollen grains of a mutant with 50% pollen lethality; the shrunken, empty grains died shortly before maturity

however, two types of pollen grains are present. One is a bit smaller than the other, but the differences in size are so small that there is not a bimodal distribution of size, only a slightly greater variability compared with other species. The differences in size only become apparent when another character, the form of the starch grains, is considered. The large pollen grains have spindle-like starch grains; in contrast, these are round in the somewhat smaller pollen grains. Both types occur with equal frequency, 50% of the pollen grains each (Renner 1919c).

On the lobes of the stigma more non-germinating pollen grains than usual are observed. The inspection of the stigmas some time after pollination shows that only the pollen grains with the oblong starch grains germinate, while those with the round ones remain unchanged and do not develop ,a pollen tube.

In cultures of pollen tubes on agar medium with a high sugar content, the two types of pollen grains behave differently. The smaller pollen grains do not react while the larger ones germinate and form a pollen tube. Again, both groups are distinguished by the form of their starch grains. Several tests prove that the non-germinating pollen grains with the round starch grains are not dead. By vital staining and other tests it has been verified that they contain a living protoplast but differ from the other pollen grains in being unable to germinate under any natural or experimental conditions.

In hybrids from *Oe. biennis* L. as the female parent, the *albata* types also have 50% of this particular type of non-germinating pollen grains which are slightly smaller than the other type and have round starch grains. In the pollen progeny of these hybrids the expected *albicans* types are missing. Similar observations have been made on other species with egg cell complexes which are never transmitted by the pollen (Renner 1919c).

The genetic explanation of this observation is that the Renner complex *albicans* contains a pollen-inactivating factor which prevents viable pollen with this Renner complex in its nucleus from germinating, thus excluding its transmission by the pollen. The factor determining germination/inactivation of the pollen grains is not only expressed at germination, as is demonstrated by the differences in starch formation during pollen development. A possible interpretation is that the last steps in pollen maturation are deleted (Noher de Halac et al. 1990). In *Oe. biennis* L., the pollen-inactivating system of *albicans* contributes very effectively to the maintenance of heterozygosity in the species. As one type of pollen grain is unable to develop a pollen tube, all the microgametes which perform fertilization have the same genotype. Recent investigations have revealed that the pollen-inactivating system is not a single gene. It is the result of incompatibility between nuclear genes in the Renner complex *albicans* and all known plastomes of *Oenothera* which makes *albicans* strictly an egg cell complex. The same holds for all cases of complete inactivation of a Renner complex in the pollen (Stubbe 1989a).

More details on the metabolism in inactivated pollen are presented by Göpel (1976). Pollen grains with the Renner complexes *flavens* or *gaudens* in combination with plastids from *Oe. parviflora* L. are inactive. The starch grains are round as in the inactive *albicans* pollen. Quantitative differences are observed in, for example, respiration rate, starch degradation and lipid metabolism. The inhibition is not caused by a single metabolic step but is correlated to a pattern of differences. This can be seen in relation to the importance of carbohydrate and lipid metabolism in pollen development as discussed by Noher de Halac et al. (1992).

In both cases, gametic lethals and pollen inactivation, 50% of the zygotes are still lethal because the egg cells contain both Renner complexes. But, in addition, 50% of the microspores are sacrificed and their genome cannot be transmitted by the pollen to the next generation. This effect of creating a homogeneous population of male gametes may be desirable for maintenance of the heterozygotic species but, at the same time, the transmission of this part of the genome to hybrids by the male gametes is made impossible. A gene-controlled developmental mechanism that could be an advantage for the species is also a disadvantage for the evolution of the genus. To transform it into an evolutionary advantage, it must be combined with mechanisms on the side of the female gametophytes which ensure a similar homogeneous population of egg cells but containing the other Renner complex of the hybrid species.

6.4.3.7 Differential Growth Rate of Pollen Tubes and Competition Between Pollen Tubes

Some observations made in the genetic experiments led to the conclusion that in populations of viable and active pollen grains of different genotypes the probability of achieving fertilization is not equal for all of them, but depends on the action of genes which influence the growth rate of the pollen tubes. When differences in growth rate exist, then the genetic composition of the population of pollen tubes arriving first at the ovary is not identical with that of the population of pollen grains on the stigma. In fact, the populations of early- and late-arriving pollen tubes from one pollination are genetically different. The pollen tubes arriving first at the ovary are not a random sample of the pollen grains on the stigma. Following generous pollination, when there are more pollen tubes than fertilizable ovules, a competition occurs in the style and the pollen tubes with the highest growth rate are favoured.

Two different types of experiments are used to elucidate the phenomenon. The first is the certation experiment, introduced to genetics by Correns. The progeny of pollinations with a generous amount of pollen, leading to fertilization exclusively by the fast-growing pollen tubes, is compared with the progeny from sparse pollination, which gives a chance for fertilization to every pollen tube reaching the ovary. Following poor pollination, the numerical relation of the genotypes in the progeny is shifted in favour of those which are rare after generous pollination. A variation of this experiment is to cut off the styles at different times after pollination, leading to a comparison between the progeny from undisturbed pollen tube growth with those in which only the pollen tubes arriving at the ovules within a given time after pollination could perform fertilization. The second method is the direct observation of pollen tube growth in the styles.

The phenomenon of differential pollen tube growth was described by Heribert-Nilsson (1920). The first experiments with cutting off the styles were carried out (Heribert-Nisson 1923) using the gene for red veins in the leaves, present in a strain of *Oe. Lamarckiana* de Vries. When the styles were cut 20 h after pollination, only the fast-growing pollen tubes could fertilize the ovules and the progeny consisted mainly of plants with red veins. However, when the styles were cut after 22 h or later, a segregation of 3:1 for red veins vs. white veins was observed. These results demonstrated that pollen tubes carrying the allele for red veins grew faster than those with the other allele.

In *Oe. flavens*·^{h}*hookeri* and several other hybrids, the deviations of the numerical relations of genotypes in the progeny from Mendelian segregation ratios revealed non-random fertilization. The certation experiment proved the existence of genes influencing the growth rate by their linkage to the marker genes *s*, *fr* and *de* (Harte 1969a,b,c). At that time it was not possible to determine whether the gene *ga*, influencing pollen tube competition, was identical with the locus influencing female competition in the ovule. Meanwhile, it has been clarified that competition between megaspores is made possible by disturbance of polarity in the ovule (Noher de Halac and Harte 1975, 1977; Sniezko and Harte 1984a,b, 1985, 1986). However, the decision which of the competing young embryo sacs will win is determined by differences in growth rate. When the alleles at the locus *ga* influence the growth rate in the haploid generation, then one gene could be responsible for competition of both pollen tubes and embryo sacs, if the possibility for development of two embryo sacs is given in homopolar tetrads.

Direct observations on pollen tubes have been made by Renner and Preuss-Herzog (1943) and Kaienburg (1950), but the data do not give quantitative information on the competition of pollen tubes in the style.

Other experiments proved directly the occurrence of considerable differences in the growth rate of pollen tubes. In the sect. *Raimannia* styles of mature flowers were cut to a length of 3 cm, placed with the wound on a gelatine-coated slide and pollinated with pollen of different genotypes (Schwemmle and Koepchen 1953; Schneider 1956). The time needed by the pollen tube to grow through the style into the gelatine was measured. As an example, the l-pollen tubes grow faster than the B-tubes. In a mixture of the two types of pollen tubes, as is the case after self-pollination in *Oe. Berteriana*, the l-pollen tubes arrive first at the site of action. They take their chance and fertilize the B-egg cells first, leaving the l-egg cells. When, later, the B-pollen tubes arrive, they find many l-egg cells to fertilize first, and then maybe a few leftover B-egg cells which at the end can be fertilized by some late-coming B-pollen tubes which do not find unfertilized l-eggs any more, giving the rare lethal B·B embryos. Similar results are obtained with *Oe. odorata* (Renner complexes I·v) (Schwemmle 1953a,b). The differences in growth rate can be so large that in some species crosses with double pollinations the pollen tubes of the second batch can overtake those of the first pollination.

6.4.3.8 Physiology of Incompatibility

In the genetic analysis an incompatibility system with multiple alleles on the S-locus was detected. In spite of the fact that incompatibility systems are widespread in the genus *Oenothera* (Steiner 1964), all the physiological experiments are performed on only one species. In *Oe. organensis* Munz, the species in which the incompatibility system was discovered and first analyzed, an investigation of the behaviour of the pollen grains on mature styles showed clearly that they can all germinate but, following self-pollination, the pollen tubes are inhibited in the stigma after growth of ca. 1 mm and never reach the style. On the other hand, following cross-pollination between plants within the population, either all pollen grains or ca. 50% or none can form a growing pollen tube, depending on the combination of S_I-alleles. On the style of another species all pollen grains develop pollen tubes. The same behaviour of the pollen tubes is found in other homozygotic species of the genus (Cleland 1960). In the heterozygotic species of the *biennis-1* group, a similar inhibition of 50% of the pollen tubes in the stigma is observed after self-pollination or after pollination to other plants of the same population (Steiner 1956, 1957, 1964). Linder (1952a) found genetic incompatibility in *Oe. fruticosa* and *Oe. missouriensis* and could demonstrate the inhibited pollen tubes in the stigma (Linder 1952b).

The physiological basis for this type of incompatibility reaction is an interaction between two partners. The alleles at the S-locus in the cells of the stigma and the stylar tissue of the sporophyte through which the pollen tubes must pass in order to reach the ovary are the first partner. The other is the corresponding allele in the vegetative nucleus of the pollen tube which directs its growth. When the S_I-allele in the haploid vegetative nucleus of the pollen tube is identical with one of the alleles in the diploid nuclei of the cells in the transmitting tissue of the style, growth of the pollen tube is inhibited. The reaction is induced by incompatibility between cells of different origin but with identical alleles at the S-locus. Differences in the growth of compatible and incompatible pollen tubes have been described by Dickinson and Lawson (1975). The structure of the epidermal cells and the secretion on the surface of the stigma, which must be related to the incompatibility reaction of the pollen tubes inhibited in this area, have been described by Heslop-Harrison and Heslop-Harrison (1985) and Heslop-Harrison (1990), but still without discussing differences between stigmas and styles with different genotypes concerning the S_I-alleles.

To understand the incompatibility reaction of *Oe. organensis* Munz., the development of the pollen tubes has to be compared in compatible and incompatible pollinations. The investigations of Dickinson and Lawson (1975) are based on EM studies of pollen tubes growing in styles and in culture. The resting stage of the pollen grain is interrupted immediately after pollination. The first sign of activation is hydration, causing a considerable increase in volume. The mature pollen grain contains double-membraned inclusions which provide the wall material for the first stages of growth of the pollen tube. This starts at one of the lobes of the pollen grain. Later, the products of dictyosome vesicles, believed to be non-cellulosic glucans and included as tightly packed fibrils in bodies with a single membrane, are transported to the wall, released there and involved in wall formation. The first steps in the growth of incompatible pollen tubes are the same, but the starch remains in the pollen grain, much less free carbohydrates are present, so that material for growth is missing as is demonstrated by the very thin wall at the tip of the pollen tube. When the pollen tube has passed through the outer stigmatic cell layer, growth of the wall seems to proceed in the same way as in compatible pollen tubes but is limited by the amount of soluble reserves available. Growth is finished when the pollen tube reaches the conducting tissue. The incompatibility reaction starts as soon as the pollen comes into contact with the surface of the stigma. Most probably the proteins in the stigmatic secretion layer are involved. But it seems clear that the main reaction, the inhibition of the pollen tubes, occurs during growth in the viscous fluid coating the outer cell layers of the stigmas and is completed before the pollen tube reaches the style.

Lewis (1952) prepared rabbit antisera from four genotypes of *Oe. organensis* Munz. which differed in their S-alleles. The antibody reaction is in favour of the assumption that the S-alleles produce specific proteins in the pollen. Experiments with double pollination demonstrated that the reaction of the second batch of pollen is not influenced by the presence of pollen and pollen tubes of the first pollination. In the stigmas and style the incompatibility substances must be present at maturity. They are not produced as a reaction to the antigen stimulus of the pollen. According to this interpretation, in *Oe. organensis* Munz. the substances involved in the incompatibility reaction are preformed on both sides and are not induced by the partner.

In grafting experiments, using pollination on a compatible style and grafting its upper part (15 mm) on another style that would be incompatible for the pollen, the pollen tubes were observed to grow into the incompatible style for an average of 22.1 mm, and thereby become much longer than they would have been after germination on the incompatible stigma. However, they did not reach the ovary through the style of ca. 15 cm length. Even when, in this experiment, contact of the pollen grains and young pollen tubes with the excretion layer and the papillary cells of the stigmatic surface of an incompatible genotype was avoided, the pollen tubes were inhibited. The incompatibility reaction can be induced in every part of the style but it is weaker in the transmitting tissue than in the stigma where, in the case of *Oenothera*, the inhibition usually occurs (Emerson 1940; Hecht 1960, 1964).

By treatment of the style with water at 50 °C some substance which is connected with the incompatibility reaction could be removed and pollen tube growth of incompatible pollen was initiated (Hecht 1964). It seems that an age- and temperature-dependent degradation of an incompatibility substance in the style is involved. Quantitative and qualitative differences in the amino acid composition of compatible and incompatible styles were detected which could indicate a relation between incompatibility and protein synthesis (Kumar and Hecht 1967).

Investigations of Lewis (1948, 1949, 1951) on mutations in the S-locus indicated that the reaction of the style and the pollen tube are directed by different parts of the gene locus. It is a complex gene consisting of at least two parts which can mutate independently of one another. The results of experiments testing the sensitivity of pollen mother cells and pollen grains for mutagenic influences were given the interpretation that the reaction of the pollen tube was determined early in the development of the pollen grain and could not be changed after that time by mutation of the S-locus. This would exclude an interaction between the pollen tube and stylar tissue leading to the inhibition. However, an alternative interpretation, based on the development of the microspore into the pollen grain and not considered at the time, is possible.

In the mature binucleate pollen the mutation has to occur in the vegetative nucleus to influence the development of the pollen tube. Since the vegetative nucleus is not involved in fertilization, and the generative nucleus of the same pollen tube contains the original S_I-allele, the pollen tube will grow and fertilize an egg cell, but the developing plant will contain the original S_I-allele of the generative nucleus. The mutation cannot be transmitted to the next generation and will be lost. Exactly this situation, as a result of his experiments, was described by Lewis (1951). As he did not consider the binucleate condition of the irradiated pollen, the observations were interpreted as the consequences of a reversible mutation. This would change the allele in such a way that the reaction of the pollen tube would be changed once and afterwards the gene would be reversed to its original structure. The correct interpretation is that the development of the pollen tube was directed by the mutated gene in the vegetative nucleus, eliminating the incompatibility reaction, and that in the generative nucleus, which was involved in fertilization, the mutation was not reverted but did not occur at all.

To be detected in the next generation, the generative nucleus must carry the mutated allele, but in this case the mutation would never reach the egg cell and become integrated in the zygote because the incompatibility reaction of the pollen tube is governed by the vegetative nucleus of the same pollen tube, carrying the original S_I-allele. A mutation of the same locus in both the vegetative and generative nuclei of a pollen grain at the same time will only occur with such a low probability that detection cannot be expected.

From the successful experiments with irradiation of younger flower buds the conclusion was drawn that metaphase I was the stage in development most sensitive to mutagenesis of the S-locus. Here too, the development of the flower bud and the anthers was neglected and a possible cytological control was not mentioned. In the anthers of a flower bud, meiosis of the pollen mother cells is not completely synchronized. When buds, which according to their size are expected to contain PMCs "in metaphase I", are investigated cytologically, every anther can be found to be within a certain range in a different stage of meiosis. Furthermore, one bud can contain stages from late pachytene to dyads, while in another bud of apparently the same size stages from metaphase to young microspores can be found (Harte 1953a). As is known from investigations on other species, metaphase I is very susceptible to severe disturbances by irradiation, causing stickiness and structural changes of the chromosomes, leading to irregularities in the distribution of the chromosomes in the following anaphase and, as a consequence, death of the meiocyte or microspores. It can safely be assumed that the pollen grains, developing from the surviving cells in irradiated young flower buds, were treated just after meiosis, between the end of anaphase II and the stage of haploid mononucleate microspores, before DNA replication in preparation for the first pollen grain mitosis has started. In a pollen grain developing from such a cell, a mutation will be represented in both nuclei, and expressed in both the pollen tube and the next generation.

It is necessary to bear in mind the mechanism of this type of incompatibility. Because the reaction is determined by the genes in the haploid vegetative nucleus of the pollen grain, it was called "gametophytic incompatibility", in contrast to other cases in which the reaction of the pollen grain is determined by the genotype of the sporophyte on which the pollen grain develops. But this is only one half of the story. In reality two partners are involved in the reaction. The interaction is between the male gametophyte, the pollen tube, and the tissues on the stigmas and in the style of the sporophyte. Their interaction decides the fate of the pollen tube. For this reason, the proper name should be **gametophytic–sporophytic incompatibility**. This nomenclature will be required later in another context to describe other phenomena involved in fertilization.

6.5 Fertilization, Selective and Otherwise

The gametophytes have to meet in order to enable a fusion of the immobile gametes, the egg cell and polar nucleus in the embryo sac and the generative cells of the pollen tube. The development of the embryo sac and of the pollen tube up to the point of contacting the micropyle have already been described in preceding sections. The fertilization process itself has not been properly investigated in *Oenothera*. Glenk always observed porogamy in the sect. *Raimannia*. This confirms earlier observations on species of the sect. *Euoenothera* already mentioned. There exist only a few old microscopic observations on pollen tubes entering a micropyle and some nice unproven hypotheses about attraction of the pollen tubes either by the micropyle and sporophytic tissues or by the egg cell or the synergidae or the fibrillary structure, the "Fadenapparat", a structure connecting the two synergidae which is mentioned in old publications (Ishikawa 1918; Haberlandt 1927) and considered again recently as important in the process of fertilization. The wall at the tip of the pollen tube should either be dissolved from the inside or the synergidae should produce a substance for its dissolution. Pollen tubes with open ends have been observed in the synergidae of mature, complete embryo sacs but in ovules with defective embryo sacs they only entered the micropyle without opening at the tip. As no quantitative information is available about the number of ovaries and ovules inspected and about the frequency of the different stages and, furthermore, the histological methods were according to the time, which means insufficient for detailed analysis, cytochemistry was non-existent and biochemistry had just started, the conclusions are open to any amount of doubt.

A quantitative investigation of the whole fertilization process between approximately 15 and 40 h after pollination, carried out with modern histological and cytochemical methods, is badly needed. It will become interesting because even in this process *Oenothera* has its specialities. On the other hand, the result is the same as in all angiosperms, namely a fertilized egg cell and polar nucleus. Therefore, it can be concluded that at the end the fusion of cells and nuclei must occur as usual, once they have overcome all the difficulties and traps. In *Oenothera* with a four-nucleate embryo sac, both products of nuclear fusion, the zygote and the fertilized polar nucleus, are diploid and genetically identical; the only process which can give rise to a genetic difference between embryo and endosperm is mutations in the generative cells, occurring in the short time span between pollen tube mitosis and fertilization.

In the cases discussed in the preceding sections on the haploid generation, selective mechanisms were at work to select and discard gametophytes of unwanted genotypes, but fertilization between the remaining gametes was considered to occur at random. In species of the sect. *Raimannia*, however, the genetic analysis revealed that this cannot be the case in some crosses. This statement was sufficient to finish the investigations on the genetic side of the problem with a formal description and name it "selective fertilization". The question remains, what happened in the fertilization process to achieve this result of non-random fusion of the gametes. As the gametes of higher plants are completely passive, the reaction must take place between the gametophytes, the pollen tube and the embryo sac.

In the genetic analysis of *Oe. Berteriana* with the Renner complexes B·l it became evident that some mechanism must secure the fertilization of the B-egg cells with male l-gametes and of the l-egg cells with male B-gametes in order to avoid formation of lethal zygotes after self-pollination. The frequency of the rare B·B homozygotes demonstrates that the fertilization of B-egg cells by B-pollen occurs only when there is a surplus of B-pollen tubes and no other egg cells are available. Other experiments demonstrated that l-pollen tubes selectively fertilize only B-egg cells, thus avoiding fertilization of embryo sacs with a genetic constitution similar to their own (Schütz 1939; Schwemmle 1948, 1951, 1952a,b; summaries by Arnold 1958a,b; Haustein 1967; Schwemmle 1968b).

One additional observation is important. When *Oe. Berteriana*, B·l, is pollinated with pollen from the hybrid B·II, which has only the Renner complex B in the active pollen tubes, then a pollination with a small batch of pollen will give only B·l-plants, resulting from fertilization of the l-egg cells by the B-pollen nucleus. However, when the pollination was generous, than lethal B·B-embryos were observed at nearly the same frequency as the vital B·l-embryos. From this it can be concluded that there is not an active rejection of the B-pollen tubes by the B-embryo sacs. When no other ovules are available, a B-pollen tube will accept, faute de mieux (for want of anything better), an ovule with a B-embryo sac. Egg cells with the Renner complex v can attract pollen tubes with B, but not vice versa (Schwemmle 1948). This demonstrates that the inability of recognition is not based on the overall genetic similarity and is not necessarily reciprocal.

The question remains: What happened to the unused ovules? Will pollen tubes enter them but not complete fertilization, or are they neglected by the pollen tubes? The experiment to answer this question must test whether these ovules can still be fertilized by providing another type of pollen tube, whose progeny can be distinguished from the result of the pollination giving the B·l plants. The pollen tubes reach the ovary between 12 and 20 h after pollination. After 24 h they can have taken their chance to enter the ovules. Following a first pollination with pollen of the Renner complex l, the stigmas were pollinated again with pollen of *Oe. argentinea*, a homozygotic species with the Renner complexes *ha·ha*. Analysis of the progeny revealed that by the first pollination with l-pollen only the B-egg cells were fertilized, leaving the l-egg cells for the ha-pollen tubes which arrived later. Similar results were obtained with other Renner complexes. The results demonstrate that a pollen tube with the Renner complex B or l in its vegetative nucleus does not enter the micropyle of an ovule whose embryo sac has the same genotype, thus leaving it free for other pollen tubes.

Schwemmle introduced the term "affinity" for the attraction between ovules and pollen tubes. The result of self-pollination of *Oe. Berteriana* can be described as differential affinity between pollen tubes and egg cells on a genetic basis.

The conclusion is obvious: in the species of the sect. *Raimannia* which were tested in these experiments, the pollen tubes cannot fertilize every egg cell which they encounter on their way through the ovary. In order to be recognized by a certain pollen tube as containing a possible partner for mating, the ovule must contain an egg cell or embryo sac of a special genotype. In most cases, the embryo sac and pollen tube must contain different Renner complexes.

The older observations on the behaviour of pollen tubes in the ovaries demonstrated that normally a pollen tube enters the first unfertilized ovule it encounters on its way, but in the case of selective fertilization a large number of pollen tubes passes several ovules before entering a micropyle.

In experiments with pollen and ovules on a gelatine-coated slide the pollen grains germinate and the pollen tubes grow in the direction of the ovules. However, when pollen and ovules, which do not react to one another in an ovary, are combined, the frequency of pollen tubes growing towards the ovules decreases and cannot be distinguished from random orientation (Schneider 1956). The time needed by genetically different pollentubes to react to ovules of certain genotypes can also be responsible for selective fertilization (Lechner 1963, 1964).

Some observations on the chemistry of attraction of pollen tubes by ovules are available but the problem is far from being solved. In fact, there are two different problems which are not always clearly distinguished between: there is the general attraction of pollen tubes by mature ovules, and the specific attraction depending on the genotype of the two partners involved in a particular case.

In extracts from mature ovules of several species of the sect. *Raimannia* amino acids, peptides, an indole compound and five sugars were found. Some of these were considered to participate in attracting pollen tubes in vitro, but the same result was achieved with KCl. In several experiments mixtures seemed to be more effective than pure substances. At the time the possibility was discussed that the species differed not in the substances present but in the relation in which they occurred, and that this was the basis for the specific reaction of the pollen tubes to ovules of their own species (Schildknecht and Benoni 1963). From a modern point of view the substances investigated can be found in every plant tissue and are not specific for ovules. It has not been proved that they are present in a secretion by the ovules as should be the case when they are involved in guiding pollen tubes through the ovary. Under no conditions can these substances explain the differential reaction of two genotypes of pollen tubes of one species to the genetic differences between embryo sacs of their own species.

In discussing the development of the pollen tubes the conclusion was drawn that the fertilization process is directed by pheromones on the basis of two sets of genes, one for the synthesis of the pheromone and another for the receptor. When genes exist, they can mutate, and when they can, they will do so at some time in evolution. The new alleles will change the quantity or quality of either the pheromones or the receptors, thereby influencing the attraction between embryo sac and pollen tube. If this is the case, then it is not difficult to assume that the chemical composition of the fluid appearing in the micropyle depends on the genotype of the cell that produces it and, on the other hand, the genotype of the reacting cell, the pollen tube, decides whether or not the secreted substances are recognized as acceptable pheromones.

The pheromone and receptor must cooperate when the pollen tube has to find its way to the egg cell. Now we only have to assume that the receptor is specific and cannot, in some cases, react to a pheromone made by a cell of the same genotype, then we have the mechanism for selective fertilization. The system enables fertilization and, at the same time, works as an inhibitor of fertilization of isogenic gametes, making the best possible use of all gametes, avoiding spoilage of gametes and also loss of lethal zygotes.

The terms "selective fertilization" and "affinity" are derived from the result. By changing the point of view and looking for what is avoided, the same process can be interpreted as a genetically determined incompatibility reaction between the embryo sac and pollen tube. As two gametophytes are involved, we can describe the phenomenon as **gametophytic–gametophytic incompatibility**. The term "affinity" becomes superfluous and "selective fertilization" is just a formal description of the result which does not explain anything.

Here it becomes clear why we introduced the term "gametophytic–sporophytic incompatibility". To improve the nomenclature of incompatibility systems, the former term "sporophytic incompatibility" should be renamed as "sporophytic–sporophytic incompatibility".

Now we know that in *Oenothera* gene-dependent pheromones and receptors are present, as are the tissues and developmental stage in which they are active, and that the genetic differences can be used, it should not be too difficult for an experienced molecular geneticist to identify them. Or will *Oenothera* act like a traffic sign marked *Deviation*, sending them off in another direction?

6.6 The Developmental Model of *Oenothera*

Now that the genetic model of behaviour of *Oenothera* is known, the developmental model, based on the combined application of genetic and developmental investigations, can be described.

The Renner complexes differ in their alleles at many loci. Some of the loci influence the phenotypic characters of the diploid sporophyte, governing morphogenesis and growth from embryogenesis onwards. To this group belong the recessive zygotic lethals. As balanced lethals in the two Renner complexes of each species they are responsible for elimination of the homozygotes, making the species a constant heterozygote.

Many genes are responsible for flower development and fertility. Only a few genes are known to influence biochemical characters like synthesis of pigments in flowers and vegetative parts. Recently, physiological investigations on incompatibility reactions and on chloroplast development and function, mainly based on molecular biology of the interesting genome–plastome interaction, have become important new fields.

Overall in the development of many characters of the sporophyte a combination of direct influence of gene expression with processes of induction or repression of a developmental step by adjacent tissues is observed. And the plastome plays a decisive role too, probably for more characters then are yet known.

Of special interest are those genes which regulate the development and behaviour of the gametophytes in both sexes. In no other genus have similar genes been analyzed to such an extent. Moreover, the interaction of the genome and plastome is important for gametophyte development. Acting at the sporophytic level, but influencing the development of the female gametophyte, are those genes which regulate the development of polarity in the ovule. Other genes, active only in the haploid phase, influence the competitive ability of megaspores and young female gametophytes. In the mature embryo sac the secretion of pheromones forms a basis not only for attraction of pollen tubes but also for a gametophytic–gametophytic incompatibility system. To explain the reactions observed in the crossing experiments there must be two loci or gene systems, one for synthesis and secretion of the pheromone and another for the receptor on the pollen tube.

On the male side there exist loci for pollen inactivation, pollen lethality, male sterility, gametophytic–sporophytic incompatibility, growth rate of the pollen tubes and chemotropic reactions of the pollen tubes in response to general signals of the stigma and style and of mature embryo sacs, and also to others indicating the genotype of the egg cell.

Did some people really assume that structures like embryo sacs and pollen tubes, the reduced but, nevertheless, most important gametophytic generation of seed plants, could live and work without genes? *Oenothera* was the first plant to provide the answer that this part of the life cycle has its own genetic system, consisting of two groups of genes. Some are important for general cell metabolism and others are active for only a short time between meiosis and fertilization and are inactivated in all other cells during the life of the sporophyte. *Oenothera* demonstrates that this system can be analyzed too. Only very recently gene expression in pollen has been recognized as a general problem (Brown and Crouch 1990).

In both sexes cooperation and coordination of developmental processes in the sporophyte and gametophyte are observed. The two generations are separated morphologically and by ploidy of the nuclei, but are linked together by their mutual interdependence, not only as the alternating stages of the life cycle but also by metabolism and determination of development. *Oenothera* demonstrated the importance of cooperation between the generations of different ploidy for maintaining the propagation on a level that enables survival of the species and also for its evolution.

7 Propagation Systems and Evolution in *Oenothera*

In *Oenothera* several different systems exist which influence on a genetic basis developmental processes in such a way that heterozygosity and the profits of heterosis are maintained with great efficiency. These issues and the problems related to them are discussed by Cleland (1972) and Stubbe (1980).

The Renner complexes, groups of genes in nearly absolute linkage, formed by reciprocal translocations between chromosomes, are the basis for the discussion. These unfavourable conditions, limiting the genetic recombination of recessive, favourable alleles and the elimination of unfavourable alleles, are turned into a profit by mechanisms perpetuating heterozygosity and heterosis.

Balanced lethals are the simplest form, maintaining heterozygosity by elimination of the lethal homozygotes. This is compensated by excessive seed production. To avoid the drawback of loss of embryos, pollen lethality and pollen inactivation, both causing an otherwise senseless loss of 50% of the microspores, are introduced. The loss of microspores is also compensated for by the large number of pollen grains which could not be used in fertilization even if they were all fertile. These mechanisms alone will not reduce the frequency of lethal embryos, but these will now all have the same genotype. Competition phenomena in pollen tubes and embryo sacs, which both as isolated events are again unfavourable for certain genotypes, increase the efficiency of the propagation process by limiting the probability for meeting of gametes of identical genotypes, thereby avoiding the production of lethal embryos. Genetic incompatibility systems, acting on both the sporophytic and gametophytic level, contribute to the same goal of providing on a heterozygote genetically homogeneous populations of active gametes in each sex, complementary to one another. Genome–plastome incompatibility can contribute to the elimination of certain Renner complexes from one sex as a very safe mechanism.

In some of the North American species of the *biennis* group there is a balanced system of an *alpha* complex with an S_I-allele and a pollen lethal as a safeguard to eliminate it from the male gametes, and megaspore competition with preference of the *alpha* complex. This is combined with a *beta* complex with an S_f allele and a recessive zygotic lethal which will eliminate any homozygotic *beta* embryos (Steiner 1964). It is a rather elaborate but very efficient system for maintaining a heterozygotic species.

The same mechanisms which ensure the continuity of the species are active in the evolution of the genus. By high mutability of the chromosomes new chromosome complexes are created. In combination with all the other peculiarities of the genus they can be constituents of new species, and this possibility opens the way for unlimited evolution.

8 Summary

Oenothera teaches one lesson: Why make things simple when you, as a genus, have the possibility to demonstrate what you can by making things complicated?

The whole scenario looks like a major quiz problem: How can you combine the contents of a big bag full of unfavourable characters to achieve at low cost a high probability of survival for the species and, at the same time, keep the way open for evolution? *Oenothera* has solved this problem. The genus should be awarded a large prize for it. But, instead, it is given a bad name as a collector of exceptions with the sole purpose of creating difficulties for biologists.

9 Postscript: The End Is a Start

Letter to a Young Biologist

Dear young friend (f/m),

by now you have read a lot about *Oenothera* and I hope you have become interested. I started with the beginning of *Oenothera* research and intended to go as often as possible to the point where we are now and to indicate the open questions. I tried to elucidate the unlimited phantasy of this genus in inventing new problems and to demonstrate what *Oenothera* can.

As an investigator you can leave such a crazy genus to its own fate, or you can see the challenge, take your chance, work as a detective in a crime story to clarify the situation and demonstrate what *you* can. Surely *Oenothera* has a surprise in store for you too. Why not join the club? You can now see where the competition is open. You know where the stadium is, you can look for the good trainers,

but the running is something you must do yourself.

START!

My best wishes for your success!
Sincerely yours

Cornelia Harte

References

Alpino P (1627) De plantis exoticis. Venetii

Andrews FM (1910) Twin hybrids (laeta and velutina) and their anatomical distinctions. Bot Gaz 50:193–201

Arnold CG (1955) Genetisch-zytologische Untersuchungen an trisomen Mutanten der *Oenothera Berteriana*. Z Indukt Abstammungs Vererbungsl 86:622–662

Arnold CG (1958a) Selektive Befruchtung. Ergeb Biol 20:67–95

Arnold CG (1958b) Untersuchungen zur Pollenfertilität trisomer Oenotheren. Z Vererbungsl 89:161–169

Arnold CG (1959) Die Blütenöffnung bei *Oenothera* in Abhängigkeit vom Licht-Dunkelrhythmus. Planta 53:198–211

Arnold CG (1963a) Versuch einer Homologisierung der Chromosomenenden von Euoenotheren mit denen von Raimannia- und Renneria-Oenotheren. Chromosoma 14:31–44

Arnold CG (1963b) Untersuchungen an Röntgenmutationen der *Oenothera Berteriana*. Biol Zentralbl 82:185–207

Arnold CG (1967a) Genominduzierte Mutationen. Biol Zentralbl Festschr 65.Geb H. Stubbe 86(Suppl):127–146

Arnold CG (1967b) Zur Plasmavererbung bei *Oenothera*. Ber Dtsch. Bot Ges 80:124–127

Arnold CG, Bina H (1957) Über 15-chromosomige Mutanten der *Oenothera odorata*. Flora 144:537–561

Arnold CG, Fellenberg G (1955) Untersuchungen über Incompatibilität bei *Oenothera*-Kreuzungen. Z Bot 52:532–538

Arnold CG, Kressel M (1965a) Über das gehäufte Auftreten trisomer Pflanzen in der Nachkommenschaft diploider Röntgenmutanten bei *Oenothera Berteriana*. Z Vererbungsl 96:83–92

Arnold CG, Kressel M (1965b) Erhöhung der Translokationsrate in trisomen Mutanten der *Oenothera Berteriana*. Z Vererbungsl 97:255–271

Arnold CG, Kressel M (1965c) Versuche zur Auslösung von Plasmamutationen bei *Oenothera*. Z Vererbungsl 96:213–216

Arnold CG, Kressel M, Fellenberg G (1963) Über Translokationen in trisomen Mutanten der *Oe. Berteriana*. Chromosoma 14:541–548

Baerecke ML (1944) Zur Genetik und Cytologie von *Oe. ammophila* Focke, *Bauri* Boedijn, *Baeckeri* Renner, *parviflora* L., *rubricaulis* Klebahn, *silesiaca* Renner. Flora 38:57–92

Bali NP, Hecht A (1965) The genetics of self-incompatibility in *Oe. rhombipetala*. Genetica 36:159–171

Basler A (1973) Das falcifolia-Syndrom der Oenotheren. 2. Mitteilung: Untersuchungen zur Entwicklungsgeschichte. Ber Dtsch Bot Ges 86:525–528

Baumgarten G, Harte C (1964) Beiträge zur Entwicklungsphysiologie der Blätter von *Oenothera Hookeri* durch vergleichende Untersuchungen der Normalform und einiger Blattmutanten. Z Bot 52:222–244

Baumgarten G, Harte C (1965a) Quantitative Untersuchungen zur Genwirkung bei der Blattentwicklung von *Oenothera Hookeri* de Vries und einigen Mutanten. Genetica 36:11–53

Baumgarten G, Harte C (1965b) Untersuchung der Morphogenese der Blätter an *Oenothera Hookeri* de Vries und einigen ihrer Mutanten. Genetica 36:286–300

Beer R (1906) On the development of pollen grain and anther of some Onagraceae. Bot Centralbl 19 A:286–313

Binder M (1938) Genetische und zytologische Untersuchungen an Euoenotheren, Teil IV. Die Eliminierung der l·l-Embryonen mit Plastiden der *Oenothera odorata*. Z Indukt Abstammungs Vererbungsl 75:739–800

Blakeslee AF, Cleland RE (1930) Circle formation in *Datura* and *Oenothera*. Natl Acad Sci 46:177–189

Boedijn KB (1924a) Die typische und heterotypische Kernteilung der Oenotheren. Z Zellen Gewebel, Abt B Z Wiss Biol 1:267–277

Boedijn KB (1924b) Die Gigas- und Deuterogigas-Formen der Oenotheren. Biol Zentralbl 44:127–137

Boedijn KB (1925a) Mehrfache Chromosomenverdoppelungen bei *Oenothera Lamarckiana*. Z Bot 18:161–171

Boedijn KB (1925b) Der Zusammenhang zwischen den Chromosomen und Mutationen bei *Oenothera Lamarckiana*. Recl Trav Bot Neerl 22:173–259

Boedijn KB (1928) Chromosomen und Pollen der Oenotheren. Recl Trav Bot Neerl 25a:25–35

Brennecke A, Schwemmle B (1984) Inheritance of mitochondrial DNA in *Oenothera Berteriana* and *Oenothera odorata* hybrids. Z Naturforsch 39c:191–192

Brittingham WH (1931a) Variations in the evening primrose induced by radium. Science 74:463–464

Brittingham WH (1931b) *Oenothera Lamarckiana* mut. *acutifolia*, a new mutant type produced by a gene outside the first linkage group. Am Nat 65:121–133

Broekels DJ (1924) Über den Stammbaum der Onagraceae. Recl Trav Bot Neerl 21:383–510

Brown SM, Crouch ML (1990) Characterization of a gene family abundantly expressed in *Oenothera organensis* pollen that shows sequence similarity to polygalacturonase. Plant Cell 2:263–274

Buchner E (1956) Genetische und cytologische Untersuchungen an trisomen Mutanten der *Oe. odorata* I. Flora 143:543–574

Capinpin JM (1930) Meiotic behaviour of triploid Oenotheras. Am Nat 64:566–570

Capinpin JM (1933) Studies on the genetics and cytology of triploid Oenotheras. Cytologia 4:355–426

Carrillo N, Seyer P, Tyagi A, Herrmann RG (1986) Cytochrome b-559 genes from *Oenothera hookeri* and *Nicotiana tabacum* show a remarkably high degree of conservation as compared to spinach. Curr Genet 10:619–624

Catcheside DG (1930) Chromosome linkage and syndesis in *Oenothera*. Trans R Soc 56, II:467–484

Catcheside DG (1931a) Critical evidence of parasynapsis in *Oenothera*. Proc R Soc Abt B 109:165–184

Catcheside DG (1931b) Meiosis in a triploid *Oenothera*. J Genet 24:145–163

Catcheside DG (1932) The chromosomes of a new haploid *Oenothera*. Cytologia 4:68–113

Catcheside DG (1933a) Chromosome configurations in trisomic Oenotheras. Genetica 15, 1933:177–201

Catcheside DG (1933b) Chromosome catenation in some F1 *Oenothera* hybrids. J Genet 27:45–69

Catcheside DG (1935) X-ray treatment of *Oenothera* chromosomes. Genetica 17: 313–341

Catcheside DG (1936) Origin, nature and breeding behaviour of *Oenothera Lamarckiana* trisomics. J Genet 33:1–23

Catcheside DG (1937a) Recessive mutations induced in *Oenothera blandina* by X-rays. Genetica 19:134–172

Catcheside DG (1937b) The extra chromosome of *Oenothera Lamarckiana lata*. Genetics 22:564–574

Catcheside DG (1939) The position effect in *Oenothera*. J Genet 38:345–352
Catcheside DG (1940) Structural analysis of *Oenothera* complexes. Proc R Soc Lond Ser B 128:509–535
Catcheside DG (1947a) A duplication and a deficiency in *Oenothera*. J Genet 48:99–110
Catcheside DG (1947b) The P-locus position effect in *Oenothera*. J Genet 48:31–42
Catcheside DG (1954) The genetics of brevistylis in *Oenothera*. Heredity 8:125–137
Charlesworth B (1979) Selection for gamete lethals and S-alleles in complex heterozygotes. Heredity 43:159–164
Chiu W, Stubbe W, Sears BB (1988) Plastid inheritance in *Oenothera*: organelle genome modifies the extent of biparental plastid transmission. Curr Genet 13:181–189
Chrometzka P (1955) Zur Kenntnis der Morphologie und des Wuchsstoffverhaltens der *Oenothera hybrida* mut. *helix*. Z. Indukt Abstammungs Vererbungsl 87: 267–297
Chrometzka P (1956) Zur Kenntnis der Lage des Gens hel der *Oenothera hybrida* mut. *helix*. Planta 46:643–648
Cleland RE (1922) The reduction divisions in the pollen mother cells of *Oenothera franciscana*. Am J Bot 9:391–413
Cleland RE (1925) Chromosome behaviour during meiosis in pollen mother cells of some Oenotheras. Am Nat LIX:475–479
Cleland RE (1926) Meiosi in the pollen mother cells of *Oenothera biennis* and *Oenothera biennis sulfurea*. Genetics 11:127–162
Cleland RE (1929) Chromosome behaviour in the pollen mother cells of several strains of *Oenothera Lamarckiana*. Z Indukt Abstammungs Vererbungsl 51: 126–145
Cleland RE (1931) The probable origin of *Oenothera rubricalyx* "Afterglow" on the basis of the segmental interchange theory. Proc Nat Acad Sci 17:437–440
Cleland RE (1932) Further data bearing upon circle formation in *Oenothera*, its cross and its genetical effect. Genetics 17:572–602
Cleland RE (1933) Predictions as to chromosome configurations, as evidence for segmental interchange in *Oenothera*. Am Nat 67:407–418
Cleland RE (1942) The origin of haplo-decipiens from the complexes of *Oenothera Lamarckiana* and its bearing upon the phylogenetic significances of similarities in segmental arrangements. Genetics 27:55–83
Cleland RE (1944) The problem of species in *Oenothera*. Am Nat 78:5–28
Cleland RE (1949) Phylogenetic relationships in *Oenothera*. Proc 8th Int Congr genetics, Hereditas (Suppl) 1949, pp 173–188
Cleland RE (1981) Extra diminutive chromosomes in *Oenothera*. Evolution 5: 165–176
Cleland RE (1957) Chromosome structure in *Oenothera* and its effect on the evolution of the genus. Cytologia, Proc Int Genet Symp 1956, pp 5–19
Cleland RE (1960) The S-factor situation in a small sample of an *Oenothera* (*Raimannia*) heterophylla population. Z Vererbungsl 91:303–311
Cleland RE (1962a) Plastid behaviour in North American Eu-oenotheras. Planta 57:699–712
Cleland RE (1962b) The cytogenetics of *Oenothera*. Adv Genet 11:147–237
Cleland RE (1967) The origin of closed circles of five chromosomes in *Oenothera*. Am J Bot 54:993–997
Cleland RE (1970) The missing petal character in *Oenothera* and its relation to the cruciate character. Am J Bot 57:850–855
Cleland RE (1972) Oenothera, cytogenetics and evolution. Academic Press, London
Cleland RE, Blakeslee AF (1930) Interaction between complexes as evidence for segmental interchange in *Oenothera*. Proc Natl Acad Sci 16:177–189
Cleland RE, Blakeslee AF (1931) Segmental interchange, the basis of chromosomal attachments in *Oenothera*. Cytologia 2:175–233

Cleland RE, Brittingham WH (1943) A contribution to an understanding of crossing over within chromosome rings of *Oenothera*. Genetics 19:62–72

Cleland RE, Hyde BB (1963) Evidence of relationship between extra diminutive chromosomes in geographically remote races of *Oenothera*. Am J Bot 50: 179–185

Cleland RE, Oehlkers F (1929) New evidence bearing upon the problem of the cytological basis for genetical peculiarities in the Oenotheras. Am Nat 63: 497–510

Cleland RE, Oehlkers F (1930) Erblichkeit und Zytologie verschiedener Oenotheren und ihrer Kreuzungen. Jahrb Wiss Bot 73:1–124

Crowe LL (1955) The evolution of incompatibility in species of *Oenothera*. Heredity 9:293–322

Darlington CD (1929) Ring formation on *Oenothera* and other genera. J Genet 20:345–363

Darlington CD (1931) The cytological theory of inheritance in *Oenothera*. J Genet 24:405–474

Darlington CD (1933) Behaviour of interchange heterozygotes in *Oenothera*. Proc Nat Acad Sci 19:101–103

Darlington CD (1936) The limitation of crossing-over in *Oenothera*. J Genet 32: 343–352

Darlington CD, La Cour LF (1962) The handling of chromosomes, 4th edn. George Allen and Unwin, London

Dart RA (1966) A tribute to Reginald Ruggles Gates, 1882–1962. Mankind Q 7:5pp

Davis BM (1909) Cytological studies on *Oenothera*. I. Pollen development of *Oenothera grandiflora*. Ann Bot 23:551–571

Davis BM (1911) Genetical studies on Oenothera. II. Some hybrids of *Oenothera biennis* and *O. grandiflora* that resemble *Oenothera Lamarckiana*. Am Nat 45:193–233

Davis BM (1915) Professor de Vries, on the probable origin of *Oenothera Lamarkiana*. Am Nat 49:59–64

Davis BM (1918) The segregation of brevistylis from crosses with *Oe. Lamarckiana*. Genetics 3:501–533

Davis BM (1921) Malnutrition as a cause of irregularities in the segregation of *Oe. brevistylis* from crosses with *Oe. Lamarckiana*. Genetics 3:574–586

Davis BM (1922) Species, pure and impure. Science 55:108–114

Davis BM (1926a) The history of *Oenothera biennis* Linnaeus, *Oenothera grandiflora* Solander and *Oenothera Lamarckiana* de Vries in England. Proc Am Philos Soc 65:349–378

Davis BM (1926b) The segregation of *Oenothera nanella – brevistylis* from crosses with *nanella* and with *Lamarckiana*. Genetics 11:57–72

Davis BM (1927) Lamarck's evening primrose (*Oenothera Lamarckiana* Seringe) was a form of *Oenothera grandiflora* Solander. Bicent No. Am Philos Soc Proc 66:319–355

Davis BM (1931) Some attempts to obtain haploids from *Oe. Lamarckiana*. Am Nat 65:233–242

Davis BM (1933) The genetics and cytology of a tetraploid from *Oenothera franciscana* Bartlett. Genetics 18:293–323

Davis BM (1934) What is *Oenothera Hookeri* Torrey and Gray? Proc. Am Philos Soc 74:245–251

Davis BM (1937) A segregation of sulfur and dwarf from crosses involving *Oe. franciscana* and certain hybrid derivatives. Proc Am Philos Soc 77:99–160

Davis BM (1940) *Oenothera cantabrigiana*. Genetics 25:433–437

Davis BM (1943) An amphidiploid in F_1-generation of the cross *Oe. franciscana* × *Oe. biennis* and its progeny. Genetics 28:275–285

Davis BM (1947) The appearance of a balanced lethal situation in the *Oenothera* heterozygote *gaudens*·^{h}*franciscana*, following severe heat treatment of seeds. Genetics 32:185–199

Davis BM, Kulkarny ChG (1930) The cytology and genetics of a haploid sport from *Oenothera franciscana*. Genetics 15:55–80

Davy de Virville A, Obaton F (1923) Etude biologique de l'epanouissement des fleurs. Rev Gen Bot 35:161–185

Dickinson HG, Lawson J (1975) Pollen tube growth in the stigma of *Oenothera organensis* following compatible and incompatible intraspecific pollinations. Proc R Soc Lond B 188:327–344

Diers L (1963a) Elektronenmikroskopische Beobachtungen an der generativen Zelle von *Oenothera Hookeri* Torr. et Gray. Z Naturforsch 7:562–566

Diers L (1963b) Elektronenmikroskopische Beobachtungen an der vegetativen Zelle im auskeimenden Pollenkorn von *Oenothera Hookeri* Torr. et Gray. Z Naturforsch 18b:1092–1097

Dietrich W (1977) The South American species of *Oenothera* sect. Oenothera (Raimannia, Renneria, Onagraceae). Ann M Bot Gard 64:425–626

Dietrich W (1984) *Oenothera*. In: Reitz R (ed) Flora ilustrada catarinense, I parte. Fasciculo ONAG, Itajai, Santa Catarina, Brasil, pp 9–33

Dietrich W (1988) *Oenothera*. In: Corrca MN (ed) Flora patagonica (Republica Argentina), vol VIII. Coleccion cientifica del INTA, p 267

Dietrich W, Raven PH (1976) An earlier name for *Oenothera strigosa* (Onagraceae). Ann M Bot Gard 63:382–383

Dietrich W, Wagner WL (1987a) New taxa of *Oenothera* L. sect. Oenothera (Onagraceae). Ann M Bot Gard 74:144–150

Dietrich W, Wagner WL (1987b) A new combination and new subspecies in *Oenothera elata* Kunth (Onagraceae). Ann M Bot Gard 74:151–152

Dietrich W, Wagner WL (1988) Systematics of *Oenothera* Section Oenothera, Subsection Raimannia and Subsection Nutantigemma (Onagraceae). Syst Bot Monogr 24:1–91

Dietrich W, Raven PH, Wagner WL (1985) Revision of *Oenothera* sect. Oenothera subsect. Emersonia (Onagraceae). Syst Bot 10:29–48

Dittmann C (1956) Genetische und zytologische Untersuchungen an trisomen Mutanten der *Oenothera odorata* II. Flora 143:575–602

Duchartre P (1842) Observation sur la fleur et plus particulierement sur l'ovaire de l'*Oenothera suaveolens* H.P. Ann Sci Nat Tome 18:339–356

Dulfer H (1926) Die Erblichkeitserscheinungen der *Oenothera Lamarckiana* semigigas. Rec Trav Bot Neerl 23:1–71

Eichler A (1878) Blütendiagramme II. Engelmann, Leipzig

Ellstrand NC, Levin DA (1980a) Association of alleles with chromosomal complexes in the permanent translocation heterozygote, *Oenothera laciniata*. Heredity 44: 169–498

Ellstrand NC, Levin DA (1980b) Recombination system and population structure in *Oenothera*. Evolution 34:923–933

Ellstrand NC, Levin DA (1980c) Evolution of *Oenothera laciniata* (Onagraceae), a permanent translocation heterozygote. Syst Bot 5:6–16

Ellstrand NC, Levin AD (1982) Genotypic diversity in *Oenothera laciniata* (Onagraceae), a permanent translocation heterozygote. Evolution 36:63–69

Emerson StH (1929a) Chromosome configuration in a dwarf segregate from *Oenothera* "franciscana sulfurea". Pap Mich Acad Sci, Arts Lett 9:117–120

Emerson StH (1929b) Multiple crosses inheritance between *Oenothera grandiflora* and *Oenothera franciscana*. Pap Mich Acad Sci Arts Lett 9:121–138

Emerson StH (1929c) The reduction division in a haploid *Oenothera*. Cellule 39: 160–164

Emerson StH (1930) The inheritance of rubricalyx bud color in crosses with *Oenothera Lamarckiana*. Proc Nat Acad Sci 16:796–800

Emerson StH (1931a) The inheritance of certain characters in *Oenothera* hybrids of different chromosome configurations. Genetics 16:325–348

Emerson StH (1931b) Parasynapsis and apparent chiasma formation in *Oenothera*. Am Nat 65:551–555

Emerson StH (1932) Chromosome rings in *Oenothera*, *Drosophila* and maize. Proc Nat Acad Sci 18:630–632
Emerson StH (1935) The genetic nature of de Vries's mutations in *Oenothera Lamarckiana*. Am Nat 69:545–559
Emerson StH (1936) The trisomic derivatives of *Oenothera Lamarckiana*. Genetics 21:200–224
Emerson StH (1938) The genetics of incompatibility in *Oenothera organensis*. Genetics 23:190–202
Emerson StH (1939) A preliminary survey of the *Oenothera organensis* population. Genetics 24:524–537
Emerson StH (1940) Growth of incompatible pollen tubes in *Oenothera organensis*. Bot Gaz 101:890–911
Emerson StH, Sturtevant AF (1931) Genetical and cytological studies on *Oenothera*. III. The translocation interpretation. Z Indukt Abstammungs Vererbungsl 59: 395–419
Emerson StH, Sturtevant AH (1932) The linkage relations of certain genes in *Oenothera*. Genetics 17:339–412
Epp MD (1973) Nuclear gene-induced plastome mutations in *Oenothera Hookeri* I. Genetic analysis. Genetics 75:465–483
Epp MD, Parthasarathy MV (1987) Nuclear gene-induced plastome mutations in *Oenothera Hookeri*. II. Phenotypic description with electron microscopy. Am J Bot 74:143–151
Epp MD, Erion JL, Sears BB (1987) The plastome mutator of *Oenothera* continues to function as originally described. Mol Gen Genet 206:515–518
Fischer HP (1963) Die Analyse der *Oenothera curvifolia*. Dissertation, Universität Erlangen-Nürnberg
Fork CD, Heber UW (1968) Studies on the electron-transport reactions of photosynthesis in plastome mutants of *Oenothera*. Plant Physiol 43:606–612
Gates RR (1907) Pollen development in hybrids of *Oenothera lata* × *O. lamarckiana* and its relation to mutation. Bot Gaz 43:81–115
Gates RR (1908) The chromosomes of *Oenothera*. Science New Ser 27:190–195
Gates RR (1912) Parallel mutations in *Oenothera biennis*. Nature 89:659–660
Gates RR (1923a) The trisomic mutations of *Oenothera*. Ann Bot 37:543–563
Gates RR (1923b) The chromosomes of a triploid *Oenothera* hybrid. Ann Bot 37:567–569
Gates RR (1928) The cytology of *Oenothera*. Bibliogr Genet 4:401–499
Gates RR (1929) A haploid *Oenothera*. Nature 124:948
Gates RR (1931) The cytological basis of mutations. Am Nat 65:97–120
Gates RR (1958) Taxonomy and genetics of *Oenothera*. Uitgeverij Junk, Den Haag
Gates RR, Catcheside DG (1931) Origin of chromosome linkage in *Oenothera*. Nature: 1p
Gates RR, Catcheside DG (1932) Gamolysis of various new Oenotheras. J Genet 26:143–173
Gates RR, Goodwin KM (1931) Meiosis in *Oenothera purpurata* and *Oenothera blandina*. Proc R Soc B 109:149–164
Geerts JM (1909) Beiträge zur Kenntnis der Cytologie und der partiellen Sterilität von *Oenothera lamarckiana*. Rec Trav Bot Neerl 5:93–208
Gerhard K (1929) Genetische und cytologische Untersuchungen an *Oenothera grandiflora* Ait. Jena Z Naturwiss 64 NF 57:283–338
Glenk HO (1964) Untersuchungen über die sexuelle Affinität bei Oenotheren. In: Linskens HF (ed) Pollen physiology and fertilization. North Holland, Amsterdam, pp 170–181
Göpel G (1970) Untersuchungen über Plastom-abhängige Pollensterilität bei *Oenothera*. TAG 40:111–116
Göpel G (1976) Stoffwechseländerungen in plastomabhängig inaktiviertem Pollen von *Oenothera*. Z Pflanzenphysiol 78:411–415

Goldschmidt RB (1958) Genic conversion in *Oenothera*? A critical review. Am Nat 92:93–104
Godron KJH, Hildebrandt JW, Bohnert HJ, Herrmann RG, Schmitt JM (1980) Analysis of the plastid DNA in an *Oenothera* plastome mutant deficient in ribulose bisphosphate carboxylase. Theor Appl Genet 56:203–297
Gordon KJH, Crouse EJ, Bohnert HJ, Herrmann RG (1981) Restriction endonuclease cleavage site map from the Euoenothera plastome IV. Theor Appl Genet 59:281–296
Gordon KJH, Crouse EJ, Bohnert HJ, Herrmann RG (1982) Physical mapping of differences on the DNA of the five Euoenothera wild plastomes. Theor Appl Genet 61:373–384
Gregory DP (1963) Hawkmoth pollination in the genus *Oenothera*. Allisu 5:357–384
Gutte P, Rostanski K (1971) Die Oenothera-Arten Sachsens. Ber Arbeitsgem Sächs Bot NF 9:63–88
Gutte P, Rostanski K (1981) Die Oenothera-Arten Sachsens (Nachtrag). Ber Arbeitsgem Sächs Bot NF 11:185–189
Haberlandt G (1921) Über experimentelle Erzeugung von Adventivembryonen bei *Oenothera Lamarckiana*. Sitzungsber Preuss Akad Wiss 40:695–725
Haberlandt G (1927) Zur Zytologie und Physiologie des weiblichen Gametophyten von *Oenothera*. Sitzungsber preuß Akad Wiss 7:33–47
Haberlandt G (1938) Über experimentelle Adventivembryonie. Sitzungsber preuß Akad Wiss, phys-math Klasse 24:1–8
Hadorn E (1955) Letalfaktoren in ihrer Bedeutung für Erbpathologie und Genphysiologie der Entwicklung. Georg Thieme, Stuttgart
Hakansson A (1926) Über das Verhalten der Chromosomen bei der heterotypischen Teilung schwedischer *Oenothera Lamarckiana* und einiger ihrer Mutanten und Bastarde. Hereditas 8:225–304
Hakansson A (1928) Die Reduktionsteilung in den Samenanlagen einiger Oenotheren. Hereditas 11:131–181
Hakansson A (1930a) Zur Zytologie trisomischer Mutanten der *Oenothera Lamarckiana*. Hereditas 14:1–32
Hakansson A (1930b) Die Chromosomenreduktion bei einigen Mutanten und Bastarden von *Oenothera Lamarckiana*. Jahrb Wiss Bot 72:385–402
Hakansson A (1931) Beobachtungen über Chromosomenverbindungen in einer triploiden *Oenothera*. Bot Not: 339–342
Hall IV, Steiner E, Threadgill P, Jones RW (1988) The biology of Canadian weeds. 84: *Oenothera biennis* L. Can J Plant Sci 68:163–173
Hallier UW (1966) Untersuchungen des Photosyntheseapparates bei Plastommutanten von *Oenothera*. Ber Dtsch Bot Ges 79:69–71
Hallier UW (1968) Photosynthese-Reaktionen einiger Plastom-Mutanten von *Oenothera* II. Die Bildung von ATP und NADPH. Z Pflanzenphysiol 58:289–299
Hallier UW, Heber UW, Stubbe W (1968) Photosynthese-Reaktionen einiger Plastom-Mutanten von *Oenothera* I. Der reduktive Pentosephosphatzyklus. Z Pflanzenphysiol 58:222–239
Hallier UW, Schmitt JM, Heber U, Chaianova SS, Volodarsky A (1978) Ribulose-1, 5-bisphosphate carboxylase-deficient plastome mutants of *Oenothera*. Biochim Biophys Acta 504:67–83
Hance RT (1919) Variations in the number of somatic chromosomes in *Oenothera scintillans* de Vries. Genetics 3:225–275
Harte C (1942) Meiosis und crossing-over. Weitere Beiträge zur Cytogenetik von *Oenothera*. Z Bot 38:65–137
Harte C (1943) Über die Aufhebung des Gonen- und Zygotenausfalls bei *Oenothera*, Teil 2. Untersuchungen an *Oenothera strigosa*. Flora NF 36:120–124
Harte C (1948) Cytologisch-genetische Untersuchungen an spaltenden Oenothera-Bastarden. Z Vererbungsl 82:495–640

Harte C (1949a) Differenzen in der Keimungsgeschwindigkeit zwischen genetisch verschiedenen Pflanzen in Bastardnachkommenschaften. Z Vererbungsl 83: 26–35

Harte C (1949b) Mutationsauslösung durch Ultrakurzwellen. Chromosoma 3: 440–447

Harte C (1953a) Entwicklungsphysiologie und Genetik der pflanzlichen Mißbildungen in den Arbeiten von Hugo de Vries. Naturwissenschaften 40:421–427

Harte C (1953b) Untersuchungen über Gonenkonkurrenz und crossing-over bei spaltenden Oenothera-Bastarden. Z Vererbungsl 85:97–117

Harte C (1953c) Untersuchungen über die Variabilität der Chiasmenbildung bei Oenothera-Bastarden I. Die Häufigkeit der Endchiasmen bei Formen mit einem Viererring. Chromosoma 6:91–114

Harte C (1954a) Untersuchungen über die Variabilität der Chiasmenbildung bei Oenothera-Bastarden. II. Die Verteilung der Endchiasmen auf Ring- und Bivalentchromosomen in Formen mit einem Viererring. Chromosoma 6:237–276

Harte C (1954b) Die Verteilung der Endchiasmen auf Ring- und Bivalentchromosomen bei *Oenothera albicans·stringens*. Chromosoma 6:301–313

Harte C (1958a) Untersuchungen über die Gonenkonkurrenz in der Samenanlage bei *Oenothera* unter Verwendung der Letalfaktoren als Markierungsgene. I. Die individuelle Komponente der Variabilität. Z Vererbungsl 89:473–496

Harte C (1958b) Untersuchungen über die Gonenkonkurrenz in der Samenanlage bei *Oenothera* unter Verwendung der Letalfaktoren als Markierungsgene. II. Die Umweltkomponente der Variabilität. Z Vererbungsl 89:497–507

Harte C (1958c) Untersuchungen über die Gonenkonkurrenz in der Samenanlage bei *Oenothera* unter Verwendung der Letalfaktoren als Markierungsgene. III. Die genetische Komponente der Variabilität. Z Vererbungsl 89:715–728

Harte C (1961) Untersuchungen über die Gonenkonkurrenz bei *Oenothera* unter Verwendung der Testloci *fr*, *s* und *de*. Z Vererbungsl 92:142–164

Harte C (1967) Gonenkonkurrenz. In: Ruhland W (ed) Handbuch der Pflanzenphysiologie, vol 18. Springer, Berlin Heidelberg New York, pp 447–478

Harte C (1969a) Gonenkonkurrenz bei *Oenothera* unter dem Einfluß eines gametophytisch wirksamen Gens in der ersten Koppelungsgruppe sowie ein Modell für die Untersuchung verzweigter Koppelungsgruppen. TAG 39:163–178

Harte C (1969b) Gonenkonkurrenz in der Samenanlage von *Oenothera*. TAG 39: 241–250

Harte C (1969c) Certationsversuch zur Gonenkonkurrenz unter dem Einfluß des Locus *ga* bei *Oenothera*. TAG 39:320–325

Harte C (1971) Die Formbildung der Rosettenblätter von *Oenothera hookeri* und einigen Mutanten. Z Pflanzenphysiol 64:309–327

Harte C (1972a) Auslösung von Chromosomenmutationen durch Meterwellen in Pollenmutterzellen von *Oenothera*. Chromosoma 36:329–337

Harte C (1972b) Abhängigkeit der Gonenkonkurrenz im Pollen vom Entwicklungszustand des Griffels bei *Oenothera*. TAG 42:215–217

Harte C (1972c) Auftreten von Mutationen bei *Oenothera hookeri* nach Behandlung des Pollens mit Meterwellen. TAG 42:297–305

Harte C (1973a) Cytologische Untersuchungen an zwei immerspaltenden Schmalblattmutanten von *Oenothera hookeri*. Chromosoma (Berl) 40:213–220

Harte C (1973b) Mutationen bei *Oenothera hookeri* nach Dauereinwirkung von Meterwellen während einer Vegetationsperiode. TAG 43:6–12

Harte C (1973c) Genetische Mosaike in der M_2 bei *Oenothera hookeri* nach Behandlung mit Meterwellen. TAG 43:54–58

Harte C (1973d) Haploide Pflanzen bei *Oenothera*. Biol Zentralbl 92:361–364

Harte C (1975) Competition in the haploid generation in *Oenothera*. In: Mulcahy DL (ed) Gamete competition in plants and animals. North Holland, Amsterdam, pp 31–41

Harte C (1983) Phänogenetik der Blattform bei *Antirrhinum majus* L. VI. Die Endgröße der Blätter von *A. majus*, Sippe 50 und einigen Mutanten in Beziehung zur Blühinduktion. Biol Zentralbl 102:299–321
Harte C (1984) Genetic control of the development of the haploid generation. Acta Soc Bot Pol 53:279–295
Harte C (1992) Developmental genetics of sexual propagation in *Oenothera*. In: Ottaviano E, Mulcahy DL, Sari Gorla M, Bergamini-Mulcahy G (eds) Angiosperm, pollen and ovules. Springer, Berlin Heidelberg New York
Harte C, Bissinger B (1952) Entwicklungsgeschichtliche Untersuchungen der durch die Faktoren *fr* und *ster* bedingten Pollensterilität bei *Oenothera*. Z Vererbungsl 84:251–269
Harte C, Sniezko R (1988) Polyembryony in *Oenothera*. In: Cresti M, Gori P, Pacini E (eds) Sexual reproduction in higher plants. Springer, Berlin Heidelberg New York, pp 413–418
Harwood J (1987) The controversy over cytoplasmic inheritance in interwar Germany. Ber Dtsch Bot Ges 100:59–67
Haselwarter A (1937) Untersuchung zur Physiologie der Meiosis V. Z Bot 31: 273–328
Haustein E (1938) Genetische und zytologische Untersuchungen an Euoenotheren, Teil IV. Plastidenvererbung bei *Oenothera Berteriana* und *Oenothera odorata*. Z Indukt Abstammungs Vererbungsl 75:661–689
Haustein E (1939) Die Chromosomenanordnung bei einigen Bastarden der *Oenothera argentinea*. Z Indukt Abstammungs Vererbungsl 76:411–421
Haustein E (1951) Die Erblichkeit des cruciata – Merkmals bei den Oenotheren. Biol Zentralbl 70:481–516
Haustein E (1952) Die Endenbezifferung der Chromosomen einiger Oenotheren aus dem Subgenus Raimannia. Z Indukt Abstammungs Vererbungsl 84:417–453
Haustein E (1956) Zur Zytologie einiger aus der *Oenothera odorata* abgeleiteter Trisomer. Flora 143:385–394
Haustein E (1961) Eine androgene haploide *Oenothera scabra*. Planta 56:275–278
Haustein E (1967) Selektive Befruchtung. In: Ruhland W (ed) Handbuch der Pflanzenphysiologie, vol 18. Springer, Berlin Heidelberg New York, pp 479–505
Hecht A (1942) Colchicine-induced tetraploidy in *Oenothera*. Proc Indiana Acad Sci 51:87–93
Hecht A (1960) Growth of pollen tubes of *Oenothera organensis* through otherwise incompatible styles. Am J Bot 47:32–36
Hecht A (1964) Partial inactivation of an incompatibility substance in the stigmas and styles of *Oenothera*. In: Linskens HF (ed) Pollen physiology and fertilization. North Holland, Amsterdam, pp 237–243
Hecht A, Tandon SL (1953) Chromosomal interchanges as a basis for the delimitation of species in *Oenothera*. Science 118:557–558
Hedayitullah S (1933) Meiosis in *Oenothera missouriensis*. Proc Soc B 113:57–70
Hegi G (1935) Illustrierte Flora von Mitteleuropa, vol V-2. Paul Parey, Berlin, pp 857–873
Heribert-Nilsson N (1920) Zuwachsgeschwindigkeit der Pollenschläuche und gestörte Mendelzahlen bei *Oenothera Lamarckiana*. Hereditas 1:41–67
Heriber-Nilsson N (1923) Zertationsversuche mit Durchtrennung des Griffels bei *Oenothera Lamarckiana*. Hereditas 4:177–190
Heribert-Nilsson N (1939) Eine diploide Form aus der tetraploiden *Oenothera gigantea*. Hereditas 25:1–8
Herrmann RG, Possingham JV (1980) Plastid DNA – the plastome. In: Reinert J (ed) Results and problems in cell differentiation, vol X. Springer, Berlin Heidelberg New York, pp 45–96
Herrmann RG, Seyer P, Schedel R, Gordon K, Bisanz C, Winter P, Hildebrandt JW, Wlaschek M, Alt J, Sears BB (1980) The plastid chromosomes of several dicotyledons. In: Bücher Th, Sebald W, Weiß H (eds) Biological chemistry of

organelle formation. 31. Colloq Gesellschaft Biol Chemie. Springer, Berlin Heidelberg New York, pp 97–112

Herrmann RG, Bisanz C, Gordon K, Hildebrandt JW (1981) Studies on interspecific genome/plastome hybrids of *Oenothera*. In: Schwemmler W, Schenk HEA (eds) Endocytobiology, endosymbiosis and cell biology, vol 1. De Gruyter, Berlin, pp 671–672

Herzog G (1940) Genetische und zytologische Untersuchungen über 15-chromosomige Mutanten von *Oenothera biennis* und *Oenothera Lamarckiana*. Flora NF 34:379–432

Heslop-Harrison J, Heslop-Harrison Y (1985) The secretory system of the stigma of *Oenothera organensis* Munz: some developmental and quantitative features. Isr J Bot 34:187–204

Heslop-Harrison Y (1990) Stigma form and surface in relation to self-incompatibility in the Onagraceae. Nord J Bot 10:1–19

Hesse M (1983) Dissimilar pollen tetrad development in Ericaceae and Onagraceae causes family-specific viscin thread configurations. Plant Syst Evol 143:163–165

Hiesel R, Wissinger B, Schuster W, Brennicke A (1990) RNA editing in plant mitochondria. Science 246:1632–1634

Hildebrandt J, Alt J, Bottomley W, Bisanz C, Herrmann RG (1984) A plastome mutant of *Oenothera hookeri* has a lesion in the gene for ribulose bisphosphate carboxylase/oxygenase. Biochim Biophys Acta 783:67–73

Hiorth G (1926) Zur Kenntnis der Homozygoten-Eliminierung und der Pollenschlauchkonkurrenz bei *Oenothera*. Z Indukt Abstammungs Vererbungsl 43:171–237

Hoeppener E, Renner O (1928) Genetisch-zytologische Oenotherenstudien. I. Zur Kenntnis der *Oenothera ammophila* Focke. Z Indukt Abstammungs Vererbungsl 49:1–25

Hoeppener E, Renner O (1929) Genetische und zytologische Oenotherenstudien II. Zur Kenntnis von *Oe. rubrinervis*, *deserens*, *Lamarckiana-gigas*, *biennis-gigas*, *franciscana*, *hookeri*, *suaveolens*, *lutescens*. Bot Abh 15:1–86

Hoffmann A (1954) Untersuchungen über interstitielle Chiasmen bei *Oenothera*. Chromosoma 6:277–300

Illick JTh (1929) The cytological study of meiosis in the pollen mother cells of some Oenotheras. Genetics 14:591–633

Illick JTh (1932) Significance of chromosome behaviour during diakinesis in *Oenothera*. Bot Gaz 93:313–327

Ishikawa M (1918) Studies on the embryo sac and fertilization in *Oenothera*. Ann Bot 32:279–317

Jalouzot MF (1971) Aspects ultrastructuraux de la megasporogenese d'*Oenothera Lamarckiana* en rapport avec les depots callosiques observés. Ann Univ ARERS 9:36–45

Jalouzot MF (1973) Quelques aspects ultrastructuraux du noyau du megasporocyte d'*Oenothera Lamarckiana* en prophase heterotypique. Caryologia 25:67–84

Jalouzot MF, Buvat MR (1970) Cytologie vegetale – Mise en evidence de parois callosiques au cours de la megasporogenese et de l'oogenese d'*Oenothera biennis*. CR Acad Sci Paris 270:317–319

Jalouzot MF, Buvat MR (1975) Cytologie vegetale – Aspects ultrastructuraux du sac embryonaire d'*Oenothera Lamarckiana*. CR Acad Sci Paris 281:1305–1308

Japha B (1939) Die Meiosis von *Oenothera* II. Z Bot 34:321–369

Jean R (1974) La letalité pollinique dans le systeme genetique du sous-genre Eu-Oenothera. These, Universite des Sciences et Techniques de Lille

Jean R (1985) *Oenothera*, Clefs pour quelques genres de la flore francaise. Bull Soc Bot Centre-Ouest 16:168–172

Jean R (1989) Variabilité de la production d'huile et d'acide gamma-linolénique dans les graines des espece du genre *Oenothera*, section Oenothera, sous-section Euoenothera. Actes Congr Int "Chevreuil" pour l'etude des corps gras, Angers. Tome 1, pp 91–98

Jean R, Linder R, Hoffmann F, Thomas E, Wenzel G (1976) Transmise par Maurice Durchon: Comportement in vitro de microspores d'*Oenothera coronifera* Renner. CR Acad Sci Paris 283:781–784

Jehlik V, Rostanski K (1979) Beitrag zur Taxonomie, Ökologie und Chorologie der *Oenothera*-Arten in der Tschechoslowakei. Folia Geobot Phytotaxon 14:377–429

Jehlik V, Rostanski K (1980) Rozsireni *Oenothera biennis* L. v Ceskoslovensku a klic k urcovani nasich pupalek. Die Verbreitung von *Oenothera biennis* L. in der Tschechoslowakei und Bestimmungsschlüssel unserer Nachtkerzen. Cas Nar Muz 148:200–210

Johansen DA (1928) The hypostase: its presence and function in the ovule of the Onagraceae. Proc Natl Acad Sci 14:710–713

Johnson EM, Sears BB (1990) Structure and expression of cytochrome f in a Oenothera plastome mutant. Curr Genet 17:529–534

Kaienburg A (1950) Zur Kenntnis der Pollenplastiden und der Pollenschlauchleitung bei einigen Oenotheraceen. Planta 38:377–430

Kandeler O, Schötz F (1956) Untersuchungen über die photooxydative Farbstoffzerstörung und Stoffwechselhemmung bei Chlorella-Mutanten und panaschierten Oenotheren. Z Naturforsch 11b:708–718

Kappus A (1957) Wilde Oenotheren in Südwestdeutschland. Z Indukt Abstammungs Vererbungsl 88:38–55

Kappus A (1958) Untersuchung über die Verbreitung des sulfurea-Merkmals in Populationen von Wildoenotheren. Z Vererbungsl 89:647–650

Kappus A (1960) *Oenothera chicaginensis*, eine neue Adventivpflanze in Freiburg i. Breisgau. Mitt Bad Landesver Naturkd Naturschutz, NF 7:487–491

Kappus A (1966) *Oenothera Oehlkersi*, eine neue Wildart am Oberrhein. Z Vererbungsl 97:370–374

Kaul CL (1988) Male sterility in higher plants. Springer, Berlin Heidelberg New York

Khafaji SD al, Steiner E (1970) Further analysis of *Oenothera biennis* populations for incompatibility alleles. Am J Bot 57:183–189

Kihara H (1927) Über das Verhalten der "end-to-end" gebundene Chromosomen von *Rumex acetosella* und *Oenothera biennis* während der heterotypischen Kernteilung. Beitrag zur Frage der Para- und Metasyndese. Jahrb Wiss Bot 66: 429–460

Kisch R (1937) Die Bedeutung der Wasserversorgung für den Ablauf der Meiosis (Untersuchungen zur Physiologie der Meiosis VI). Jahrb Wiss Bot 85:450–484

Kistner G (1955) Über plastidenbedingtes Absterben während der Embryoentwicklung einiger Oenotheren-bastarde. Z Indukt Abstammungs Vererbungsl 86: 521–544

Kowalewicz R (1956a) Entwicklungsgeschichtliche Studien an normalen und cruciaten Blüten von *Epilobium* und *Oenothera*. Planta 46:569–603

Kowalewicz R (1956b) Zur Kenntnis von *Epilobium* und *Oenothera*. I. Über die Rhaphidenschläuche. II. Über intergenerische Transplantation. Planta 47: 501–509

Krautblatter H, Arnold CG (1967) Die unterschiedliche Strahlenempfindlichkeit der Genome von *Oenothera Berteriana*. Radiat Bot 7:283–287

Kressel M, Arnold CG (1967a) Die unterschiedliche Empfindlichkeit der Genome von *Oenothera Berteriana* gegenüber Äthylmethansulfonat. Mutat Res 4: 797–804

Kressel M, Arnold CG (1967b) Die mutagene Wirkung von Äthylmethansulfonat bei der komplexheterozygoten *Oenothera Berteriana*. Züchter 37:364–371

Krumbholz G (1925) Untersuchung über die Scheckung der Oenotherenbastarde, insbesondere über die Möglichkeit der Entstehung von Periklinalchimären. Jena Z Naturwiss 62, NF 55:187–260

Krumbholz G (1930) Über Verschiedenheiten in der Embryonengröße einiger Oenotheren und ihrer reziproken Bastarde. Z Indukt Abstammungs Vererbungsl 56:383–392

Kuhn E (1933) Verknüpfung von Zytologie und Genetik bei Oenothera. (Sammelreferat). Züchter: 181–192

Kulkarny ChG (1929a) Meiosis in *Oenothera franciscana*. Pap Mich Acad Sci Arts Lett 9:223–226

Kulkarny ChG (1929b) Meiosis in pollen mother cells of strains of *Oenothera pratincola* Bartlett. Bot Gaz 87:218–259

Kumar S, Hecht A (1965) Growth of styles and pollen tubes of *Oenothera organensis* following emasculation. J Hered 56:125–126

Kumar S, Hecht A (1967) Studies on amino acid composition of self-compatible and self-incompatible *Oenothera organensis* L. Indian J Exp Biol 5:194–195

Kuran H (1972) Callose localization in the walls of megasporocytes and megaspores in the course of development of monospore embryo sacs. Acta Soc Bot Pol 41:509–539

Kutzelnigg H (1968) Versuche zur Auslösung von Plastommutationen bei *Oenothera*. Dissertation, Universität Düsseldorf

Kutzelnigg H (1972) Mutationsversuche mit N-Methyl-N-Nitro-N-Nitrosoguanidin an *Oenothera Hookeri*. Radiat Bot 12:63–75

Kutzelnigg H, Stubbe W (1974) Investigations on plastome mutants in *Oenothera*. I. General considerations. Sub cell Biochem 3:73–89

Kutzelnigg H, Meyer B, Schötz F (1975a) Untersuchungen an Plastom-Mutanten von *Oenothera*. II. Überblick über die Ultrastrucktur der mutierten Plastiden. Biol Zentralbl 94:513–526

Kutzelnigg H, Meyer B, Schötz F (1975b) Untersuchungen an Plastom-Mutanten von Oenothera. III. Vergleichende ultrastrukturelle Charakterisierung der Mutanten. Biol Zentralbl 94:527–538

Kwack B (1965) Stylar culture of pollen and physiological studies of selfincompatibility in *Oenothera organensis*. Physiol Plant 18:297–305

Langendorff J (1930) Zur Kenntnis der Genetik und Entwicklungsgeschichte der *Oenothera fallax*, *rubirigida* und *hookeri-albata*. Bot Arch 29:474–530

Lechner K (1963) Selektive Befruchtung aufgrund unterschiedlicher Reaktionsgeschwindigkeiten zwischen Pollenschlauch und Eizellsorten. Naturwissenschaften 50:695

Lechner K (1964) Selektive Befruchtung aufgrund unterschiedlicher Reaktionsgeschwindigkeiten zwischen Pollenschlauch- und Eizellsorten. Biol Zentralbl 83: 541–560

Lelieveld JA (1930) The heterotypic division in the genus *Oenothera*. Proc K Akad Wet Amsterdam 33:1–7

Lelieveld JA (1931) Cytological studies in some species of the genus *Oenothera*. Cellule 40:195–256

Lelieveld JA (1932) Cytological observations on the diploid off-spring of a haploid *Oenothera franciscana*. Cellule 41:281–287

Lepouse J, Romain MF (1967) L'etude de l'ultrastructure des enveloppes polliniques chez *Oenothera biennis*. Pollen Spores 9:403–413

Levy M, Levin DA (1975) Genetic heterozygosity and variation in permanent translocation heterozygotes of the *Oenothera biennis*-complex. Genetics 79: 493–512

Levy M, Winternheimer PL (1977) Allozyme linkage disequilibria among chromosome complexes in the permanent translocation heterozygote, *Oenothera biennis*. Evolution 31:465–476

Lewis D (1948) Structure of the incompatibility gene. I. Spontaneous mutation rate. Heredity 2:219–236

Lewis D (1949) Structure of the incompatibility gene. II. Induced mutation rate. Heredity 3:339–355

Lewis D (1951) Structure of the incompatibility gene. III. Types of spontaneous and induced mutation. Heredity 5:399–414

Lewis D (1952) Serological reactions of pollen incompatibility substances. Proc R Soc B 140:127–135

Linder R (1952a) Observation de l'incompatibilité dans l'*Oenothera fruticosa* L. CR Acad Sci 235:516–517

Linder R (1952b) L'incompatibilité dans *Oenothera missouriensis* Sims. Ann Amelior Plant 2:189–213

Linder R (1953) Une anomalie hereditaire de l'ovaire conduisant à la sterilité. CR Acad Sci 236:851–853

Linder R (1954) Etude genetique de mechanisme qui limite la fertilité dans l'*Oenothera missouriensis* et *Oenothera fructicosa*. Annee Biol 30:501–518

Linder R (1957a) Les Oenothères recemment reconnue en France. Bull Soc Bot Fr 104:515–525

Linder R (1957b) Clé provisoire des Oenothères reconnue en France. Bull CERS 4:571–575

Linder R (1957c) Apercu des Oenotheres rencontré dans le sud-ouest. Bull CERS 4:575–576

Linder R (1961) Faktorielle Pollenletalität. Z Vererbungsl 92:1–7

Linder R (1967) Zum Problem der leeren Pollenkörner bei Eu-Oenotheren. Ber Dtsch Bot Ges 80:539–544

Linder R (1970) Determination of pollen lethality. Proc XII Int Congr Genet 1:189

Linder R, Brun J (1987) L'incompatibilité dans *Oenothera Serrulata* Nutt. Experientia 13:23

Linder R, Jean R (1968) La letalité pollinique chez les Eu-Oenotheras. Bull Soc Bot Nord Fr 21:209–219

Linder R, Jean R (1969) *Oenothera ersteinensis*, espèce nouvelle. Bull Soc Bot Fr 116:523–529

Linder R, Jean R, Butantin M (1957) Etude des Oenothères en Alsace. Bull Soc Histoire Nat Colmar 48:21–49

Linnert G (1948) Untersuchungen über die Zytologie polyploider Pflanzen I. Chromosoma 3:328–356

Linnert G (1951) Die Bestimmung des Zeitpunktes für das Auftreten von Chromosomenmutationen in der Meiosis von *Oenothera* nach experimenteller Einwirkung. Z Indukt Abstammungs Vererbungsl 83:414–421

Linnert G (1953) Der Einfluß von Chemikalien auf Chiasmenbildung und Mutationsauslösung bei *Oenothera*. Chromosoma 5:428–453

Linnert G (1962) Untersuchungen an hemiploiden Nachkommen Autotetraploider. I. Der Verteilungsmodus einer reziproken Translokation und die daraus folgende Erhöhung der Heterozygotenfrequenz in der Nachkommenschaft von Duplex-Heterozygoten. Z Vererbungsl 93:389–398

Linnert G (1965) Untersuchungen an polyhaploiden Nachkommen Autotetraploider. II. Die spontane Entstehung von Genom- und Chromosomen- und Genmutationen während des tetraploiden Zustandes bei *Oenothera Hookeri* de Vries. Z Vererbungsl 96:190–212

Lutz AM (1907) A preliminary note on the chromosomes of *Oenothera Lamarckiana* and one of its mutants, *Oenothera gigas*. Science 26:151–152

Lutz AM (1908) Chromosomes of somatic cells of the Oenotheras. Science 27:335

Marquardt H (1937) Die Meiosis von *Oenothera* I. Z Zellforsch Mikrosk Anat 27:159–210

Marquardt H (1943) Über eine röntgeninduzierte Mutation mit abweichendem Erbverhalten bei *Oenothera hookeri* (vorläufige Mitteilung). Flora NF 37: 152–165

Marquardt H (1948) Das Verhalten röntgeninduzierter Viererringe mit großen interstitiellen Segmenten bei *Oenothera hookeri*. Z Vererbungsl 82:415–429

Marquardt H (1951) Die Auslösung von Chromosomenmutationen durch Röntgenstrahlen und durch die Invertseife Zefirol in der Meiosis von *Oenothera hookeri*. Z Indukt Abstammungs Vererbungsl 83:513–530

Marquardt H (1952) Die Auslösung von Chromosomenmutationen in der Meiosis durch tiefe Temperatur. Z Indukt Abstammungs Vererbungsl 84:169–181

Meyer B, Stubbe W (1974) Das Zahlenverhältnis von mütterlichen und väterlichen Plastiden in den Zygoten von *Oenothera erythrosepala* Borbas (syn. *Oe. lamarckiana*). Ber Dtsch Bot Ges 87:29–38

Michaelis P (1930a) Über einige experimentell erzeugte Chromosomenmutanten bei *Oenothera hookeri*. Z Indukt Abstammungs Vererbungsl 54:245

Michaelis P (1930b) Über experimentell erzeugte heteroploide Pflanzen von *Oenothera hookeri*. Z Bot 23:288–308

Mühlethaler K, Bell PR (1962) Untersuchungen über die Kontinuität von Plastiden und Mitochondrien in der Eizelle von *Pteridium aquilinum* (L.) Kuhn. Naturwissenschaften 49:63–64

Munz PA (1935) Studies in the Onagraceae. IX. Subgenus Raimannia. Am J Bot 22:645–663

Munz PA (1949) The *Oenothera hookeri* group. Aliso 2:1–47

Munz PA (1965) Onagraceae. North American Flora Ser II, pt. 5, 1–231. NY Botanical Garden

Noher de Halac I (1980a) Fine structure of nucellar cells during development of the embryo sac in *Oenothera biennis* L. Ann Bot 45:515–521

Noher de Halac I (1980b) Callose deposition during megagametogenesis in two species of *Oenothera*. Ann Bot 46:473–477

Noher de Halac I (1980c) Expression genica en las megaspores y durante la megagametogenesis en especies de *Oenothera* (Onagraceae). Actas IV. Congr Latinoam Genet vol 2:161–169

Noher de Halac I (1985a) Sterility of microspores in *Oenothera* ^{h}*hookeri·velans*. In: Willemse MTM, van Went JL (eds) Sexual reproduction in seed plants, ferns and mosses. PUDOC, Wageningen, pp 47–49

Noher de Halac I (1985b) Stages of tapetal cell development in sterile anthers of *Oenothera* ^{h}*hookeri·velans*. In: Willemse MTM, van Went JL (eds) Sexual reproduction of seed plants, ferns and mosses. PUDOC, Wageningen, p 50

Noher de Halac I, Cismondi IA (1992) Gametophyte ontogeny. An overview based on *Oenothera*. In: Iobal M (ed) Growth patterns in vascular plants. CRC Press, Boca Raton (in press)

Noher de Halac I, Harte C (1975) Female gametophyte competition in relation to polarization phenomena during megasporogenesis and development of the embryo sac in the genus *Oenothera*. In: Mulcahy DL (ed) Gamete competition in plants and animals. North Holland/American Elsevier, Amsterdam, pp 43–56

Noher de Halac I, Harte C (1977) Different patterns of callose wall formation during megasporogenesis in two species of *Oenothera* (Onagraceae). Plant Syst Evol 127:23–38

Noher de Halac I, Harte C (1985) Cell differentiation during megasporogenesis and megagametogenesis. Phytomorphology 35:189–200

Noher de Halac I, Cismondi IA, Harte C (1990) Pollen ontogenesis in *Oenothera*: a comparison of genotypically normal anthers with the male-sterile mutant *sterilis*. Sex Plant Reprod 3:41–53

Noher de Halac I, Fama G, Cismondi IA (1992) Changes in lipids and polysaccharides during pollen ontogeny in Oenothera anthers. Sex Plant Reprod 5:110–116

Oehlkers F (1921) Vererbungsversuche an Oenotheren I. Z Indukt Abstammungs Vererbungsl 26:1–31

Oehlkers F (1923) Vererbungsversuche an Oenotheren II. Z Indukt Abstammungs Vererbungsl 31:202–206

Oehlkers F (1924) Vererbungsversuche an Oenotheren III. Das Sulfurea-Merkmal bei den Oenotheren. Biol Zentralbl 44:1–9

Oehlkers F (1926) Erblichkeit und Zytologie einiger Kreuzungen mit *Oenothera strigosa*. Jahrb Wiss Bot 65:401–446

Oehlkers F (1927) Entwicklungsgeschichte der Pollensterilität einiger Oenotheren. Z Indukt Abstammungs Vererbungsl 43:265–284

Oehlkers F (1930) Studien zum Problem der Polymerie und des multiplen Allelomorphismus I. Z Bot 22:473–537

Oehlkers F (1933) Crossing-over bei *Oenothera*. (Vererbungsversuche an Oenotheren V). Z Bot 26:385–430

Oehlkers F (1935a) Die Erblichkeit der Sepalodie bei *Oenothera* und *Epilobium*. Studien zum Problem der Polymerie und des multiplen Allelomorphismus III. Z Bot 28:161–222

Oehlkers F (1935b) Untersuchungen zur Physiologie der Meiosis I. Z Bot 29:1–53

Oehlkers F (1936) Untersuchungen zur Physiologie der Meiosis III. Z Bot 30: 253–276

Oehlkers F (1937) Neue Versuche über zytologisch-genetische Probleme (Physiologie der Meiosis). Biol Zentralbl 57:126–149

Oehlkers F (1938) Über die Erblichkeit des cruciata-Merkmals bei den Oenotheren: eine Erwiderung. Z Indukt Abstammungs Vererbungsl 75:277–297

Oehlkers F (1940a) Der Einfluß der Plastiden auf den Ablauf der Meiosis. Naturwissenschaften 28:219–222

Oehlkers F (1940b) Meiosis und crossing-over. Biol Zentralbl 60:337–348

Oehlkers F (1940c) Meiosis und crossing-over. Zytogenetische Untersuchungen an Oenotheren. Z Indukt Abstammungs Vererbungsl 78:157–186

Oehlkers F (1940d) Über Chromosomenfärbungen mit Gentianaviolett. Z Bot 36: 55–62

Oehlkers F (1943a) Die Auslösung von Chromosomenmutationen in der Meiosis durch Einwirkung von Chemikalien. Z Indukt Abstammungs Vererbungsl 81: 313–341

Oehlkers F (1943b) Über die Aufhebung des Gonen- und Zygotenausfalls bei *Oenothera*, 1. Teil. Untersuchungen an *Oenothers Cockerelli*. Flora NF 36:106–119

Oehlkers F (1946) Weitere Versuche zur Mutationsauslösung durch Chemikalien. Biol Zentralbl 65:176–186

Oehlkers F (1949) Mutationsauslösung durch Chemikalien. Sitzungsber Heidelb Akad Wiss, Math-Naturwiss Klasse, 149, 9. Abh

Oehlkers F (1952) Chromosome breaks influenced by chemicals. Heredity 6:95–105

Oehlkers F (1956) Das Leben der Gewächse. Ein Lehrbuch der Botanik. Springer, Berlin Göttingen Heidelberg

Oehlkers F, Bergfeld-Gärtner H (1962) Mutationsauslösung durch Radioisotope. Z Vererbungsl 93:264–279

Oehlkers F, Linnert G (1949) Neue Versuche über die Wirkungsweise von Chemikalien bei der Auslösung von Chromosomenmutationen. Z Indukt Abstammungs Vererbungsl 83:136–156

Oehlkers F, Linnert G (1951) Weitere Untersuchungen über die Wirkungsweise von Chemikalien bei der Auslösung von Chromosomenmutationen. Z Indukt Abstammungs Vererbungsl 83:429–438

Oehlkers F, Linnert G, Stange L (1951) Die Häufigkeit experimentell durch Röntgenstrahlen und Chemikalien induzierter Chromosomenumbauten auf den unterscheidbaren Anteilen der Oenothera-Genome in der Meiosis. Z Vererbungsl 83:479–484

Oelkrug K (1934) Genetische und zytologische Untersuchung an Euoenotheren. Zytologie der Komplexverbindungen zwischen *Oenothera Berteriana* und *Oenothera mollissima* und die Frage der Homozygoteneliminerung bei Euoenotheren. Z Indukt Abstammungs Vererbungsl 68:22–93

Overeem van C (1921) Über Formen mit abweichender Chromosomenzahl bei *Oenothera*. Beih Bot Zentralbl 38, Abt I:1–80

Planchon LM (1896) Observation et experience sur l'ouverture des fleurs de l'*Oenothera Lamarckiana* Ser. Bull Soc Bot Fr 43:455–476

Pohl J (1895) Über Variationsweite der *Oenothera lamarckiana*. Österr Bot Z 45:166–169, 205–212

Przywara L, Rostanski K (1980) Chromosome numbers in the genus *Oenothera* L. from Poland. Acta Bio Cracov Ser Bot XXII/2:155–162

Raven PH (1964) The generic subdivision of Onagraceae, tribe Onagreae. Brittonia 16:276–288

Raven PH, Dietrich W, Stubbe W (1979) An outline of the systematics of *Oenothera* subsect. Euoenothera (Onagraceae). Syst Bot 4:242–252

Renner O (1914) Befruchtung und Embryobildung bei *Oenothera Lamarckiana* und einigen verwandten Arten. Flora 107:115–150

Renner O (1916) Die tauben Samen der Oenotheren. Ber Dtsch Bot Ges 34:858–869

Renner O (1917) Versuche über die gametische Konstitution der Oenotheren. Z Indukt Abstammungs Vererbungsl 18:121–294

Renner O (1918) Weitere Vererbungsstudien an Oenotheren. Flora NF 10:641–667

Renner O (1919a) *Oenothera Lamarckiana* und ihre Bedeutung für die Mutationstheorie und für die Bastardforschung. Sitzungsber Ges Morphol Physiol: 1–6

Renner O (1919b) Zur Biologie und Morphologie der männlichen Haplonten einiger Oenotheren. Z Bot 11:305–380

Renner O (1919c) Über das Sichtbarwerden der mendelschen Spaltung im Pollen von Oenotherabastarden. Ber Dtsch Bot Ges 37:129–135

Renner O (1921) Heterogamie im weiblichen Geschlecht und Embryosack-Entwicklung bei den Oenotheren. Z Bot 13:609–621

Renner O (1922) Eiplasma und Pollenschlauchplasma als Vererbungsträger bei den Oenotheren. Z Indukt Abstammungs Vererbungsl 27:235–237

Renner O (1924) Die Scheckung der Oenotherenbastarde. Biol Zentralbl 44:309–336

Renner O (1925) Untersuchungen über die faktorielle Konstitution einiger komplex-heterozygotischer Oenotheren. Bibl Genet 9:1–168

Renner O (1928) Über Koppelungswechsel bei *Oenothera*. Verh 5. Int Kongr Vererbungswissenschaft Berlin 1927. Z Indukt Abstammungs Vererbungsl Suppl II:1216–1220

Renner O (1933a) Zur Kenntnis der Letalfaktoren und des Koppelungswechsels der Oenotheren. Flora NF 27:215–250

Renner O (1933b) Zur Kenntnis der *gigas*- und *hemigigas*-Typen bei *Oenothera*. Flora 28:123–144

Renner O (1934) Die pflanzlichen Plastiden als selbständige Elemente der genetischen Konstitution. Ber Math-Phys Klasse Sächs Akad Wiss Leipzig 86:241–266

Renner O (1936a) Zur Kenntnis der nicht-mendelnden Buntheit der Laubblätter. Flora 30:218–290

Renner O (1936b) Kurze Mitteilungen über *Oenothera*. I. Über eine dominante, semiletale Mutation (taeniata) im *flavens*-Komplex der *Oenothera suaveolens*. Flora 30:441–453

Renner O (1937a) Wilde Oenotheren in Norddeutschland. Flora 31:182–226

Renner O (1937b) Über *Oenothera atrovirens* Sh. et Bartl. und über somatische Konversion im Erbgang des cruciata-Merkmals der Oenotheren. Z Vererbungsl 74:93–124

Renner O (1937c) Zur Kenntnis der Plastiden- und Plasmavererbung. Cytologia Fujii-Jubiläumsbd: 644–653

Renner O (1938a) Alte und neue Oenotheren in Norddeutschland. Feddes Reper Beih C:94–105

Renner O (1938b) Kurze Mitteilungen über *Oenothera*. II. Zu den Chromosomenformeln der Komplexe *albicans*, *curvans*, *flectens*, *gaudens* (*rubens*), *rigens*. Flora 32:319–324

Renner O (1939) Kurze Mitteilungen über *Oenothera*. III. Über *gigas*- und *hemigigas*-Formen und ihre Verwendung zur Untersuchung des cruciata-Merkmals. Flora 33:215–238

Renner O (1940a) Kurze Mitteilungen über *Oenothera*. IV. Über die Beziehungen zwischen Heterogamie und Embryosackentwicklung und über diplarrhene Verbindungen. Flora NF 34:145–158

Renner O (1940b) Zur Kenntnis der 15-chromosomigen Mutanten von *Oenothera Lamarckiana*. Flora NF 34:257–310

Renner O (1941) Über die Entstehung homozygotischer Formen aus komplex-heterozygotischen Oenotheren. Flora NF 35:201–238

Renner O (1942a) Über das crossing-over bei *Oenothera*. Flora NF 36:117–214

Renner O (1942b) Europäische Wildarten von *Oenothera*. Ber Dtsch Bot Ges 60:448–466

Renner O (1942c) Beiträge zur Kenntnis des cruciata-Merkmals der Oenotheren. IV. Gigas-Bastarde. Labilität und Konversibilität der Cr-Gene. Z Indukt Abstammungs Vererbungsl 80:590–611

Renner O (1943a) Kurze Mitteilungen über *Oenothera*. V. Zur Kenntnis von *O. silesiaca* N.Sp., *O. parviflora* L., *ammophila* Focke, *rubricaulis* Kleb. Flora 36:325–334

Renner O (1943b) Die Variabilität der *Oenothera biennis* L. Mitt Thür Bot Ver NF 50:14–19

Renner O (1943c) Zur Kenntnis des Pollenkomplexes *flectens* der *Oenothera atrovirens* Sh. et Bartl. Z Indukt Abstammungs Vererbungsl 81:391–483

Renner O (1943d) Über die Entstehung homozygotischer Formen aus komplex-heterozygotischen Oenotheren II. Die Translokationshomozygoten. Z Bot 39: 49–105

Renner O (1946) Artbildung in der Gattung *Oenothera*. Naturwissenschaften 33:211–218

Renner O (1948a) Die cytologischen Grundlagen des crossing-over bei *Oenothera*. Z Naturforsch 3b:186–196

Renner O (1948b) Über die Modifizierbarkeit der Oenotheren. Biol Zentralbl 67: 52–60

Renner O (1949) Die 15-chromosomigen Mutanten der *Oenothera Lamarckiana* und ihrer Verwandten. Z Vererbungsl 83:1–25

Renner O (1950) Europäische Wildarten von *Oenothera* II. Ber Dtsch Bot Ges 63:129–138

Renner O (1952) Plastidenvererbung bei *Oenothera*. Bayer Akad Wiss, Sitzung

Renner O (1953) Über *Oenothera hybrida* mut. *helix*. Planta 42:30–41

Renner O (1956) Europäische Wildarten von *Oenothera* III. Planta 47:219–254

Renner O (1957) Über den Erbgang des cruciata-Merkmals der Oenotheren V. Planta 48:343–392

Renner (1958a) Bemerkung zu der Mitteilung von Th.J. Stomps: *Oenothera biennis* war hemi-kleistogam. Bot Neerl 7:59–60

Renner O (1958b) Über den Erbgang des cruciata-Merkmals der Oenotheren. VI. Mitteilung. Verbindungen der *Oenothera Lamarckiana* mut. *blandina*. Z Indukt Abstammungs Vererbungs 89:14–35

Renner O (1958c) Über den Erbgang des cruciata-Merkmals der Oenotheren. VII. Verbindungen der *Oenothera hookeri*, *Oe. franciscana* und *Oe. purpurata*. Flora 145:339–373

Renner O (1958d) Über den Erbgang des cruciata-Merkmals der Oenotheren. VIII. Verbindungen der *Oenothera atrovirens* und Rückblick. Z Vererbungsl 89: 377–396

Renner O (1958e) Konversion labil gewordener Gene. Bayer Akad Wiss, Math-Naturwiss Klasse

Renner O (1959a) Paralipomena zur Genetik von *Oenothera*. Mutanten von *Oenothera hookeri*, reziproke Bastarde, Samengröße, subletale Kombinationen, Analyse von *hookeri-curva*. Z Vererbungsl 90:132–147

Renner O (1959b) Somatic conversion and the heredity of the *cruciata* character in *Oenothera*. Heredity 13:283–288

Renner O, Cleland RE (1933) Zur Genetik und Zytologie der *Oenothera chicaginensis* und ihrer Abkömmlinge. Z Indukt Abstammungs Vererbungsl 56:275–318

Renner O, Hirmer U (1956) Zur Kenntnis von *Oenothera*. I. Über *Oenothera conferta* n.sp. Biol Zentralbl 75:513–531

Renner O, Preuss-Herzog G (1943) Der Weg der Pollenschläuche im Fruchtknoten der Oenotheren. Flora 36:215–222

Renner O, Sensenhauer R (1942) Versuche über den Erbgang des cruciata-Merkmals der Oenotheren. III. Weitere Belege für somatische Konversion. Z Indukt Abstammungs Vererbungsl 80:570–589

Renner O, Schmidt H, Braun M, Veith K, Kowalewicz R (1952) Zur Entwicklungsgeschichte von *Oenothera* und *Epilobium*. Z Naturforsch 7b:420–424

Rodkiewicz B (1970) Callose in cell walls during megasporogenesis in angiosperms. Planta 90:39–47

Rodkiewicz B, Kuran H (1971) The specificity of callose distribution in cell walls during megasporogenesis in Onagraceae, Caryophyllaceae and Scrophulariaceae. Ann Univ ARERS 9:31–35

Rodkiewicz B, Bednara J, Pora H (1971) Alternative localization of the active megaspore in tetrads in *Oenothera muricata*. Bull Acad Pol Cl II 19:691–694

Rodkiewicz B, Bednara J, Sniezko R (1976) Polarization of cells during megasporogenesis in Onagraceae. Akad Hayk CCP Tom 61:905–909

Roscher A (1973) Untersuchung des Gehalts an Gibberellinsäure bei der Normalform von *Oenothera Hookeri* und einer frühblühenden Mutante. Staatsexamensarbeit, Universität Köln

Rossmann G (1961) Über den Erbgang des Sepalodie- und des cruciata-Merkmals. 2. Mitteilung. Die Verbindung von purpurata sepaloid mit normalkronigen Arten. Dissertation, Universität München

Rossmann G (1963a) Über den Erbgang des Sepalodie- und des Cruciata-Merkmals der Oenotheren. 1. Mitteilung: Die Entstehung des Sepalodie-Merkmals und sein Verhalten bei Kreuzung mit normalkronigen Arten. Z Vererbungsl 94:19–28

Rossmann G (1963b) Analyse der *Oenothera coronifera* Renner. Flora 153:415–468

Rostanski K (1965a) Gatunki z rodzaju *Oenothera* L. na Slasku. (The species of the genus *Oenothera* L. in Silesia.) Fragm Florist Geobot XI, pars 4:491–497

Rostanski K (1965b) Kilka nowych taksonow rodzaju *Oenothera* L. z podrodzaju Oenothera. (Some new taxa in the genus *Oenothera* L., subgenus Oenothera.) Frag Florist Geobot XI, pars 4

Rostanski K (1966) Die Arten der Gattung *Oenothera* L. in Ungarn. Acta Bot Acad Sci Hung 12:337–349

Rostanski K (1968a) Proba zastosowania metody dendrytowej w systematyce gatunkow rodzaju *Oenothera* L. (The example of the application of the dendrite method to the classification of some *Oenothera* species.) Acta Soc Bot Pol XXXVII: 235–244

Rostanski K (1968b) Neofityzm wystepujacych w europie gatunkow rodzaju *Oenothera* L. Mater Zakladu fitosoc stosowanej UW 25:67–79

Rostanski K (1968c) Kilka nowych taksonow rodzaju *Oenothera* L. z podrodzaju Oenothera. Czecsc II. (Some new taxa in the genus *Oenothera* L., subgenus Oenothera. Part II.) Frag Florist Geobotanica XIV, pars 2:189–195

Rostanski K (1969) Die Anwendung der Dendriten-Methode (Wroclawer Taxonomie) in der Systematik der Pflanzen am Beispiel der Gattung *Oenothera* L. Feddes Repert 80:373–381

Rostanski K (1975) Rozmieszczenie gatunkow rodzaju *Oenothera* L. z podrodzaju Oenothera w ZSRR. (The distribution of the species of the genus *Oenothera* L. subgenus Oenothera in the USSR.) Acta Biol I Katowice Pr Nauk Uniw Slask 99:7–35

Rostanski K (1977) Some new taxa in the genus *Oenothera* L. subgenus Oenothera. Part III. Kilka nowych taksonow rodzaju Oenothera 1. z podrodzaju Oenothera. Czesc III. Frag Florist Geobot XXIII, pars 3–4: 285–293

Rostanski K (1982) The species of *Oenothera* L. in Britain. Warsonia 14:1–34

Rostanski K (1985) Zur Gliederung der Subsektion Oenothera (Sektion Oenothera, *Oenothera* L., Onagraceae). Feddes Repert 96:3–14

Rostanski K (1988) *Oenothera*. In: Schubert R, Vent W (eds) Exkursionsflora für die Gebiete der DDR und BRD, vol 4, Kritischer Band. Volk and Wissen Volkseigener Verlag, Berlin, pp 371–376

Rostanski K, Ellis G (1979) Evening primroses (*Oenothera* L.) in Wales. Nat Wales 16:237–249

Rostanski K, Forstner W (1982) Die Gattung *Oenothera* (Onagraceae) in Österreich. Phyton (Austria) 22:87–113

Rostanski K, Glowacki Z (1977) Rozmieszczenie gatunkow rodzaju *Oenothera* L. na Wysoczyznie Siedleckiej i terenach przyleglych. (The distribution of the species of the genus *Oenothera* L. in the central part of eastern Poland.) Frag Florist Geobot XXIII, pars 3–4:309–316

Rostanski K, Kloss K (1965) Die gegenwärtige Verbreitung von *Oenothera silesiaca* Renner. Feddes Repert 72:36–45

Rostanski K, Serwatka J (1968) *Oenothera cruciata* Nutt., nowy gatunek we florze Polski. (*Oenothera cruciata* Nutt., a species new to the Polish flora.) Frag Florist Geobot XIV, pars 2:169–172

Rostanski K, Soldano A (1981) A new *Oenothera* species (Sectio Oenothera, Onagraceae) in Central Italy. Una nuova specie di *Oenothera* (Sezione Oenothera, Onagraceae) in Italia Centrale. Frag Florist Geobot XXVII:375–380

Rostanski K, Stachurska A (1964) Proby wyrozniania niektorych gatunkow rodzaju *Oenothera* L. na podstawie morfologicznych cech pylku. (The attempts to differentiate some species *Oenothera* L. on the basis of morphological features of the pollen grains.) Opolskie towarzystwo przyjaciol nauk, Zeszyty przyrodnicze 4:177–203

Rostanski K, Szotkowski P (1973) Gatunki rodzaju *Oenothera* L. w portach rzecznych gornej Odry. (The *Oenothera* species in the river ports of the Upper Oder.) Frag Florist Geobot XIX, pars 4:33–40

Rostanski K, Wika S (1988) Materialy do rozmieszczenia gatunkow rodzaju *Oenothera* L. w srodkowej czesci Wyzyny Krakowko-Wielunskiej (srodkowo-poludniowa Polska). (Materials to the distribution of the *Oenothera* species in the central part of the Krakow-Wielun Upland (South-central Poland).) Frag Florist Geobot XXXIII, pars 1–2

Rostanski K, Felisiak J, Sokol S (1989a) Analiza morfologiczna nasion gatunkow z podsekcji *Oenothera* (Sectio Oenothera, *Oenothera* L., Onagraceae). Acta Biol Siles 11(28):7–25

Rostanski K, Grzegorzek P, Rostanski A, Tokarska-Guzik B (1989b) Nowe stanowiska gatunkow z rodzaju *Oenothera* L. (wiesiolek) w wojewodztwie katowickim. Acta Biol siles 11(28):26–39

Roze ME (1895) L'epanouissement de la fleur de l'*Onothera suaveolens* Desf. Bull Soc Bot Fr 42:574–582

Roze ME (1899) Note sur l'epanouissement des fleurs des trois espèces d'Onagres (*Onothera suaveolens*, *biennis* et *Lamarckiana*). J Soc Nat Hortic Fr Ser III 21:849–853

Rudloff F (1929) Zur Kenntnis der *Oenothera purpurata* Klebahn und *Oenothera rubricaulis* Klebahn. Z Indukt Abstammungs Vererbungsl 52:191–235

Rudloff CF, Schmidt M (1932) Untersuchungen über den Einfluß ungünstiger Witterungverhältnisse auf die Reduktionsteilung und die Embryosackentwicklung bei verschiedenen Oenotheren. Planta 18:104–167

Russell J, Levin DA (1988) Competitive relationships of *Oenothera* species with different recombination systems. Am J Bot 75:1175–1180

Salisbury FB (1963) The flowering process. Pergamon Press, Oxford

Schildknecht H, Benoni H (1963) Über die Chemie der Anziehung von Pollenschläuchen durch die Samenanlagen von *Oenothera*. Z Naturforsch 18b:45–54

Schmitt JM (1978) Zur Funktion des genetischen Apparates der Plastiden: Nukleinsäuresynthese und plastomcodierte Proteine. Dissertation, Universität Düsseldorf

Schneider G (1956) Wachstum und Geotropismus von Pollenschläuchen. Z Bot 44:175–205

Schötz F (1954) Über Plastidenkonkurrenz bei *Oenothera*. Planta 43:182–240

Schötz F (1955) Untersuchungen an panaschierten Oenotheren. I. Über die photosynthetische Leistungsfähigkeit der Chloroplasten in den blassen Arealen einiger Plastidenmutanten und Bastardschecken. Z Naturforsch 10b:100–108

Schötz F (1956) Über die photosynthetische Leistungsfähigkeit einiger panaschierter Oenotheren. Photogr Forsch 7:12–16

Schötz F (1958a) Beobachtungen zur Plastidenkonkurrenz bei *Oenothera* und Beiträge zum Problem der Plastidenvererbung. Planta 51:173–185

Schötz F (1958b) Periodische Ausbleichungserscheinungen des Laubes bei *Oenothera*. Planta 52:351–392

Schötz F (1959) Periodische Veränderungen im Zusammenwirken von Zellkern und Plastiden bei Oenotheren. Photogr Forsch 8:65–72

Schötz F (1962a) Zur Kontinuität der Plastiden. Planta 58:333–336

Schötz F (1962b) Pigmentanalytische Untersuchungen an *Oenothera*. I. Vorversuche und Analyse der Blätter und Blüten von *Oenothera suaveolens* Desf., Mutante "Weißherz". Planta 58:411–434

Schötz F (1964) Elektronenmikroskopische Untersuchungen an den Plastiden eines Oenotherenbastards mit disharmonischer Genom-Plastom-Kombination. Ber Dtsch Bot Ges 77:372–378

Schötz F (1965) Zur Frage der Vermehrung der Thylakoidschichten in den Chloroplasten. Planta 64:376–380

Schötz F (1968) Über Plastidenkonkurrenz bei *Oenothera* II. Biol Zentralbl 87:33–61

Schötz F (1970) Effects of the disharmony between genome and plastome on the differentiation of the thylakoid system in *Oenothera*. Symp Soc Exp Biol XXIV. University Press, Cambridge, pp 39–54

Schötz F (1971) Untersuchungen über die Plastidenkonkurrenz bei *Oenothera*. Biol Zentralbl 93:41–64

Schötz F (1974) Untersuchungen über die Plastidenkonkurrenz bei *Oenothera*. IV. Der Einfluß des Genoms auf die Durchsetzungsfähigkeit der Plastiden. Biol Zentralbl 93:41–64

Schötz F (1975) Untersuchungen über die Plastidenkonkurrenz bei *Oenothera*. V. Die Stabilität der Konkurrenzfähigkeit bei Verwendung verschiedenartiger mutierter Testplastiden. Biol Zentralbl 94:17–26

Schötz F, Batheld H (1964) Pigmentanalytische Untersuchungen an *Oenothera*. II. Der *albivelutina*-Typ, 1. Teil. Planta 63:213–232

Schötz F, Batheld H (1965a) Pigmentanalytische Untersuchungen an *Oenothera*. III. Der *albivelutina*-Typ, 2. Teil. Planta 62:233–252

Schötz F, Batheld H (1965b) Pigmentanalytische Untersuchungen an *Oenothera*. IV. Der *pictirubata*-Typ. Planta 64:330–362

Schötz F, Diers L (1966) Über die Bildung von Stromasubstanz und Thylakoiden in dem Raum zwischen dem inneren und äußeren Raum der Plastiden-Doppelmembran. Planta 69:259–287

Schötz F, Diers L (1968) Beeinflussung der Thylakoidbildung durch Disharmonie zwischen Genom und Plastom. Protoplasma 65:335–348

Schötz F, Heiser F (1969) Über Plastidenkonkurrenz bei *Oenothera* (Zahlenverhältnisse in Mischzellen). Wiss Z PH Potsdam 13:65–89

Schötz F, Reiche GA (1957) Untersuchungen an panaschierten Oenotheren. II. Über die Lebensfähigkeit von Bastarden, die ausschließlich die durch den Bastardkern geschädigte Plastidensorte enthalten. Z Naturforsch 12b:757–764

Schötz F, Senser F (1961) Reversible Plastidenumwandlung bei der Mutante "Weißherz" von *Oenothera* suaveolens Desf. Planta 57:235–238

Schötz F, Senser F (1964) Untersuchung über die Chloroplastenentwicklung bei *Oenothera*. III. Der *pictirubata*-Typ. Planta 63:191–212

Schötz F, Senser F (1965) Untersuchungen über die Plastidenentwicklung und Pigmentausstattung der Oenotheren. Der *rubiflexa*-Typ. Z Vererbungsl 96:250–266

Schötz F, Stubee W (1962) Die Kontinuität der Plastiden. Naturwissenschaften 49:211

Schötz F, Senser F, Batheld H (1965) Untersuchungen über die Plastidenentwicklung und Pigmentausstattung der Oenotheren. II. Mutierte *biennis*-Plastiden. Planta 70:121–190

Schötz F, Diers L, Bathelt H (1968) Über den Einfluß einer Genom-Plastom-Disharmonie auf die Thylakoidanordnung. Z Naturforsch 23b, H 9:1248–1252

Schötz F, Diers L, Bathelt H (1970) Zur Feinstruktur der Raphidenzellen. 1. Die Entwicklung der Vakuolen und der Raphiden. Z Pflanzenphysiol 63:91–113

Schröder KH (1962) Mikroskopische Untersuchungen über die Vermehrung der Plastiden im Scheitelmeristem und in den Blättern von *Oenothera albilaeta*. Z Bot 50:348–367

Schütz G (1939) Genetische und zytologische Untersuchungen an Euoenotheren. Selektive Befruchtung bei der Komplexheterozygoten *Oenothera Berteriana*. Z Bot 33:481–525

Schumacher E, Steiner E, Stubbe W (1992) The complex-heterozygotes of *Oenothera grandiflora* L'Her. Bot Acta 105:375–381

Schwemmle B (1962) Über den Einfluß von Kern, Plasma und Plastiden auf die photoperiodische Reaktion einiger Raimannia-Oenotheren. Planta 58:619–646

Schwemmle J (1927) Der Bastard *Oenothera Berteriana* × Onagra (*muricata*) und seine Zytologie. Jahrb Wiss Bot 66:579–595

Schwemmle J (1929) Genetische und zytologische Untersuchungen an Euoenotheren I. Jahrb Wiss Bot 67:814–876

Schwemmle J (1932) Die Beziehung zwischen Zytologie und Genetik in der Oenotherenforschung. Z Indukt Abstammungs Vererbungsl 61:36–61

Schwemmle J (1938) Genetische und zytologische Untersuchungen an Euoenotheren, Teil I. Die Analyse der *Oenothera Berteriana* und *odorata*. Z Indukt Abstammungs Vererbungsl 75:358–468

Schwemmle J (1940a) Plastidenmutationen bei Euoenotheren. Z Indukt Abstammungs Vererbungsl 79:179–187

Schwemmle J (1940b) Keimversuche mit alten Samen. Z Bot 36:225–261

Schwemmle J (1948) Gibt es eine selektive Befruchtung? Biol Zentralbl 68:195–231

Schwemmle J (1951) Gibt es eine selektive Befruchtung? II. Biol Zentralbl 70:193–252

Schwemmle J (1952a) Selektive Befruchtung als Erklärung unerwarteter Kreuzungsergebnisse. Biol Zentralbl 71:353–384

Schwemmle J (1952b) Der Einfluß des Plasmas auf die Affinität zwischen Samenanlagen und Pollenschläuchen. Biol Zentralbl 71:487–499

Schwemmle J (1952c) Prädetermination des Keimverlaufs bei *Oenothera Berteriana*. Z Indukt Abstammungs Vererbungsl 84:568–574

Schwemmle J (1953a) Selektive Befruchtung bei *Oenothera Berteriana*. Biol Zentralbl 72:129–146

Schwemmle J (1953b) Selektive Befruchtung bei *Oenothera odorata*. Biol Zentralbl 72:405–424

Schwemmle J (1957a) Der Samenansatz bei Kreuzungen mit der *Oenothera Berteriana* nach Früh- und Spätbestäubung. Planta 49:135–167

Schwemmle J (1957b) Das unterschiedliche Reifen und Altern der Samenanlagen in den verschiedenen Abschnitten des Fruchtknotens der *Oenothera Berteriana*. Planta 49:168–202

Schwemmle J (1958) Über Kreuzungen mit verschiedenen Formen der *Oenothera odorata* bei Früh- und Spätbestäubung I. Planta 51:223–248

Schwemmle J (1960) Keimversuche auf genetischer Grundlage. I. Die Lichtabhängigkeit. Planta 54:294–313

Schwemmle J (1968a) Versuche zum Nachweis der Konstanz von Plasmon und Plastom. Ber Dtsch Bot Ges 81:22–27
Schwemmle J (1968b) Selective fertilization in *Oenothera*. Adv Genet 14:225–324
Schwemmle J (1970) Versuche zum Nachweis der Konstanz von Plasmon und Plastom. II. Ber Dtsch Bot Ges 83:75–77
Schwemmle J, Koepchen W (1953) Weitere Untersuchungen zur selektiven Befruchtung. Z Indukt Abstammungs Vererbungsl 85:307–346
Schwemmle J, Lohmeyer H (1963) Über Versuche mit *Oenothera campylocalyx*. Z Vererbungsl 94:412–426
Schwemmle J, Zintl W (1939) Genetische und zytologische Untersuchungen an Euoenotheren. Die Analyse der Oenothera argentinea. Z Indukt Abstammungs Vererbungsl 76:353–410
Sears BB, Herrmann RG (1985) Plastome mutation affecting the chloroplast ATP synthase involves a post-transcriptional defect. Curr Genet 9:521–528
Seitz FW (1935) Zytologische Untersuchungen an tetraploiden Oenotheren. Z Bot 28:481–542
Senser F, Schötz F (1964) Untersuchungen über die Chloroplastenentwicklung bei *Oenothera*. II. Der *albivelutina*-Typ. Planta 62:171–190
Sheffield FML (1927) Cytological studies of certain meiotic stages in *Oenothera*. Ann Bot 41:779–816
Sheffield FML (1929) Chromosome linkage in *Oenothera*, with special reference to some F1 hybrids. Proc R Soc B 102:207–230
Shull GH (1921) Three new mutations in *Oe. Lamarckiana*. J Hered XII:354–363
Shull GH (1923) Linkage with lethal factors, the solution of the *Oenothera* problem. Eugen Genet Fam I:86–99
Shull GH (1925) The third linkage group in *Oenothera*. Proc Nat Acad Sci 11:715–718
Shull GH (1926) "Old-gold" flower color, the second case of independent inheritance in *Oenothera*. Genetics 11:201–234
Shull GH (1927) Crossing-over in the third linkage group in *Oenothera*. Proc Nat Acad Sci 13:21–24
Shull GH (1928a) A new gene mutation (mut. *bullata*) in *Oenothera Lamarckiana* and its linkage relations. Verh V. Int Kongr Vererbungswissenschaft, Berlin 1927. Z Indukt Abstammungs Vererbungsl Suppl II:1322–1342
Shull GH (1928b) *Oenothera* cytology in relation to genetics. Am Nat 62:97–114
Shull GH (1928c) The "outside-in" *Oenothera* flower, a new morphological type produced by the interaction of two recessive Mendelian factors. Proc Nat Acad Sci 14:142–146
Shull GH (1930) The first two cases of crossing-over between old-gold and bullata factors in the third linkage group of *Oenothera*. Proc Nat Acad Sci 16:106–109
Shull GH (1934) *Oenothera* mut *pollicata*, an interesting new mutation. Am Nat 58:481–490
Shull GH (1937) New experiences with *Oenothera* mut. *pollicata*. Am Nat 71:69–82
Shull GH (1942) New mutational segregations from *Oenothera* mut. *erythrina* de Vries. Proc Am Philos Soc 85:183–214
Siebs W (1939) Neue Beiträge zur Kenntnis des cruciata-Merkmals der Oenotheren. Z Indukt Abstammungs Vererbunsgsl 57:239–290
Sigmond H (1930a) Die Entfaltung der Blütenknospen zweier *Oenothera*-Arten, I. Teil. Bot Zentralbl Beih 46:480–488
Sigmond H (1930b) Die Entfaltung der Blütenknospen zweier *Oenothera*-Arten, II. Teil. Bot Zentralbl Beih 47:69–138
Skvarla JJ, Raven PH, Chissoe WF, Sharp M (1978) An ultrastructural study of viscin threads in Onagraceae pollen. Pollen Spores 20:5–143
Sniezko R (1976) Distribution of polysaccharides during megasporogenesis in Onagraceae. Ann Univ Mariae Curie-Sklodowska 31, 16 Sect C:183–191
Sniezko R, Harte C (1984a) Polarity and competition between megaspores in the ovule of *Oenothera*-hybrids. Plant Syst Evol 144:183–197

Sniezko R, Harte C (1984b) Callose pattern and polarization phenomena in the ovules in the F_2-hybrids between *Oenothera hookeri* and *Oe. suaveolens*. Plant Syst Evol 147:79–90

Sniezko R, Harte C (1985) Genetic basis of callose pattern, megaspore degeneration and polarity in the ovules of *Oenothera*. In: Sexual reproduction in seed plants, ferns and mosses. PUDOC, Wageningen, p 136

Sniezko R, Harte C (1986) Development of polarity in the ovules of the F_2-progeny of the *Oenothera* hybrid *albicans*·^{h}*hookeri*. Plant Syst Evol 154:89–101

Sniezko R, Harte C (1993) Influence of the gene brevistylis on development of the flower in *Oenothera*. (Sex. Plant Repr., in press)

Sokol S, Rostanski K (1989) Komplexanalyse von Oenothera novae-scotiae Gates, *Oe. paradoxa* Hudziok und *Oe. pseudochicaginensis* Rostanski. Acta Biol Siles 11(28):40–47

Stange L (1951a) Zur Frage nach dem Einfluß der Temperatur bei der Auslösung von Chromosomenmutationen in der Meisosis durch Röntgenstrahlen. Z Indukt Abstammungs Vererbungsl 83:431–438

Stange L (1951b) Untersuchungen über den Einfluß von Begleitfaktoren auf die mutationsauslösende Wirkung der Röntgenstrahlen. Z Indukt Abstammungs Vererbungsl 83:485–512

Steiner E (1953) The results of hybridization between *Oenothera elata* and certain South American Eu-Oenotheras. Pap Mich Acad Sci 38:89–96

Steiner E (1956) New aspects of the balanced lethal mechanism in *Oenothera*. Genetics 41:486–500

Steiner E (1957) Further evidence of an incompatibility allele system in the complex-heterozygotes of *Oenothera*. Am J Bot 44:582–585

Steiner E (1961a) Incompatibility in the complex-heterozygotes of *Oenothera*. Genetics 46:301–315

Steiner E (1961b) Incompatibility studies in *Oenothera*. Z Vererbungsl 92:205–212

Steiner E (1964) Incompatibility studies in *Oenothera*: the distribution of S_1 alleles in *biennis*-1 populations. Evolution 18:370–378

Steiner E (1975) *Oenothera*. In: King RC (ed) Handbook of genetics, vol 2. Alan Liss, New York, pp 223–245

Steiner E, Levin DA (1977) Allozyme, SI gene, cytological, and morphological polymorphisms in a population of *Oenothera biennis*. Evolution 31:127–133

Steiner E, Schultz MS (1958) Distribution of incompatibility alleles among the complex-heterozygotes of *Oenothera*. Science 127:516–517

Steiner E, Stubbe W (1984) A contribution to the population biology of *Oenothera grandiflora* L.Her. Am J Bot 71:1293–1301

Steiner E, Stubbe W (1986) *Oenothera grandiflora* revisited: a new view of its population structure. Bull Torrey Bot Club 113:406–412

Stern C (1930) Konversionstheorie und Austauschtheorie. Biol Zentralbl 50:608–623

Stern C (1932) Über die Konversionstheorie. Biol Zentralbl 52:367–379

Stinson HT (1953) Cytogenetics and phylogeny for *Oenothera argillicola* Mackenz. Genetics 38:389–406

Stinson HT jr (1960) Extranuclear barriers to interspecific hybridization between *Oenothera Hookeri* and *Oe. argillicola*. Genetics 45:815–838

Stinson HT, Steiner E (1955) Phylogenetic studies in *Oenothera*: further analysis of plants from the southeastern United States. Am J Bot 42:905–911

Stomps TJ (1912) Mutation bei *Oenothera biennis* L. Biol Zentralbl 32:521–535

Stomps TJ (1913) Das cruciata-Merkmal. Ber Dtsch Bot Ges 31:166–172

Stomps TJ (1957) *Oenothera biennis* L. var. *hemikleistogama*, eine neue Mutation von *O. biennis*. Acta Bot Neerl 6:378–380

Stubbe W (1953) Genetische und zytologische Untersuchungen an verschiedenen Sippen von *Oenothera suaveolens*. Z Indukt Abstammungs Vererbungsl 85:180–209

Stubbe W (1955) Erbliche Chlorophylldefekte bei *Oenothera*. Photogr Wiss 4:3–8

Stubbe W (1957) Dreifarbenpanaschierung bei *Oenothera*. I. Entmischung von drei in der Zygote vereinigten Plastidomen. Ber Dtsch Bot Ges 70:221–226

Stubbe W (1958) Dreifarbenpanaschierung bei *Oenothera*. II. Wechselwirkung zwischen Geweben mit zwei erblich verschiedenen Plastidensorten. Z Vererbungsl 89:189–203

Stubbe W (1959) Genetische Analyse des Zusammenwirkens von Genom und Plastom bei *Oenothera*. Z Vererbungsl 90:288–298

Stubbe W (1960) Untersuchungen zur genetischen Analyse des Plastoms von *Oenothera*. Z Bot 48:191–218

Stubbe W (1962) Sind Zweifel an der genetischen Kontinuität der Plastiden berechtigt? Eine Stellungnahme zu den Ansichten von Mühlethaler u. Bell. Z Vererbungsl 93:175–176

Stubbe W (1963a) Extrem disharmonische Genom-Plastom-Kombinationen und väterliche Plastidenvererbung bei *Oenothera*. Z Vererbungsl 94:392–411

Stubbe W (1963b) Die Rolle des Plastoms in der Evolution der Oenotheren. Ber Dtsch Bot Ges 76:154–167

Stubbe W (1964) The role of the plastome in evolution of the genus *Oenothera*. Genetica 35:28–35

Stubbe W (1970) Das Falcifolia-Syndrom der Oenotheren. 1. Mitteilung. Mol Gen Genet 106:213–227

Stubbe W (1979) Genetic self-incompatibility in *Oenothera* subsect. Euoenothera. Science 204:327

Stubbe W (1980) Über die Bedingungen der Komplexheterozygotie und die beiden Wege der Evolution komplexheterozygotischer Arten bei *Oenothera*. Ber Dtsch Bot Ges 93:441–447

Stubbe W (1984) Biparental plastid inheritance in *Oenothera organensis* Munz. Acta Bot Neerl 33:229

Stubbe W (1989a) The falcifolia syndrome of *Oenothera*. III. The general pattern of its non-Mendelian inheritance. Mol Gen Genet 218:499–510

Stubbe W (1989b) The falcifolia syndrome of *Oenothera*. IV. Loss of falcifolia-determining factors. Mol Gen Genet 218:511–515

Stubbe W (1989c) *Oenothera* – An ideal system for studying the interactions of genome and plastome. Plant Mol Biol Rep 7:245–257

Stubbe W, Diers L (1958) Die Chromosomenformel des *albicans*-Komplexes der Sippe Grado von *Oenothera suaveolens*. Z Vererbungsl 89:320–322

Stubbe W, Herrmann RG (1982) Selection and maintenance of plastome mutants and interspecific genome/plastome hybrids from *Oenothera*. In: Edelmann M, Hallick R, Chua NH (eds) Methods in chloroplast molecular biology. Elsevier, Amsterdam, pp 149–165

Stubbe W, Raven PH (1979a) A genetic contribution to the taxonomy of *Oenothera* Sect. Oenothera (including subsections Euoenothera, Emersonia, Raimannia and Munzia). Plant Syst Evol 133:39–59

Stubbe W, Raven PH (1979b) Genetic self-incompatibility in *Oenothera* subsect. Euoenothera. Science 204:327

Stubbe W, von Wettstein D (1955) Zur Struktur erblich verschiedener Chloroplasten von *Oenothera*. Protoplasma 45:241–250

Stubbe W, Pietsch B, Kowallik KV (1978) Cytologische Untersuchungen über väterliche Plastidenvererbung und plastom-abhängige Degeneration der Samenanlagen bei einem Oenothera-Bastard. Biol Zentralbl 97:30–52

Sturm A (1938) Genetische und zytologische Untersuchungen an Euoenotheren. Teil V. Anatomische und zytologische Untersuchung an bleichen und gescheckten Komplexheterozygoten. Z Indukt Abstammungs Vererbungsl 75:690–738

Sturtevant AH (1931) Genetic and cytological studies on *Oenothera*. I. Nobska, Oakesiana, Ostrea, Shulliana and the inheritance of old-gold flower color. Z Indukt Abstammungs Vererbungsl 59:365–380

Taeckholm WG (1915) Beobachtungen über die Samenentwicklung einiger Onagraceen. Sven Bot Tidskr 9:294–361

Takahashi M, Skvarla JJ (1990) Pollen development in *Oenothera biennis* (Onagraceae). Am J Bot 77:1142–1148

Tandon SL, Hecht A (1953) Cytogenetic studies of *Oenothera affinis* Camb. Cytologia 18:133–145

Tandon SL, Hecht A (1955) Cytogenetic studies of *Oenothera molissima* L. Cytologia 20:199–210

Tandon SL, Hecht A (1956) Cytogenetic evidence for the inclusion of *Oenothera affinis* Camb. under *Oenothera molissima* L. Cytologia 21:252–271

Tiele-Winckler von EB (1956) Codein als hemmende Substanz bei der Auslösung von Chromosomenmutationen von *Oenothera*. Z Indukt Abstammungs Vererbungsl 87:338–355

Tyagi AK, Herrmann RG (1986) Location and nucleotide sequence of the pre-apocytochrome f gene on the *Oenothera hookeri* plastid chromosome (Euoenothera plastome I). Curr Genet 10:481–486

Veer PHWAM de (1969) Leven en werk van Hugo de Vries. Wolters-Noordhoff n.v. Groningen

Vries H de (1889) Intracelluläre Pangenesis. G Fischer, Jena

Vries H de (1900a) Sur la loi de disjonction des hybrides. CR Acad Sci 130:845–847

Vries H de (1900b) Das Spaltungsgesetz der Bastarde. Ber Dtsch Bot Ges 18:83–90

Vries H de (1901) Die Mutationstheorie, vol I. Verlag von Veit, Leipzig

Vries H de (1903) Die Mutationstheorie, vol II. Verlag von Veit, Leipzig

Vries H de (1905) Species and varieties, their origin by mutation. In: McDougal DT (ed) Lectures delivered at the University of California. Open Court Chicago

Vries H de (1907) On twin hybrids. Bot Gaz 44:401–407

Vries H de (1913) Gruppenweise Artbildung unter spezieller Berücksichtigung der Gattung *Oenothera*. Borntraeger, Berlin

Vries H de (1915) The coefficient of mutation in *Oenothera biennis*. Bot Gaz 59:169–196

Vries H de (1918) Mutations of *Oenothera suaveolens* Desf. Genetics 3:1–26

Vries H de (1925) Latente Mutabilität von *Oenothera biennis* L. Z Indukt Abstammungs Vererbungsl 38:141–199

Vries H de (1927) Operata et periodicis collata.

Vries H de (1929) Über das Auftreten von Mutanten aus *Oenothera Lamarckiana*. Z Indukt Abstammungs Vererbungsl 52:121–190

Vries H de (1935) Über semirezessive Anlagen in *Oenothera Lamarckiana*. Z Indukt Abstammungs Vererbungsl 70:222–256

Wagner WL (1983) New species and combinations in the genus *Oenothera* (Onagraceae). Ann Mo Bot Gard 70:194–196

Wagner WL (1986) New taxa in *Oenothera* (Onagraceae). Ann Mo Bot Gard 73:475–480

Wasmund O, Stubbe W (1986) Cytogenetic investigations on *Oenothera wolfii* (Munz) Raven, Dietrich and Stubbe. Plant Syst Evol 154:79–88

Wasmund O, Stubbe W, Michaelis G (1987) Molecular and genetic studies on the falcifolia syndrome in *Oenothera*. XIV Intern. Bot Congress Abstracts, p 180

Weidner-Rauh E (1939) Untersuchungen über die partielle Sterilität der Oenotheren: Das Pulver bei Eu-Oenotheren. Z Indukt Abstammungs Verebungsl 76:422–486

Wein K (1930) Die älteste Einführungs- und Einbürgerungsgeschichte der nordamerikanischen Vertreter der Gattung *Oenothera*. Feddes Repert Beih LX: 27–64

Weinland H (1941) Das Wachstum der Hypanthien bei den Oenotheren. Z Bot 36:401–430

Weiße A (1899) Beitrag zur Entwicklungsgeschichte der Onagraceenblüte, mit besonderer Berücksichtigung des unterständigen Fruchtknotens. Schwendener Festschr, pp 231–250

Wiebalck U (1940) Untersuchungen zur Physiologie der Meiosis. XI. Reifeteilung und Kohlehydratspiegel der Pflanze. Z Bot 36:161–212

Winter P, Herrmann RG (1988) A five-base-pair-deletion in the gene for the large subunit causes the lesion in the ribulose-bisphosphate carboxylase/oxygenase-deficient plastome mutant *sigma* of *Oenothera hookeri*. Bot Acta 101:68–75

Wisniewska E (1935) Badania cytologiczne nad *Oenothera Hookeri* de Vries. (Zytologische Untersuchungen an *Oenothera Hookeri* de Vries). Acta Soc Bot Pol 12:113–164

Zenke U (1950) Untersuchungen an aneuploiden Nachkommen von *Oenothera franciscana* de Vries. Dissertation, Universität Freiburg i Br

Zürn K (1936) Untersuchungen zur Physiologie der Meiosis. IV. Z Bot 30:577–603

Zürn K (1937) Die Bedeutung der Plastiden für den Ablauf der Meiosis. (Untersuchungen zur Physiologie der Meiosis. IX). Jahrb Wiss Bot 85:706–731

Zürn K (1939) Untersuchungen zur Physiologie der Meiosis. X. Z Bot 34:273–310

Author Index

only first authors, no authors of taxonomic names

Alpino **15**, 16
Andrews 22
Arnold 52, 93, 96, 103, 107, 113, 127–128, 153, 197, 216

Baerecke 92
Bali 47
Basler 131, 148
Baumgarten 146–149
Baur 113
Beer 154
Binder 127
Blakeslee 6, 68–69
Boedijn 56, 65, 86–87, 96, 194
Brennecke 130
Brittingham 26, 113, 147
Broekels 54
Brown 201, 220
Buchner 104

Candolle de 16
Capinpin 86, 88
Carrillo 129
Catcheside 14, 56, 69, 76, 85–87, 92, 95–97, 103, 113, 159–160
Charlesworth 49
Chapman 109–110
Chiu 128, 164, 168
Chrometzka 148
Cleland 6–7, **8**, 9–12, 14, 27–28, 39, 47, 49–50, 54, 56–57, 60, 65, **65**, 67–69, 72–73, 88, 90, 92–94, 96–99, 108, 112, 132–133, 135–136, 146, 151–152, 157, 207, 212, 221
Commelin 15
Correns 210
Crowe 47, 50

Darlington 57, 69, 72, 86, 95
Dart 102
Darwin 100–101
Davis 6, 11–12, 14, 21, 26, 39, 54, 86, 103, 108, 132, 135–136, 141–142, 146, 159
Davy de Virville 153
Dickinson 204, 212
Diers 194–195
Dietrich 10–13, 133
Dittmann 106
Duchartre 156
Dulfer 103

Eichler 18–19
Ellstrand 133
Emerson 14, 26, 39–40, 47, 65, 69, 86, 90, 92, 96–97, 107, 146, 213
Epp 127

Fischer 10
Fork 128

Gates 10, 14, 16, 25, 56, 67, 69, 92, 96, 102–103, 134, 141, 178, 184
Geerts 178
Gerhard 90, 108, 141
Glenk 205, 215
Göpel 126, 209
Goldschmidt 112, 122
Gordon 128
Gregory 162
Gutte 139

Haberlandt 142, 178, 184, 215

Hadorn 33
Hakansson 59–60, 69, 86, 96, 106
Hall 18, 134, 141
Hallier 128
Hance 97–98
Harte 3, 14, 36–38, 40, 42–46, 76, 85–86, 93–94, 96, 113, 118, 131, 141–142, 145–148, 150–151, 154–156, 178, 191, 196–197, 205, 210, 214
Harwood 123
Haselwarter 85
Haustein 52, 88, 92, 99, 112, 127, 141, 216
Hecht 14, 87, 93, 99, 213
Hedayitullah 72
Hegi 18, 134
Heribert-Nilsson 14, 103, 124, 210
Herrmann 14, 128
Herzog 87–88, 97
Heslop-Harrison, J 162, 212
Heslop-Harrison, Y 161, 204, 212
Hesse 201
Hiesel 130
Hildebrandt 129
Hiorth 140–142, 144
Hoeppener 36, 90, 134, 136, 147, 165
Hoffmann 59

Illick 65, 69
Ishikawa 215

Jalouzot 179, 185–186
Japha 58, 60, 68, 76, 86
Jean 60, 141, 193–195, 200, 207
Jehlik 139
Johansen 157
Johnson 129

Kaienburg 156, 204–205, 211
Kandeler 166
Kappus 94, 124, 135
Kaul 197
Khafaji 49
Kihara 56
Kisch 85
Kistner 142
Kowalewicz 19, 146, 151–153, 156–157
Krautblatter 113
Kressel 113
Krumbholz 144, 165
Kuhn 69
Kulkarny 65, 86
Kumar 154, 213
Kuran 186
Kutzelnigg 121, 128, 175, **176**, 177
Kwack 205–206

Langendorf 90, **143**
Lechner 217
Lelieveld 65, 141
Lepouse 195
Levy 109
Lewis 48, 213–214
Linnaeus 100
Linder 47, 136, 157, 159, 197, 200, 207, 212
Linnert 60, 85, 87, 113, **119**, 121
Lutz 55

Marquardt 57–58, 68, 76, 86, 113, 118, 120
Mendel 3–4, 8, 31, 101
Meyer 129, 165
Michaelis 113
Mühlethaler 126
Mulcahy 130
Muller 113
Munz 10, 133

Noher de Halac 155, 157, 178–180, 183–184, **185**, 186, 191, 193, 195, **195–196**, 196–197, 199–202, **203**, 209, 211

Oehlkers 6–7, **8**, 9, 11, 14, 36–39, 41–42, 45, 47, 55, **65**, 69, 82, 84–85, 94–96, 109, 111, 118, **119**, 120, 122, 146, 154–155, 163, 196–197
Oelkrug 142, 144
Overeem 86–87, 96

Planchon 19
Pohl 197
Przywara 139

Raven 9–11, 133
Renner 6, **8**, 9, 11–14, 16–17, 21–22, 25–28, 30–31, 35–41, 43, 45, 47, 54, 59, 69, 72, 86–88, 90, 92–97, 99, 101, 103, 107–108, 111–112, 121–123, 126, 132, 134–135, 140–141, 144–149, 151, 153, 160, 164, 183–184, 188, 190, 192–193, 200, 204, 206, 208–209, 211
Rodkiewicz 179, 186
Roscher 150
Rossmann 112, 134, 152
Rostanski **15**, 16, 134–136, **137**, 139
Roze 153
Rudloff 134, 178, 184, 188
Rudloff E **70**, 77

Salisbury 149
Schildknecht 218
Schmitt 129
Schneider 206, 211, 217
Schötz 13–14, 18, 126, 146, 164–165, 168–169, 171–172, **175**
Schröder 164
Schütz 50–51, 216
Schumacher 12, 133
Schwemmle B 130, 150
Schwemmle J 12, 14, 28, 52, 69, 88, 127, 130, 142, 145, 157, 180, 197, 206, 211, 216–217
Sears 129
Seitz 86
Senser 171
Sheffield 56, 69
Shull 6, 14, 21, 26, 31, 39–40, 65, 67, 69, 108, 147–148, 159–160
Siebs 111
Sigmond 153, 162
Skvarla 201
Sniezko 157, **159**, 160, 162, 178–179, **181–182**, 183–184, 186, **186**, 188–189, **189**, 191, 211
Sokol 139
Stange 118
Steiner 12, 14, 47, 49, 133, 200, 211–212, 221
Stern 112
Stinson 14
Stomps 111, 153
Stubbe 14, 47, 49, 123–124, **125**, 126–129, 131, 133, 136, 141, 148, 164–166, **166**, 168, **170**, 175, 181, 209, 221
Sturm 127
Sturtevant 13–14

Taeckholm 144
Takahashi 201
Tandon 12, 14, 93, 98
Thiele-Winckler 120
Troll 18
Tyagi 129

Vries, de 1, 3–7, **8**, 9, 11, 13–14, 21–22, 25, 30, 39, 96, 101–103, 106, 108, 111, 159

Wagner 133
Wasmund 130, 133
Weidner-Rauh 140
Wein 14, 16
Weinland 144
Weiße 156
Wiebalck 85
Winter 129
Wisniewska 57–59

Zenke 96, 98
Zürn 85

Species Index

Euoenothera 10–11, 18, 22, 49–50, 93, 132, 142, 145–146, 166, 204, 206, 215
Raimannia 10–14, 18, 22, 28, 50, 52, 54, 88, 92–93, 96, 99, 103, 106, 109–110, 121, 127, 130, 142, 144–145, 150, 157, 166, 180, 205–206, 211, 215–218
Renneria 10, 12, 206

Oenothera argillicola Mackenzie 123–124, **170**
affinis 12–13, 93
ammophila **170**
argentinea 51, 217
atrovirens 47, 111–112, 146, 152–153, 171–172, **175**

Berteriana 51–52, 88, 107, 109, **110**, 113, 122, 145, 150, 153, 197, 211, 216–217
berteriana „Erlangen" Haustein 12
biennis L. 11, 15–16, 19, 22, 25–29, 35, 38, 67, 94, 96–97, 102–103, 109–112, 123–124, 126, 133–136, **137**, 143, 146, 148, 152–153, 157, 171–172, **175**, 179–180, 183, 186, 190, 207, 209
biennis-1 group 47–48, 200, 212
biennis-2 136
blandina 95–97, 99, 103, 108, 113

campylocalyx Koch et Bouche 12, 206
cantabrigiana Davis 136
chicaginensis 47, 92, 135
conferta Renner 135

dulcis 108

elata H.K.B. 11–12, 123–124
elata H.K.B. strain Johannsen 124
elata ssp. hookeri 133
erythrosepala Borbas 11, **138**

fallax **143**
franciscana 16, 87, 103, 108, 135–136, 146
fruticosa 197, 200, 212

gigantea 103
glazioviana Michelli 11–12
grandiflora Ait. 12, 108, 141
grandiflora l'Her. 12, 49, 123–124, 133, 135–136
grandiflora Solander 11–12, 16, 108

Hookeri de Vries 12–13, 19, 22, 24, **24**, 28, 56, 58–59, **61**, 87–88, 90–91, 111–113, 118, 120–121, 124, 142, 144–148, 150, **170**, 179, 180, 183–184, 186, 188, 205–206
Hookeri Torr. and Gray 11–12, 123
Hookeri strain Johannsen 127, 164

Johannsen 87

laciniata 133
Lamarckiana de Vries 4–7, 11, 13–14, 16, 19, 23–31, 33, 35–36, 39–40, 50, 54, **64**, 67, 86–88, 96–97, 101–103, 108–109, 111, 113, 121, 124, 131, 134–136, 141–142, **143**, 144, 146–147, 155–156, 159, 165, **166**, 171, 179, 186, 197, 210
Lamarckiana Seringe 11–12
latifrons 108
lutescens 140, 144

missouriensis 157, 159, 212
mollissima L. 12, 50, 93
muricata Aut. 15–16, 22, 24, 27, 35, 88, 121, 134, 146, 183, 190
muricata L. s.str. nom.conf. **137**

nutans Atkinson 124

oakesiana (A. Gray) Robbins. 123–124
odorata 13, 104, 107, 109, **110**, 113, 122, 145, 150, 197, 211
odorata „Erlangen" Haustein 12
Oehlkersii Kappus 135
organensis Munz 14, 18, 47, 154, 161, 201, 211–213

parviflora L. 123–124, 126, **138**, 209
picensis Philippi ssp. picensis Dietrich 12, 109, **110**
pratincola 147
purpurata 188

recurva Dietrich 195
rhombipetala 46
rubricalyx 95, 108
rubricaulis Klebahn 137
rubrinervis 108
rubrisepala 108

silesiaca Renner **138**
strigosa Rybd. 11
strigosa (Rydb.) Mack and Busch 11
strigosa de Vries 11, 13, 24, 27–28, 32, 35–36, 41, 45–46, 145–146, 149
suaveolens Desf. 11, 16, **24**, 25–29, 35–36, 38, 45, 108, 121, 123, 126–127, 131, 134–136, **138**, 141, 145–146, 156, 163, 165, 169, **170**, **175**, 179, 188–189
subterminalis Gates 138
syrticola Bart. **15**, 15–16, 22, 24

versicolor Lehm. 12
villarica Dietrich 12, 109, **110**
villosa Thunb. 11, 123–124

wolfii 133

Subject Index

In the Index bold type refers to figures.

acutifolia 147
adventitious embryos 142
affinity 217–219
Afterglow 108
alpha-complex 28, 48–49, 221
anthocyanin 25–26, 36–37, 39–40, 152, 162–163
apetala 110, 152–153
apertural chambers 194
archespore 151, 154–155, 177, 180, 193, 202
auxin, s. IAA 147–148, 154

balanced lethals 31, 40, 50, 52–53, 100, 107–109, 133, 136, 219, 221
beta-complex 28, 48–49, 221
brevistylis 4–7, 21, 39, 95, 97, 101–102, 109, 111, 157, 159–161, 178, **182**, 197
bullata 39, 148

callose 60, **62**, 178–179, 183–186, **186**, 188, **189**, 194, 198, 201, **203**
centromere 57–58, 60, 70, 75–76, **77**, 82, 89, 98, 107
certation 205, 210
chiasma 9, 59, 66, 75, **77**, 84–88, 93–97, **116**, 118–119, 121
chiasma, terminalization 59, 75–76, 88
chimaera 122, 126, 165, **166**, 169, **170**, 172, 175, **175**
chlorophyll 85, 121–123, 126, 163, 165–166, **166**, 168, 170–172, **175**
chloroplast 152, 175
chromocenter 57–58, 60, **61**
chromosome arrangement 66, 72–73, 75, **80**, 107, **116**
chromosome break **77**, 97–98, 107, 113, **114**, **116**, 118, **119**, 120
chromosome catenation 9, 67–68, 84, 89, 94, 98–100, 111
chromosome chains 56, 59, 67, **70**, **77**, **81**, 88, 105, **116**
chromosome circles, incl. chromosome rings 56, 59–60, **64–65**, 67–68, 72, 74, 76, **77**, 85, 86, 89–90, 93–97, 107, 120, 133
chromosome complex 9, 70, 72–73, 75, **77**, **81**, 89–91, 93, 99–100
chromosome configuration **64**, 68, 70, 72–76, **77**, 81, 84, 87, 85–86, 89–93, 96, 105–107, **116**, **119**
chromosome formula 72–73, 75, 90–93, 96–100, 106–107, 133
chromosome pairing 56, 58, 65, 67–70, 76, **81**, 82, 84–89, 96–97, 107
chromosome, supernumerary 88, 96–97, 103, 105–107, 200
Co/co (flower size) 39, 42, 95–96, 111
competition 43–44, 46, 53, 94, 129, 164, 181, 183, 188, 190, 192–193, 210–211, 221
competitive ability 44, 46, 192–193
complex-heterozygote 6, 49, **77**, 191
constant hybrid, incl. constant heterozygosity 25, 30–32, 38, 40–42, 50, 53–54, 100, 102, 109, 132–133, 179, 193, 219
conversion 111–112
crossing over 9, 34–35, 37–38, 40–42, 45, 48, 84, 94–97, 100, 107–109, 111, 135
cross-pairing 68, 75
cross-sterility 47
cruciata 110–112, 152–153, 157, 161
cytochrome 129

decondensation 57–58, 62
deplanata (de) 45, 210
determinant 131
developmental function 148
differential (segment) 76, **77**, 89, 99
diplarrhenic hybrids 45, 192
ditypic 23–25, 30–31, 34, 47, 49–50, 191, 207
DNA-pattern 109
duplication 95, 104–106, 128
dwarf 36, 146

egg cell complex 27, 29, 34, 43–44, 46–47, 106, 126, 209
ektexine 194–195, **195–196**, 197–198, 203, **203**
embryogenesis 31, 102, 140–142, **143**, 144, 157, 164, 166, 207, 219
embryo sac 20, 23, 32, 43, 50, 140–142, 157, **159**, 178–180, **182**, 183–186, **185**, 188, 190–193, 205–206, 211, 215, 217–218, 220–221
end chiasma **64**, 70, 75, **81**, 84–85, 87–88, 94, 97, 100
endexine 194–195, **195–196**, 197–198, **203**
end-to-end connection 56, 59, 67–68, 84
erythrina 108
euchromatin 57–58, 60, 76, 87–88, 94–95, 113, **114**, **116**
evolution 7, 14, 49–50, 53–54, 75, 88, 98–102, 109, 129, 132–136, 209, 218, 220, 222

falcifolia 130–131, 148
fasciation, incl. ring fasciation 3, 147, 154, 157
formosa 147
fr/+^{fr} (male sterile) 38–39, 41–42, 45, 155, 197–198, 210
fragmentation 118–119
funifolia 147

ga (gametophytic active gene) 45, 210
gamma-linolenic acid 141
gametophyte 1, 7, 32, 43, 46, 52, 157, 177–178, 180, 191–195, 203–204, 206–207, 209, 215–216, 218–220
gamolysis 25
gene complex 33–35, 50, 100
generative cell, incl. generative nucleus 32, 124, 126, 180, 194, 204, 214–216
gibberellin 147, 150
gigas 86, 102–103
golden center (vetaurea-sulfurea) 2

half-mutants 106–108
haploids, androgenic 141
gynogenic 141
helix 148
hemigigas 86
hemikleistogama 153
heterochromatin 57–58, 60, **62**, **70**, 76, 77, 82, 86–87, 89, 94–95, 107, 113, **114**
heterogametic 24–25, 34, 43, 181, 184, 190
heteropolar **181–182**, 183–194, **185–186**, 188, 191
heterosis 110, 221
homogametic 184
homopolar **181–182**, 183–184, **185–186**, 186, 188–190, **189**, 192–193, 211
homotypic 51
homozygous species 13, 28
hypanthium 5, 19, 153–154, 156–157, 159–162
hypocotyl 145
hypostase 157, **159**

IAA, s. auxin 146
idioplasma 123
incompatibility 57–50, 54, 132, 211–215, 218–219, 221
incompatibility, gametophytic-gametophytic 219–220
incompatibility, gametophytic-sporophytic 215, 219–220
incompatibility, sporophytic-sporophytic 219
incompatibility, plastome-genome 122, 126–127, 130, 142, 163, 165,

166, 168–169, **170**, 172, 175, 181, 209, 219
instability 111–112
interstitial (segment) **70**, 75–76, **77**, 86, 89, 99, 107
intine 195, **196**, 198, **203**, 204
irregularis 166
isogametic 24–25, 27, 30–31, 34, 50, 178, 183

Knirps kr/+^{kr} 36–37, 42, 146
Koppelungswechsel 39

lata 97, 102–103, 155, 197
lethal 40, 43, 46, 50, 53, 73, 93–94, 99, 107–109, 133, 135, 141–142, 144, 163, 168, 207, 209, 218
lethal embryo 46, 50, 53, 142, 221
lethals, sporophytic 126, 141
lethals, zygotic 218–219, 221
linkage 9, 26, 33–34, 37–41, 43, 45–46, 48, 53–55, 90, 93–94, 210
linkage, absolute 33–34, 38, 40–41, 53, 221

megasporangia 177
megaspore 43, 94, 177–179, 180, **182**, 183–186, **185–186**, 188, 190–193, 211, 220–221
megaspore mother cell (incl. MMC) 19, 157, **159**, 177, 179–180, **182**, 183, 194
megaspore tetrad 157, **159**, 178, 180, **181**, 183–184, **185**, 186–188, **189**, 190, 192–193
meiocyte 58, 60, **62**, 65–66, 72, 87–89, **116**, 157, **159**, 177–180, **181–182**, 183, **185–186**, **189**, 189–190, 192, 194, 200, 215
meiosis, duration 60, 113
meiospores 60, 89, 100, 186
metasyndesis 57, 67
Mendelian 4, 21, 26, 30, 32–33, 36–38, 43, 52–53, 95, 97, 100, 102, 109, 121, 131, 181, 210
microgamete 209
microsporangia 58, 177
microspore 19, 46, 50, 60, **62**, 103, 152, 155–156, 177–178, 194–197, 199–203, 207, 209, 214–215, 221
microspore mother cell 19, 150–151, 155, 193
microwave 118
missing petals (mp) 151
mitochondria 123, 130–131
monosomic 113
monotypic 23–25, 27, 34, 43–47, 49, 53, 191, 193, 207

nanella 109
non-disjunction 96, 103–104, **104**
non-homologous pairing 107
nucellar embryo 142
nuclear sap 65–66

ovary 5, 10, 19, 26, 36–37, 40, 43, 50, 142, 149, 150, 154, 156–157, 160–162, 179–180, 203–207, 210, 215, 217–218

parasynapsis 67
parasyndesis 67
pairing cross 75
parallel mutation 96, 102–103
pheromone 205–206, 218–219, 220
pitys 146
placenta 19, 156–157, 159–160, 205–206
plasmatic inheritance 7, 9, 123
plasmon 122–124, 130, 141–142, 160
plastid 122–124, **125**, 126–127, 129–131, 142, 155–156, 162, 164–166, **166**, 168–169, **170**, 171–172, **175–176**, 177, 179–181
plastid DNA, incl. ptDNA 126, 128–129, 175
plastidome 122, 160
plastome 122–124, **125**, 126–131, 136, 163–165, **166**, 168, 171–172, 175, 180, 209, 219
plastome mutator (pm) 127–129
polarity, incl. polarization 179–180, **182**, 183–184, **185–186**, 186, 188–193, 211, 219
porogamy 215

pollen complex 27, 34, 43–45, 47, 121, 192
pollen inactivation 54, 120, 207, 209, 220–221
pollen lethality 50–51, 53, 106, 200, **208**, 220–221
pollen mother cells (including PMC) 19, 48, 56, 60, **61–64**, 82, 84–86, 98, 113, **116**, 118, 120, 151, 155, 177, 193–194, 196, 198, **199**, **203**
pollen sac 155, 177
pollen tube 32, 48–50, 124, 178, 180, 201, 204–208, 210–218, 220–221
pollicata 159–160
position effect 95
Punctata $P/+^P$ 39–40, 42, 94–95, 162

quadrivalent 87–88

raphid 18, 146
receptor 206, 218–220
recondensation 58, **62**

SAT-chromosome 97
second contraction 59, 64, 67
selection 43, 49, 101, 110–111
self-sterility 47
self-incompatibility 5, 48–50
semi-gigas 103
semi-heterogametic 24–25, 31, 34, 207
sepaloid 112, 152
S-locus, incl. *S*-allele 47–50, 212–214, 221
species, good, incl. pure 5, 54, 100–101, 132
Spitzblatt, $Sp/+^{Sp}$ 148
sporoderm 194–196, **196**, 203
sporopollenin 155–156, 197–198, 203, **203**
sporophyte 1, 32, 163, 177–178, 191, 193, 195–197, 200–201, 203, 212, 215, 219–220
sterilis 155, 197–198, **199**
sterility, female 102, 159–160
  male 36, 38, 97, 102–103, 151, 155–156, 193, 196–197, **199**, 200, 203, 220
  pollen 41, 45, 113
stigma 4–5, 19, 48, 154, 156, 159–162, 201, 203–205, 206, 210–213, 217, 220
style 4–5, 19, 47–49, 54, 154, 156, 159–162, 201, 204–205, 210–215, 220
sulfurea, $s/+^s$ 2, 28, 36, 38–39, 42, 45, 48, 94, 109, 135, 146, 151, 162, 171, 210
supplena 39, 159–160
synergide 20, 164, 180, 184, 206, 215
synizesis 59, **62**, 64–66

tapetum 58, 155, 196, 198–199, 202–203
telesynapsis 67
tenella 197
tetraploid 87, 102–103, 194
translocation, incl. reciprocal translocation **9**, 11, 68–69, **70–71**, 72, 74–76, 77, **80–81**, 82, 84, 86, 88–93, 95, 97–99, 106–109, 113, **114**, **116**, 118–121, 132–133, 135–136, 221
translocation point 60, **70**, 75–76, **77**, 87, 89, 94, 107
transposable element 131
triploid 86–88, 96, 103, 194
trisomic 87, 96–97, 103–107, 113, 197, 200
trivalent 86–88, 105
twin embryo sacs **182**, 184, **185**, 188–189, 191
twin hybrids 22, **24**, 28, 30, 32–34, 38
thylacoid 128, 171, **175–176**

univalent 86–88, 93, 103
urethane 85, 118, 120

variegation 95, 127
variegation, three-colour **165**

vegetation point 141, 146–152, 154, 156–157, 159–161
vegetative cell, incl. vegetative nucleus 49, 195, 200–201, **203**, 204, 212, 214
vetaurea 2, 28, 39, 109, 162, 171
viscin threads 194, **195–196**, 201, **203**

Weißherz **166**, 169, 171, **175**

Zwerg 36, 146
zygote 31, 46, 52, 54, 142, 165, 201, 209, 214, 216, 218

SPRINGER NATURE

GPSR Compliance

The European Union's (EU) General Product Safety Regulation (GPSR) is a set of rules that requires consumer products to be safe and our obligations to ensure this.

If you have any concerns about our products, you can contact us on ProductSafety@springernature.com

In case Publisher is established outside the EU, the EU authorized representative is:

Springer Nature Customer Service Center GmbH
Europaplatz 3
69115 Heidelberg, Germany

The manufacturer's authorised representative in the EU is Springer Nature Customer Service Centre GmbH, Europaplatz 3, 69115 Heidelberg, Germany. If you have any concerns regarding our products, please contact ProductSafety@springernature.com

Printed and bound by CPI Group (UK) Ltd, Croydon, CR0 4YY
22/07/2026
02174178-0006